GAME THEORY RELAUNCHED

Edited by **Hardy Hanappi**

Game Theory Relaunched
http://dx.doi.org/10.5772/2563
Edited by Hardy Hanappi

Contributors
Hardy Hanappi, Riccardo Alberti, Atulya K. Nagar, Eizo Akiyama, Ryuichiro Ishikawa, Mamoru Kaneko, J. Jude Kline, Naima Saeed, Odd I. Larsen, Sheng Zeng, Emmanuel Fernandez, Alberto Garcia-Diaz, Dong-Ju Lee, Mohamed Baslam, Rachid El-Azouzi, Essaid Sabir, Loubna Echabbi, El-Houssine Bouyakhf, Senka Hadzic, Shahid Mumtaz, Jonathan Rodriguez, Hisao Kameda, Anjiao Wang, Zhongxing Ye, Yucan Liu, Dariusz G. Mikulski, Antoniou Josephina, Lesta Papadopoulou Vicky, Libman Lavy, Pitsillides Andreas, David Schneider, A.M. Kowalski, A. Plastino, M. Casas, Alireza Chakeri, Nasser Sadati, Guy A. Dumont, Paul A. Wagner

Published by InTech
Janeza Trdine 9, 51000 Rijeka, Croatia

Publishing Process Manager Oliver Kurelic
Typesetting InTech Prepress, Novi Sad
Cover InTech Design Team

Additional hard copies can be obtained from orders@intechopen.com

Game Theory Relaunched, Edited by Hardy Hanappi
p. cm.
ISBN 978-953-51-1078-1

For EU product safety concerns:
IN TECH d.o.o., Prolaz Marije Krucifikse Kozulić 3, 51000 Rijeka, Croatia,
info@intechopen.com or visit our website at intechopen.com.

Meet the editor

Hardy Hanappi was born in Vienna (1951) and studied economics and Informatics at the University of Vienna and the Technical University of Vienna, respectively. Before returning to university he worked as econometrician for OPEC, and as CEO for the consulting firm ECON GmbH. Afterwards, he became a university teacher and researcher at the TU-Vienna and concentrated on macroeconomics, political economy, simulation methods, and game theory. From 1992 till 1997, he was the deputy director of socioeconomics at the Austrian Academy of Sciences. After his return to the TU Vienna he became head of economics. From 2001 till 2005, he was also the director of the Institute for Monetary Economics (Ludwig Boltzmann-Institute). He is currently ad personam chair for political economy of European integration, a position granted by the European Commission. He is a professorial research associate at SOAS (University of London), where he spent a sabbatical in 2011. Hardy Hanappi now lives in Vienna.

Contents

Preface

The scientific discipline called game theory has now reached the age of 71, the age which reminds ordinary human individuals of retirement. Fortunately enough, profound scientific achievements do not lose their powers at the same speed as their human creators. If they provide exciting new ideas and are able to stimulate a broader scientific community, then they can become a public good, a precious intellectual heritage, which is passed on from generation to generation of researchers. And even if such theories are proven to be only special cases of a more general approach, or are even falsified, their preliminary impasses or mistakes are usually acknowledged as necessary steps towards state-of-the art knowledge. There is no need for a pension system; such a theory lives on as an ingredient of ever expanding human knowledge.

When game theory entered the scientific arena in the 40s of the last century, it was a highly acclaimed new star. A star launched by one of the most admired mathematical geniuses of the century, John von Neumann. In contrast to these high aspirations, the name of the new discipline - 'Game Theory' – at first sight suggested that it might be not really a serious scientific project. It reminded of playing games, of card games like poker, or board games like chess, or even of children not being able to understand how to spend time in serious activities and thus having to train their abilities in metaphorical game play. The tension between the most stern and most abstract scientific discipline, mathematics, on the one hand, and on the other hand, the reference to the *homo ludens*, the fun of exploring reality by acting in sheltered simulated contexts, explains to some extent why game theory immediately was attractive to many young scientists on both sides of the Atlantic. It also explains why any attempt to change this name (some more 'serious' researchers have proposed to call it *Theory of Strategic Behavior*) should be avoided. Game theory still thrives on its ability to be sufficiently rigorous and open to a wide range of metaphorical game play at the same time. This is why it is so charming.

Additional profile came from the science onto which its two creators, John von Neumann and Oskar Morgenstern, decided to graft the new apparatus: economic theory. After the Great Depression of the world economy in the 30s, few scientific disciplines were visibly in a more disastrous state than economics. In tight cooperation with the trained economist Morgenstern, the extremely ambitious John von Neumann jumped on this subject to revolutionize it. But in doing so, the two underestimated the resistance against deep changes, which is an unavoidable characteristic of established researchers trying to defend their own, with life-long great difficulties acquired, intellectual capital. The relatively sudden dismissal of game theory as a scientific fashion, immediately after John von Neumann's death in the 50s, certainly has another root in the long-run upswing of a reconstructing world economy, which seemed to need no revolutionized economic theory. Some piece-meal social engineering based on Keynes ideas, plus the mathematical framework of Paul Samuelson's 'Neoclassical Synthesis', seemed to be good enough to run the show. In economic theory, game theory was out, and left to some mathematical 'nerds' who were mainly elaborating refinements along the lines of John Nash's equilibrium approach to game-theoretic questions.

But ignorance of the economic mainstream could not kill the beast. The last three decades saw a slowly starting, but exponentially rising influence of game theory in surprisingly different fields of science. In its old domain, the social sciences (mainly political economy), it is now hard to imagine that an innovative paper can succeed without at least apologizing for why it doesn't use a more appropriate game-theoretic approach. But in many other areas – from biology via abstract network theory to ICT engineering – game-theoretic modeling has reappeared as an indispensable tool.

This is the very reason why this book is called '**Game Theory Relaunched**'. Which parts of game theory are used, and which kind of further development is contributed to game theory by the respective research, does, of course, assume different forms, which in turn depends on the respective discipline. Game theory is still not a finalized body of knowledge – and (as chapter 1 argues) will not be for a very long time to come. Many existing textbooks on game theory therefore spend most of their pages on describing the very specific history of refinements of Nash's equilibrium concept – and mention the actually happening renaissance of game-theoretic thought across different disciplines and engineering activities as an exotic outlier at best. The collection in this book takes a different perspective; it proceeds along the lines of currently developed game-theoretic work. At the current stage of the renaissance of game theory, all that can be done is to collect and to structure the diverse contributions: The book thus consists of four parts containing (1) social science oriented chapters, (2) chapters related to engineering problems, (3) chapters enhancing the mathematics of game theory, and (4) chapters that stress the transdisciplinary character of game theory. By working through this mosaic of building blocks of game theory, the reader can hopefully get an impression of the breadth and depth of the intellectual potential, which the founders of this theory had envisaged.

All the chapters have been written by different authors spread all over the globe, and thus showing how international the game theoretic community already is these days. It has been my honor to collect and edit their contributions, but the intellectual surplus they produced is completely their own merit. Special thanks have to also go to InTech Publisher, and in particular to publishing process manager Oliver Kurelic, whose patience and wise consultation made this book possible. This type of publishing, provided by publishers like InTech, will hopefully not only help young and innovative researchers in many countries, but also benefit students around the world. This book is dedicated to them.

Hardy Hanappi
University of Technology Vienna,
Austria

Section 1

Game Theory in the Social Sciences

Chapter 1

The Neumann-Morgenstern Project – Game Theory as a Formal Language for the Social Sciences

Hardy Hanappi

Additional information is available at the end of the chapter

http://dx.doi.org/10.5772/56106

1. Introduction

In 1942 John von Neumann and Oskar Morgenstern published their book 'Theory of Games and Economic Behavior'. The impact of this book on the scientific community and in particular on the further development of the social sciences was tremendous. John von Neumann's reputation as a mathematical genius and Oskar Morgenstern's ability to contribute truly innovative and original ideas to economic theory helped to spread the fame of their monumental masterpiece. Since this first wave of excitement in the last 70 years the theory of games has experienced a rather mixed fate, periods of ignorance changing with periods of redirection towards new fields of interest, or even new re-definition of its basic aims. There is no doubt that in each of these emerging sideways towards which specific scientific communities modified the original formal framework tremendous scientific progress was stimulated. The range of the diversity of the affected fields can hardly be exaggerated; it reaches from political economy via sociology and psychology to pure mathematics. But the price paid for these wide-spread singular success stories was another effect accompanying it: an increasing disintegration of the original project. Moreover the incredible swelling of research papers in each area during the last decades made it impossible - even for large research teams – to survey what was going on with the use of the theory of games in science. This is the starting point for the line of argument presented in this chapter.

The need for a re-integrative attempt of the basic tenets of a theory of games probably currently is felt most urgently in the area of political economy. In this area the mainstream theory of economic policy seems to be particularly helpless when confronted with questions arising in times of global economic crisis. To answer most of these questions would require to formulate, non-linear strategic behavior in situations of disequilibrium, a task which the equilibrium centered part of economic game theory hardly can tackle. Therefore in this chapter the modest attempt is made to return to the original Neumann-Morgenstern project to learn from it how to frame a formal language that is able to capture the essence of such a situation. In doing so it can be shown that the development of algorithmically oriented evolutionary economics (e.g. agent-based simulation) can play an important role to approach the Neumann-Morgenstern project. From the opposite side, namely the most advanced formalization attempts of disequilibrium economics, economists like Vela Velupillai (compare (Velupillai, 2011)) are aiming to bridge the distance to algorithmic considerations too.

The result of the chapter will be the formulation of an updated version of the Neumann-Morgenstern project. On the basis of this research program the most recent structural crisis in economic theory building, and its possible future merits will appear in a new, more comprehensive light.

2. The origins of game theory – Personalities and milieus

John von Neumann, probably the most influential scientist of the 20th century, for many researchers in the structural sciences has been the unique personality, the reference point, from which the theory of games has been developed. Indeed John von Neumann's lifelong work, his intellectual trajectory leading him through a whole range of different disciplines, is an excellent starting point for a better understanding of the *logical* origin of the theory of strategic games. But before using von Neumann's biography to develop this argument it is useful, even necessary, to recognize that his contribution can easily be interpreted as a rediscovery and a more general redesign and unification of older theory fragments, which can be traced back in history almost two thousand years.

Strategic considerations are implicitly enabled by the characteristic feature, which distinguishes the human species from earlier forms of living organisms: the capacity to use internal model building as an instrument for survival and growth. In this context the adjective 'internal' evidently means internal to the species, to the respective tribe. Model building therefore is congruent to the existence of a communication system of the tribe. And this in turn implies that the constituent parts of a communication system have emerged. As there are:

- The ability of the members of the tribe to send, to perceive, and to store signals inside and outside their brains.
- The ability of the members of the tribe to interpret signals as representations of dynamics going on outside the world of signals.
- An environment of the tribe which is favorable enough for the tribe to allow for a fast enough adaption of the communication system to adjust to (possibly) deteriorating conditions.

As soon as these features are added to a purely animal species - and some sophistication of their evolution has occurred – systems of spoken and written language will serve as the major constituting element of the tribe. Members of a tribe will recognize other members as mirror images of themselves, and the division of labor within the tribe will open up the road to the interplay of cooperation and conflict. In other words, with the emergence of language all the ingredients necessary for setting up a strategic game in terms of modern game theory are given.

First consider cooperation. Division of types of activity within the tribe, division of labor, needs coordination, needs an internal image of this division in the mind of each member. Moreover this common image only exists since it is maintained via an externalized common language. To control the self-consciousness of the tribe perpetuated by the common communication system is itself a task carried out by a specialized group of tribe members. In game theoretic terms it works by producing images in the minds of tribe members which align their behavior by predicting individual disasters in case of breaking the rules of traditional cooperative behavior. The most archaic model of such an internalized game in strategic form is shown in figure 1.

		T	
		G	P
M_i	TB	C^M, C^T	F^M, F^T
	NTB	E^M, E^T	D^M, D^T

Figure 1. Cooperation enforcing mental model of M_i

A tribe member M_i can either follow its *traditional behavior* choosing actions TB, or it can decide to deviate from choosing NTB (not choosing TB). But a look inside its model shows that these two options lead to results which also depend on the reaction of the entity tribe, called T. In this mental model the column player T chooses between gratification (G) and punishment (P) depending on conformity of the member's behavior. In a functioning tribe each member chooses traditional cooperative behavior by predicting that it is preferable using the comparison of possible payoffs in figure 1. This is how the concept of free will in a cooperative tribe emerges – there is a choice. The subgroup of tribe members controlling cooperation works on implementing these mental models in all diverse groups of other tribe members. Control typically concerns two levels, an ideological level (e.g. religion) which aims at directly implanting a certain game (including payoff) structure, and a directly coercive level (e.g. police) which provides actual examples of punishment to reinforce the believe in the mental model. In game theoretic terms the task simply is to guarantee that

$$p^C \cdot C^M + p^F \cdot F^M > p^E \cdot E^M + p^D \cdot D^M ,$$

with p^C, p^F, p^E and p^D being the respective probabilities ascribed by member M_i to each of the four possible events. Expected utility from deviating from cooperative behavior must be smaller than sticking to the norm, and all measures which either concern the probabilities or the predicted utilities can be used by the tribes control instances to maintain self-control of its members[1].

[1] In the Middle Ages particularly cruel forms of punishment of non-conform members could compensate for a lower probability of detection due to increasing empires; the necessary decrease in p^D was counteracted by an increase in D^M.

Turn to conflict now. As tribes expand across areas with different fertility conflicts between tribes cannot be avoided. Again reoccurrence of battles will lead to division of labor, to a specialized subgroup within each tribe, warriors[2]. For most of human history, till the times of Napoleon, the force of a group of warriors could be directly related to the number of warriors drafted. Napoleon's rule famously stated that the higher number of soldiers in a battle between two armies will decide who wins. Complications and even possibilities for a weaker army to win a battle enter the picture as soon as internal model building of army leaders is introduced. From earliest historical sources onward military theory has emphasized the importance of knowledge of the expected battleground. Knowing where to fight and how to position the own warriors is based on the anticipation of the moves of the approaching enemy. The core of game theoretic reasoning - namely the fact that the own final outcome depends not only on my own choices but as well on the actions of another conscious player, both anticipating each other's actions – this essence of strategic decision-making immediately emerges as soon as a more detailed environment for conflicts is taken into consideration. Larger conflicts, wars, are usually spit into a set of smaller battles taking place at different locations forcing the two generals to split their armies according to these predetermined battlefields. The art of warship for several thousand years consisted to a large extent of informal game-theoretic considerations on how to deal with this issue. More than a decade before Neumann and Morgenstern published their path-breaking book (Neumann and Morgenstern, 1944) the mathematician Borel had already formalized this basic military problem as what today is known as Colonel Blotto game (Borel, 1921). In figure 2 the strategic form of a simple Colonel Blotto Game is presented. It is assumed that each of the two armies consists of six units of warriors, all with the same number of soldiers. There are three battlefields and a battle at a battlefield is won by the army which has sent more soldiers to this location. If the amount of soldiers is equal, then this battle is a draw. The war is won by the army which wins more battles. The task for a General A thus is to distribute his units over the three locations to win as many battles as possible against General B. Consider the strategic form of this game in figure 2.

The six strategies in the table only indicate the first decision on how to split the troops; they do not concern battlefields and do not include any anticipation of the opponent's plan. Assuming that every strategy of the enemy has the same probability – this is the famous assumption about 'insufficient reason' in cases of no information – the payoff matrix of figure 2 can be constructed. The first payoff in each cell relates to general A, the second to general B. If the number is positive it shows the probability to win the war, if it is negative the probability to lose it – zeros indicate draws. A cell is shaded in grey if players can improve their chances if they know the allocations of the enemy's troops (espionage – to discover the mental model of the opponent - makes sense). Despite this oversimplified setting strategic choice is already a sophisticated enterprise[3]. The centuries' old art of

[2] In earlier societies specialists in exerting coercive power were just one subgroup guaranteeing internal cooperation (today police) as well as success in external conflict (today military). Till today some overlapping can be observed.

[3] Colonel Blotto Games are still a flourishing area of game theory. Its mathematical treatment was already introduced during the first seminars with John von Neumann by John Tukey (Tukey, 1949), further enhanced by eminent scholars

warfare has produced a large amount of insights, refinements and partial solutions to the strategic games of conflict. In the last century Hitler's early success with Blitzkrieg (fast moves of tank regiments between locations) as well as terrorist and counter-terrorist strategy building by consideration of hubs in social networks, see [Barabási, 2002, pp. 109-122], are examples of extensive use of formalizations of game theoretic ideas about conflict.

		General B				
	Strategy	6, 0, 0	5, 1, 0	4, 1, 1	3, 2, 1	2, 2, 2
General A	6, 0, 0	0, 0	$-\frac{1}{3}, \frac{1}{3}$	-1, 1	-1, 1	-1, 1
	5, 1, 0	$\frac{1}{3}, -\frac{1}{3}$	$\frac{1}{6}, \frac{1}{6}$	$-\frac{1}{3}, \frac{1}{3}$	$-\frac{1}{3}, \frac{1}{3}$	-1, 1
	4, 1, 1	1, -1	$\frac{1}{3}, \frac{1}{3}$	0, 0	$-\frac{1}{3}, \frac{1}{3}$	-1, 1
	3, 2, 1	1, -1	$\frac{1}{3}, -\frac{1}{3}$	$\frac{1}{3}, -\frac{1}{3}$	$\frac{1}{6}, \frac{1}{6}$	0, 0
	2, 2, 2	1, -1	1, -1	1, -1	0, 0	0, 0

Figure 2. Conflict anticipating mental model of A

Back in history, the mental models guiding actions in a game theoretic sense (i.e. by taking into account that other tribe members also act on the basis of their mental models) could only use languages available to the respective culture. But with the scientific revolution of the natural sciences a quantum jump in formalization techniques had taken place. In 1900, three years before John von Neumann was born, David Hilbert proposed his famous list of 23 problems of mathematics, the most abstract form of human language. It was due to the use of this language, of mathematics, that the continuing success of the natural sciences had been possible. Hilbert's program was thought to be a pathfinder to reach the highest zenith of mathematical analysis – a point where most abstract theoretical results coincide with insights into the actual physical structure of nature. It was this presumably triumphant phase of mathematical research during which the young Hungarian mathematician John von Neumann, who later became a collaborator of Hilbert, was socialized. In the first two decades of the 20th century the vision of an ultimately correct language, which has to be cleaned from all semantic references and resides only in the sphere of logic, was extremely attractive for talented young mathematicians. Alfred North Whitehead and Bertrand Russell wrote their Principia Mathematica (Whitehead and Russell, 1910), Wittgenstein produced his Tractatus (Wittgenstein, 1921), and Albert Einstein's two papers from 1905 first remained

like Bellman (Bellman, 1969), and in certain mathematical dimensions even finalized (Roberson, 2006). But as (Kovenock and Roberson, 2008) show the interpretative power of this type of conflict models has not been exhausted at all.

almost completely unknown (Einstein, 1905a, 1905b). This intellectual milieu within the scientific community of mathematicians for John von Neumann was additionally amplified by the outstanding historical record of the successes of Hungarian mathematicians[4]. Two other important elements should be mentioned to better understand the milieu which contributed to John von Neumann's early socialization.

Born as the son of a successful Jewish banker in Budapest, who received the title of nobility, the 'von', when John was 10 years old, he was very aware of business practice and the more subtle advantages of having a higher income. Following the advice of his father John started an academic career as engineer, a down-to-earth study of chemistry. The extraordinarily talented young genius developed his mathematics as his hobby; perhaps it therefore was even more fascinating and original. Despite his outstanding ability to work on highest levels of mathematical abstraction all his life John von Neumann never shied away from using mathematics for engineering problems, used it as a tool for practical problems. This attitude seems to stem from his formative years as pupil and young student[5].

The second special characteristic of von Neumann's formation came from the peculiar cultural milieu of intellectuals interested in analysis and logic in central Europe after World War 1. After the breakdown of the feudal empires freewheeling intellectual exchange of opinions flourished, not just via research papers[6] but also in the coffeehouses of Vienna and Budapest. Debates often resembled games, challenges for competing brains looking for solutions to abstract problems, usually extremely difficult games, but nevertheless still intellectually highly rewarding competitions for the players. And parallel to serious science there was a real board game which everybody in central Europe played: chess[7]. When later in his life John von Neumann lived in the USA he still cultivated this highly cooperative central European culture for which intellectual property rights simply not existed. Knowledge was freely exchanged, voluntarily shared, sometimes copied, in principle considered as public good. The reward for outstanding achievements was mainly the admiration of the other members of the scientific circle, the authority gained within the scientific community. Of course, this authority hopefully in the end would win a position at a university. Since von Neumann never had a problem in this respect, all his life he completely neglected intellectual property rights, cooperation of all scientists was the name of his game.

John von Neumann's reputation was surging. But then Hilbert's research program received a lethal blow: Kurt Gödel proved - using Hilbert's analytical apparatus – that any such apparatus can contain statements which necessarily cannot be evaluated as correct or incorrect (Gödel, 1931). Today 15 of Hilbert's problems from 1900 have been solved, 3 are

---

[4] The legend has it that for every important mathematical proof, there seems to exist a less known Hungarian mathematician who had proved the theorem one year earlier.

[5] The sharpest contrast to von Neumann probably is the personality of a contemporary; the British mathematician G. Hardy, who insisted on the purity of the discipline (Hardy, 1940).

[6] An extremely concise reconstruction of Neumann's early years in Viennese circles can be found in (Punzo, 1989).

[7] An extremely impressive recently published book shedding light on these historical roots of game theory comes from Robert Leonard (Leonard, 2010).

still unsolved, but for 5 of them it is certain that they cannot be decided. The formidable research program of logical apriorism suddenly imploded. Wittgenstein turned to the concept of a diversity of 'Sprachspiele'; Russell gave up to investigate how the consistency of the mathematical apparatus can be saved from the contradictions he himself had discovered – and devoted his time to political activism. And John von Neumann turned away from pure mathematics to advance theoretical physics and economics[8]. Of course, he took his extraordinary mathematical skills with him when he directed his attention to these new fields thus becoming the prime example for successful transdisciplinarity.

In his introduction to the sixtieth-anniversary edition of the famous book by Neumann and Morgenstern Harold Kuhn notes that he agrees with Robert Leonard that 'had von Neumann and Morgenstern never met, it seems unlikely that game theory would have been developed.' The personality of Oskar Morgenstern therefore is the second, equally essential ingredient to the Neumann-Morgenstern project.

Oskar Morgenstern's career is in many respects remarkable. He often is considered to have belonged to the school of Austrian Economics, though only few economists have a clear picture of what characterizes Austrian Economics[9], or what has been produced by Morgenstern – apart of having been the co-author of John von Neumann. As many other economists socialized in central Europe during the first two decades of the 19th century, Morgenstern's vita shows a high volatility of his views, which often changed according to the intellectual milieu he just experienced. Like Joseph Schumpeter, Ludwig von Mises, Friedrich Hayek and several other less known young scientists he never really could settle down intellectually in the established circle of Vienna's economists dominated mainly by Böhm-Bawerk. The Vienna Circle, collecting so many outstanding scientists from diverse disciplines, for a short time also was a home for some of the economists whose careers in Vienna were blocked. The smallest common denominator of this group perhaps was the emphasis which they put on the combination of underlining the importance of disequilibrium and the insistence on clarity and logic. It might be speculated that this strange mix reflects the turmoil several of them had experienced in their own individual lives.

Indeed it is again the game of chess which can serve as a metaphor explaining this type of fascination. It is a game of perfect recall with all desirable clarity necessary for logical analysis. It is immediately amenable to complete analysis: both players before they start to play could agree that for all finite games with perfect information there exists a Nash equilibrium in mixed strategies, a result later provided by (Kuhn, 1953) generalizing (Zermelo, 1913). Of course, such a game would be extremely boring if it indeed would be possible to play it that way[10]. The reason why chess proved to be so fascinating in the milieu

---

[8] Leonard seems to be correct to reject Philip Mirowski's claim that von Neumann's turn to game theory was a direct reaction on Gödel's proof (Mirowski, 2002). Indeed it is much more plausible that John von Neumann felt the immediate needs of modeling warfare from 1939 onwards were the main motivation.

[9] Moreover Austria Economics in Oskar Morgenstern's time in Vienna was radically different to what today the so-called 'Austrians' in economic circles of the USA are representing as 'Austrian Economics'.

[10] The number of possible constellations on the chess board is about 10^{47} and exceeds the number of atoms in the universe. And without a carrier system for the symbols of internal modelling no operational strategy is possible.

of Viennese economists like Morgenstern was that the intelligence of the players was not reduced to being good in deductive reasoning or guessing the opponents secrets – intelligence of a player consisted in the ability to produce internal models of the not-too-far future *in time*[11]. This was similar to the still valid critique of the 'Austrians' (e.g. von Wieser, Mayer, and Mises. Morgenstern's teachers) concerning Walras: The result that with certain assumptions on functional forms and market rules used by agents in a perfectly competitive society is compatible with the existence of a unique and stable price vector is formally interesting, but the real challenge clearly is to model what agents and institutions do if they build expectations based on models using limits of perception (of total complexity) and have to result in actions taken before full enumeration of consequences is possible. The quantitative overload of atomistic agents with such an extremely interdependent network leads to structuring and summarizing certain specialized features. In real economic life – Schrödinger stated this quantitative overkill as one of the characteristics of life itself (Schrödinger, 1928) – division of actions breeds further division of labor resulting in social classes. But to synthesize this divided world all internal models not only have to be built, they have to be kept consistent by communication. Morgenstern, after some 'nihilistic' years in Vienna, where in the face of these methodological difficulties he doubted any possibility of predicting overall socio-economic development at all, went to Britain where he met Edgeworth. The special twist in Oskar Morgenstern's vita is his insistence on the use of abstract language, of mathematics, to overcome mostly useless results put forward in this same language. He admired Edgeworth's ability to clean his abstract arguments from any 'normative' reference, while at the same time these models escaped the rigid framework of Walrasian economics. Already from 1928 onwards Morgenstern's desire to produce a new abstract formal language for economics became visible. It just needed his encounter with the mathematical genius of John von Neumann to take serious steps towards this goal.

3. The context of the further evolution of game theory

In his paper "Zur Theorie der Gesellschaftsspiele" (Neumann, 1928) John von Neumann had already developed a blueprint of what was later to become known as game theory. But for more than a decade he did not further consider the topic. Only after some fundamental changes - in his life as well as in the general state of the world - he returned to this seemingly mundane theme. In 1938 his first wife had left him, he had left Europe and had settled in Princeton, and Hitler's armies were successfully conquering Europe[12]. An enormous amount of intellectual capacity was driven out of Europe and almost exclusively found its exile in the United States. John von Neumann's world-wide reputation as a mathematical genius made him, together with Albert Einstein and Kurt Gödel to one of the key personalities at the epicenter of this exodus from Europe – Princeton University and the Institute of Advanced Studies in Princeton.

---

[11] It is interesting that many contemporaries of John von Neumann mention his *speed of thought* as the most impressive feature of his genius.

[12] For a detailed account of these crisis years in von Neumann's creativity – he only published one paper during 1938 and 1939 - see (Leonard, 2012, pp. 195-223).

Oskar Morgenstern was attracted by John von Neumann's genius too, and during a visit to the USA managed to meet him. During the 30-ties Morgenstern had been able to achieve a well-respected status as economist in Vienna. With Mises helping him to get into contact with the Rockefeller foundation he had become director of the "Trade Cycle Research Institute" in Vienna, but in March 1938 (while Morgenstern was in the USA) Hitler's academic collaborators took over and installed Ernst Wagemann and Reinhard Kamitz as new directors cleaning the staff from all Jewish and "politically unbearable" elements. Morgenstern had never been a socialist though; quite to the contrary his early formation has had a strong anti-socialist, during school days even anti-Semitic, tendency. Like John von Neumann's family background his family background rather was characterized by a flight from the threatening communist regimes in Eastern Europe. In his student days in Vienna he had been closer to the conservative groups around Mises, Mayer, and Hayek, and he always stayed in distance to the social-democrats Otto Neurath and Otto Bauer. Only when he met Karl Menger, son of the famous father of marginalism Carl Menger, he developed an interest beyond the difficulties of modeling of complex individual decisions, an interest in theories of social justice and fairness. Karl Menger and Hans Hahn had been those members of the school of Austrian Economics who were closest to socialist thought. By the putsch of Austro-fascism in 1934 the carriers of the tradition of liberal thought of the Austrian School were forced to take a decision: Either they could transfer liberalism into a political agenda that (due to its anti-socialism) was hopefully compatible with the new political rulers, or they could take the problems of modeling liberalism to some higher methodological grounds, carefully separating theory building from normative political judgments[13]. Friedrich Hayek took the first option while Karl Menger and Oskar Morgenstern took the second alternative. As a consequence Morgenstern during the 30-ties worked through an extensive list of readings in mathematics and philosophy to acquire more knowledge on the state of the art of formal methods across all disciplines. It probably was this extraordinary broad aspiration, which made Morgenstern an ideal partner – a "necessary interlocutor" as Leonard calls it – for John von Neumann.

When Neumann and Morgenstern met the excitement they both experienced during the relatively short time it needed to produce their common book had immediately emerged. They set out to produce a new formal language for the social sciences; the deficiency of mathematical economics is best expressed in von Neumann's words: "You know, Oskar, if these books (on mathematical economics, G.H.) are unearthed sometime a few hundred years hence, people will not believe that they were written in our time. Rather they will think they are about contemporary with Newton, so primitive is their mathematics. Economics is simply still a million miles away from the state in which an advanced science is, such as physics." (Morgenstern, 1976).

A detailed discussion of the content of the masterpiece of John von Neumann and Oskar Morgenstern goes beyond the scope of this chapter; a few remarks have to suffice. First, it is

[13] This strict distinction between objective knowledge and subjectively determined normative issues stemmed from Max Weber and was highly influential in the interwar period (Weber, 1904).

important to realize that the book to a considerable extent has to be understood as a critique of political economy, of prevailing economic theory existing in 1942. It is evident that the difference to be made between models of non-living atoms and models of agents in social settings consists in the necessity to consider internal model-building of the agents. Since they use these internal models to identify variables they want to optimize (goals), variables they can control (instruments), and relations between these variables, which they have to observe (rules plus auxiliary variables) it is necessary to characterize these subsets. Chapter 1 and the appendix of the book thus contain a new theory on how to formalize goal variables, i.e. the famous Neumann-Morgenstern utility theory. As a side product of this formalization the notion of rationality is given a clear definition. Then, turning to instruments and rules instead of using the already formalized metaphor of Newtonian mechanics the authors rather take a look at social learning of goal directed action as it occurs in human societies. And they discover it as learning via games, games with which children learn, card games with which adults entertain themselves and explore their psychological interaction, games like chess where masterminds encounter the problem of 'infinite' regress[14]. As chapter 2 of the book demonstrates goals, instruments and rules can neatly be packed in a formal definition of a strategic game built on the archetype of a simple board game. In this chapter another important feature of the new formalization had to appear: Internal model-building needs assumptions describing the processing of information. Chess again proves to be a good starting point for an analysis, since the only information kept secret by each player is his or her internal model. But even this last hide-away of secret personal knowledge is hard to capture because it contains all memories and pattern recognition capabilities of a player. Neumann and Morgenstern react to this difficulty by restricting their attention to most simple settings and the structure of the theory they imply: Structure implied by the number of players, by payoffs being constant-sum or not, by decomposability. The detailed treatment of all of these cases constitutes the core of their book.

As Morgenstern later wrote their work was not just intended to show the capabilities of a modern mathematical treatment, neither was it just an alternative spotlight on economic processes: "The theory … deals in a new manner even with such things as substitution, complementarity, superadditivity of value, exploitation, discrimination, social 'stratification', symmetry in organizations, power and privilege of players, etc. Thus the scope of the book extends far beyond economics, reaching into political science and sociology … "(Morgenstern, 1976). When the United States were entering the war - German submarines were already near its East coast - John von Neumann became involved in war activities. Economic, political, and military strategic interaction could not be properly disentangled anymore.

Interestingly enough a certain inversion of emphasis of the two authors of the path-breaking book after its publication can be observed. John von Neumann took a turn towards

[14] "If he thinks that I think, what he thinks that I think …" is just the imagined mirror image of a very large decision tree of possible moves in chess. It actually is finite, but so large that it cannot be used to derive the best next move, thus providing a new meaning for "infinite", namely inoperative.

operationalization, supporting the development weapons technology[15] and in the realm of modeling living agents he designed the first modern computers[16]. Oskar Morgenstern, who already had turned his back to 'normative' political economy, pushed the elaboration of the mathematical generalizations of the new theory[17].

In retrospective the genesis of the Neumann-Morgenstern project to a considerable part can be understood as a coincidence of the particular biographies of the two protagonists and the conditions of a world thrown into a global war, a war the roots of which could hardly be understood by traditional political economy – not to speak of a Walrasian equilibrium theory. What is most significant for this project is that it first treats the contradictions that occurred in the real world as well as in the formalizations with an utmost extension of the existing formal apparatus. But if this apparatus proves to be insufficient then the language of this apparatus might have to be changed - at least this seems to be the implicit message of John von Neumann's own vita leading to the invention of modern computer technology. In the end the project today appears as an enormous attempt to redesign formal modeling of human societies by including what usually is summarized by the notion of *communication*. During the last 60 years actually used communication technologies have profoundly changed our lives often in surprising ways, while the Neumann-Morgenstern project of an adequate theoretical correlate still seems to be far away from catching up with reality. Though the trigger event of their published book seemed to be a satisfactory round-up of the project at least for von Neumann, it nevertheless remained less influential for the social sciences than the immediate euphoric reviews it experienced would have indicated. Von Neumann and Morgenstern only produced one more paper together after its publication, and the further development of the theory fell completely into the hands of mathematicians.

4. Advantages and dangers of narrowing the focus of game theory

In his introduction to the sixtieth anniversary edition of "Theory of Games and Economic Behavior" the mathematician Harold Kuhn contemplating the decades after its publication in 1944 writes: "A crucial fact was that von Neumann's theory was too mathematical for the economists. ... As a consequence, the theory of games was developed almost exclusively by mathematicians in this period." (Kuhn, 2004). And then he refers to the entry on "Game Theory" written by Nobel Prize Winner Robert Aumann in the *New Palgrave Dictionary of*

[15] This chapter ignores von Neumann's eventually occurring and irritating transformations into 'Dr. Strangelove mode' (see (Strathern, 2002)): He sometimes put forward extremely naïve and politically unacceptable suggestions. Comparable to other strange behavioral traits of his personal life (see (Macrae, 1992)) this should be delegated to a discussion on his psychology and is largely independent of the Neumann-Morgenstern project.

[16] Towards the end of his life he tried to isolate the essential elements of internal model-building by comparison with the human brain (Neumann, 1958), and worked out the necessary general rules for self-reproducing automata. Evidently returning to his early engineering background he was in a kind of search for a material correlate of a game-theoretic player's internal structure. His earlier interest in describing hydrodynamic turbulence at that time already had initiated his work on computer simulations, compare (Ulam, 1958).

[17] Only late he had discovered John von Neumann's growth model (Neumann, 1937) and was rather excited about its links to game theory. In a book published together with Gerald Thompson (Morgenstern and Thompson, 1976) he contributed to the development of the Kemeny-Morgenstern-Thompson model, a generalization of Neumann's growth model.

Economics, where the latter enumerates the following success story of mathematical results in game theory in these years: Nash, Shapley, Gillies, Milnor, Tucker, and Kuhn himself. Since Kuhn was cooperating with Neumann and Morgenstern as a young researcher, he is also truthful enough to admit that these theoretical developments ran counter the theoretical aspirations of the two original authors. He writes:

"It is important to recognize that the results that Aumann enumerated to not respond to some suggestion of von Neumann; rather they were new ideas that ran counter to von Neumann's preferred version of the theory. In almost every instance, it was a repair of some inadequacy of the theory as presented in the TGEB (The Theory of Games and Economic Behavior, G.H.). Indeed von Neumann and Morgenstern criticized Nash's non-cooperative theory on a number of occasions." (Kuhn, 2004).

This statement allows for an interesting interpretation of what had happened to game theory in the after war period. As many introductory textbooks on game theory today proudly state on the first pages, this discipline can be considered as a proper branch of mathematics. Having developed during the years of rapid formalization of mainstream economic theories of all sorts, during the "golden years of high theory" as some feel inclined to call this period, these mathematical improvements of many ideas of John von Neumann certainly can be considered as progress within the realm of mathematics. To some extent several seemingly new contributions of game theory could be shown to be isomorph to already existing parts of mathematics[18], in other cases game theory did provide a new vista on an already existing mathematical technique. This should not be too surprising since the formal game theoretic framework was built by a traditionally trained mathematician, and since progress in this discipline is brought about by a world-wide community of similarly educated scientists it can be expected that any seemingly new development pops up with limited time delays at different places. Till this is discovered by the scientific community some small idiosyncratic frameworks can take off and certain astonishment occurs as soon as somebody strips off the disguising nomenclature and shows that the essence of two approaches is equivalent. In a scientific discipline like mathematics, which sets itself the goal to be as free as possible from any reference outside its own language, it is not always easy to discover such an isomorphism. There is no physical outside object on which the language is applied and where different language perspectives on this same object are held together by the structure of the object as it is reflected in the different perspectives. Perhaps this is even a more general point than just a characteristic difficulty of progress in highly abstract structural disciplines: Having lost an outside point of reference and being thrown back to a self-defined criterion of consistency such sciences are prone to become quasi-religious believe systems. Standard microeconomic theory is an outstanding example of such a development. The mathematical framework that is used is a non-stochastic version of the mathematical framework that was so successfully applied in thermodynamics. And once the

[18] Kuhn mentions the early success he and his young colleagues had when they showed that mathematically linear programming and the theory of two-person zero-sum games are equivalent. In other cases a difference was just a matter of naming, e.g. 'dynamic programming' used in operations research is the same as 'backward induction' in game theory.

outside physical origin of this framework was forgotten (or even consciously deleted) the independent sprachspiel used in the new domain could sway freely[19] with an aura of eternal validity. Of course, there is always room for additional insight by deduction within the same language, the case of simple syllogisms is telling in this respect. But as Gödel to some embarrassment of von Neumann had proved, there are limits to such success in every possible analytical language[20].

At this point it has to be remembered that von Neumann – and the turn towards an interpretation of game theory as only a branch of mathematics is mainly directed against him, Morgenstern with his non-mathematical socialization was a much less critical admirer of mathematics – was also an engineer, he had studied chemistry at the ETH Zürich in Switzerland. For engineers there always exists a more or less physical object of investigation, and this feature also characterizes some of the turns in John von Neumann's life. To complement any theoretical result by work referring and using the finite tools of the physical world was important. When Neumann encountered difficulties in the (theoretical) mathematical description of hydromechanics he turned to (practical) simulation of partial differential equations, which in turn became the catalyst for his important (theoretical) work on computers. Later, in his last book 'The Computer and the Brain', (Neumann, 2000), he (practically) compares biological processes to (theoretical) problems of the necessary elements of self-reproducing automata. Practice and theory are always intrinsically interwoven. The importance of the engineering perspective is even more pervasive if one looks beyond the monolithic contribution of the Neumann-Morgenstern project. World War 2 has not only lead to a mass emigration of Jewish intelligentsia from Europe to the USA, it also had forced all scientific workers involved in the resistance against the Nazi forces to combine theoretical and practical insights to derive operational devices. Within a short time a collective of researchers was organized not just by military leaders but also by organizations like Bell Laboratories. The most outstanding – all to some extend contributing to the Neumann-Morgenstern project of a new language for the social sciences – have to be briefly mentioned.

One of the most profound innovations for the development of the project came from an American mathematician, Claude Shannon, who surprisingly enough concentrated on stripping communication theory from any reference to semantic content. Shifting the focus from the search for a fundamental framework of essential communication concepts to the engineering perspective of quantification of goal-oriented symbol transmission opened up the exploration of a brand-new set of definitions[21]. Shannon's aspiration was to provide a far-reaching and deep general theory of communication, a historical fact often ignored by

---

[19] Wittgenstein's idea of a parallelism of sprachspiele is just the over-pessimistic downside of his exaggerated and euphoric rigidity in the Tractatus. Languages used as tools can play on both pianos.

[20] Quick as he always was, John von Neumann managed to incorporate contradictions like the wave-particle problem of light, a challenge that had emerged with quantum theory, by providing an additional level of generalization (Neumann, 1996 (1932)). He also quickly had accepted Gödel's point, though in this case he saw no contradiction there to be repaired; note again the engineering perspective.

[21] See (Shannon, 1949).

authors concentrating on his importance for practical engineering tasks, e.g. adding redundancy to overcome noise in channels[22]. In a paper presented in 1950 at the Conference of Cybernetics (in front of Von Neumann and Norbert Wiener) Shannon[23] proposed to quantify what he called 'information content', H, by a formula, which used the total number of symbols used in communication, n, and the (usually smaller) number, s, sent or received in a particular communicative signal:

$$H = n \cdot \log s$$

This formula assumed that all symbols occur with the same probability, and Shannon showed how to generalize it for a vector of different probabilities p_i :

$$H = -\sum_i p_i \cdot log_2 p_i$$

In this form the information encapsulated in a message could be measured in bits and H was dubbed Shannon entropy.

Despite the omnipresent engineering jargon, it is evident that this quantification constructs a context of purposeful transmission of selected subsets between agents sharing the same alphabet. Shannon's ignorance with respect to any further (semantic) reference leading from a signal to an element outside the symbol set should not entrap to overlook that this type of signal transmission, i.e. communication using an alphabet that is present in the consciousness of sender and receiver, is a pivotal characteristic of tribes of living entities. Shannon's extremely rigid formalization cannot escape from being a linguistic foundation for the social sciences. These living entities might use tools to change the form of the signals they use, e.g. telegraphs or computers sending or receiving transcriptions of everyday language, nevertheless outside the context of human societies these entities would not exist – just like a hammer would not be a hammer if it were not a tool with a specified function in human work[24]. Improving the encoding for given capacity limits – bandwidth and time constraints – was the typical engineering task Shannon was trying to accomplish. And under the same stressful wartime conditions as Shannon, von Neumann and Morgenstern in parallel work were developing another piece of the puzzle of decision-making in human societies. Looking at Neumann-Morgenstern utility from the opposite perspective, namely by starting with the view that it is a theory about human individuals[25] and deriving from it the mathematical engineering problem of finding the coincidence of optimal responses explains why the mathematician John von Neumann in his famous paper from 1928 acted as an engineer for parlour games. Only more than a decade later, confronted with the harsh necessities of global warfare and after meeting Morgenstern, the Neumann-Morgenstern project took on shape. For Shannon too, wartime spurred his efforts, his first – secret and

[22] James Gleick's recent book can take the credit to readjust the image of its hero Claude Shannon. (Gleick, 2011)

[23] The formula had been developed shortly before by Nyquist and Hartley.

[24] In this perspective, the long-standing dispute concerning communication between computers burns down to the need to use proper definitions for ''communication' and 'tool'.

[25] Methodological individualism might be a typical inheritance brought into the cooperation by Morgenstern, who was still partially rooted in some Jevons-oriented Austrian economics.

unpublished – research paper concerned cryptographic military methods and precluded many results of his famous later work (Shannon, 1949). After the war Shannon and Neumann met at a series of conferences and again the engineering perspective seemed to be the common denominator of the two mathematicians[26].

Another famous scientist present at this series of conferences was Norbert Wiener. Some of his theoretical conclusions were similar to those elaborated by Shannon, but Wiener had less modest views on the implications of his theory of cybernetics on the social evolution of mankind (Wiener, 1948, 1954). While it used to be popular to consider some processes with rigid engineering attitude as a 'black box' - to take the behaviorist position that only inputs and outputs of this black box need to be considered to understand what's going on – Wiener proposed to look into black boxes to turn them into 'white boxes'[27]. In a sense the Neumann-Morgenstern project proposes to take internal model building processes of living entities serious, to open the black box of a stimulus-response reaction pattern and to substitute it by a white box, i.e. the explicit statement of a full-fledged equation system or program[28]. Wartime needs again had been an important motive for Norbert Wiener but there also was an intrinsic methodological imperative, which enlivened Wiener. Like Morgenstern he was deeply opposed to subordinate formalization in the social sciences to the ready-made apparatus used for non-living matter. Primacy of equilibrating forces and increasing entropy had to be challenged:

"We are swimming upstream in a torrent of disorganization, which tends to reduce everything to the heat death of equilibrium and sameness. ... The heat death in physics has a counterpart in the ethics of Kierkegaard, who pointed out that we live in a chaotic moral universe. In this, our main obligation is to establish arbitrary enclaves of order and system. ... Like the Red Queen, we cannot stay where we are without running as fast as we can." (Wiener cited in (Gleick, 2011, p.237)).

Norbert Wiener originally had been working in mathematics and probability theory, applying his knowledge – like Alan Turing - during wartime to cryptography. He also had graduated in zoology at Harvard and after the war became more interested in the biological foundations of cybernetics, a scientific research area he had created earlier. Like the mathematician John von Neumann the mathematician Wiener late in his life looked for inspiration in biology to see how human thought processes are conditioned by the physical constraints present in the human brain.

How much the question 'What is Life?' was in the air during the first half of the 20th century can also be seen by taking a look at Erwin Schrödinger's book with exactly that title

[26] An example is the second Conference on Cybernetics: While the electro-mechanical device labeled 'Shannon's Mouse' initiated a discourse on cognition and learning on agent- and system-level, Neumann took the discourse to a broader understanding of the primacy of discrete phenomena in the context of empirically observed 'continuous' biological phenomena.

[27] For a more detailed discussion of Wiener's work see (Hanappi H. and Hanappi-Egger E., 1999).

[28] The important step from dynamic equation systems towards programs was explored by Alan Turing (Turing, 1936) and will be sketched below.

(Schrödinger, 1944). Schrödinger too was an outstanding mathematician and physicist and was already famous for having reframed Einstein's theory in wave equations. In this book he provided an interesting answer as to why it is possible that 'arbitrary enclaves of order and system' (see the citation of Wiener above) can be established at all. It is the sheer amount, the mass of tiny atoms in random motion, which can produce its opposite, namely order that is describable by rules[29], by (always stochastic) natural laws. According to Schrödinger order has to occur on both sides of the perception process if understanding shall be possible: on the side of the observed phenomenon as well as on the side of the observer, the human brain. Ideas how order can emerge out of randomness, a topic made prominent much later by Ilya Prigogine (Prigogine, 1984) and Stuart Kaufman (Kaufman, 1993), can be traced back to the first half of the 20th century. For the Neumann-Morgenstern project this implies that the rule set for a formal language of social interaction might mimic a large amount of heterogeneous internal models, only partly stratified by communication and mass media, which nevertheless can lead to an aggregate behavior of the system that exhibits law-like features. A theory of such emergent properties thus is possible; indeed Schrödinger proposes that all theory even in the natural sciences is of precisely this type.

The last personality of particular importance for the Neumann-Morgenstern project is Alan Turing. He had taken up a scientific research program that had been almost forgotten: The work of Charles Babbage and Ada Lovelace aiming at a machine that can carry out complicated human thought processes. Turing, a logician and mathematician, had met Claude Shannon in 1943 at Bell Laboratories when both were successful cryptanalysts, and the two men exchanged some ideas on the possibility of 'thinking machines'[30] – again a sign how important a common global political environment can be. Ten years earlier Turing had started to work on the development of what is called a 'universal machine', a thought model of a device, which should be able to encompass all possible logical deductions. His blueprint, the so-called Turing machine, decades later became famous[31]. Long before programming became ubiquitous Turing's thought machine already enabled sets of instructions[32], which transformed the state of the machine in discrete steps by a finite set of pre-defined actions. To write down programs, instead of using equation systems as a metaphor for internal model-building is an important ingredient of the Neumann-Morgenstern project, despite the fact that John von Neumann only late in his life seemed to

---

[29] This basic idea had been introduced by Ludwig Boltzmann several decades ago, though not really understood by Boltzmann's contemporaries. It constituted not only a pivotal step in theoretical physics but also advanced probability theory proper. In 1906 - feeling completely misunderstood by his contemporary researchers - Ludwig Boltzmann had committed suicide.

[30] Andrew Hodges, Turing's biographer, writes: "They (Shannon and Turing) found their outlook to be the same: there was nothing sacred about the brain, and that if a machine could do as well as a brain, then it *would* be thinking ... This was a back-room Casablanca, planning an assault not on Europe, but on inner space." (Hodgson, 1983, p. 251).

[31] Turing, after being prosecuted and convicted for homosexuality in 1952, committed suicide in 1954 at the age of 42. Like Boltzmann's death, this was an enormous loss for science.

[32] Turing used different words for programming: In a (finite) state table each state was related to instructions that lead to another state. Like Babbage's machine his thought model had memory (a 'tape') and used symbols. With this setting he was able to reproduce, and even to generalize, Kurt Gödel's famous answer to the 'Entscheidungsproblem' mentioned earlier.

recognize the significance of this turn of style. Even much later, when cellular automata (CA) like 'game of life' became popular, many mathematicians underlined the fact that it has been proven that any CA is equivalent to a traditional dynamic system – the new gadget CA thus being of minor importance for the development of formalisms. John von Neumann, on the one hand turning to computer science and on the other hand recognizing the severe impact, which quantum theory has on the prevailing mathematical apparatus (Neumann, 1996 (1932)) was looking for the evolution of formalisms.

Perhaps this daring great leap towards the future of scientific development can explain somewhat why after the generation of the founding fathers of game theory had disappeared a period of stagnation set in. What had been envisaged entailed a fundamental overhaul of how to do science in the area of social sciences. Since doing science always predominantly involves the use of a scientific language (often including rigid formal elements) the reformulation of this language – the Neumann-Morgenstern project – was amidst a broad and radical scientific reformation project. Though the brightest minds of this wave of scientific revolution produced prophetic vistas on what might be its future, there was no mass movement in 'normal science' (see (Kuhn T., 1962, chapter 2)) that could backup this burst of intellectual energy once its leaders were gone.

The breakpoint to the following epoch of oblivion of the Neumann-Morgenstern project seems to be close to the occupation of the intellectual terrain of game theory by a new cohort of devoted – but differently inspired – young mathematicians. As Robert Leonard insightfully reports, Neumann disliked John Nash (Leonard, 1994). Not just as a matter of personal antipathy but due to a profoundly different methodological approach to the tenets of game theory:

'By the same token (Neumann's refusal to Kuhn's proposal of experimental study of stable sets[33], H.H.), we can understand von Neumann's dismissal of John Nash's 1950 proof of existence of an equilibrium point in a game without coalitions. Given everything we have observed about him, it seems that to von Neumann, the formation of alliances and coalitions was *sine qua non* in any theory of social organization. It is easy to understand why the idea of noncooperation would have appeared artificial to him, ... At a Princeton conference in 1955, he defended, against the criticism of Nash himself, the multiplicity of solutions permitted by the stable set: "[T]his result", he said, "was not surprising in view of the correspondingly enormous variety of observed stable social structures; many different conventions can endure, existing today for no better reason than that they were here yesterday.' (Leonard, 2010, p. 245)

The project to produce a theory that was able to reveal and to understand the structures existing outside the realm of the formal language (game theory for social structures[34]) was

---

[33] Neumann's argument against Kuhn's attempt of what today is called *experimental economics* is interesting – and still true: 'I think that nothing smaller than a complete social system will give a reasonable "empirical" picture [of the stable set solution].' (Leonard, 2010, p. 244)

[34] For an engineer – von Neumann's alter ego - this outside reference of theoretical work might be any material object of investigation.

different from what Nash, Kuhn, Shapely, and their colleagues were focusing on. Indeed the following decades saw the development of a type of game theory, which in general immunized itself from all impure influences of empirically observed phenomena[35]. To establish this newly defined discipline as a proper branch of mathematics implied that the two age-old ethical tenets of mathematics - namely to clean its language from any reference outside itself and to reduce its core to the smallest number of statements – became the goals of this type of game theory too.

A comparable development took place in economic theory. Starting with Paul Samuelson's PhD thesis that appeared as a book with the modest title 'Foundations of Economic Analysis' (Samuelson, 1947) the style of mathematics for Newtonian physics, i.e. calculus, started to reign over economic content – Neumann's cynic statement ' There's no sense in being precise when you don't even know what you're talking about.' was forgotten. What the proponents of the new era later proudly proclaimed as the Golden Age overcoming the Keynesian 'years of high theory' (see (Shackle, 1967)), for economic theory from the perspective of the Neumann-Morgenstern project has been a dark age. But it has to be noted that the retreat of economic theory-building into an ivory tower of mislead, self-referring dream-worlds had been possible – even necessary – because of the pragmatic take-over of the decision-making process in Western economies by political business men. In the tremendous capitalist growth process, the reconstruction possible after WW2, not much advice from outside the business community was needed. At best, economic theory should legitimize ex post what was in the interest of the business community anyway, or at least it should involve bright but potentially critical social scientists in tedious – but economically void – theoretical disputes. In hindsight it is thus not surprising that until the late 70-ties John von Neumann's and Oskar Morgenstern's epochal project became a Sleeping Beauty[36].

5. Renaissance of the Neumann - Morgenstern – project

It is a rather revealing coincidence that the renaissance of the Neumann-Morgenstern project started just a few years after the first severe and synchronized crisis in Western economies since the end of WW2 – at the beginning of the 80-ties. With the breakdown of the fixed-exchange rate system in 1971 and the two oil price crises induced by this event the world economy went into troubled waters again. A more pronounced economic policy stance was needed, and as the largest Western countries just had elected conservative leaders – Margaret Thatcher, Helmut Kohl, and Ronald Reagan – economic theory mainly should prove that policy has only to assure that the free interplay of market forces suffices to

---

[35] The history of game theory was re-interpreted as the history of theories resembling the Nash-equilibrium. Till today the respective parts of most textbooks use this distorted perspective, e.g. (McCain, 2009). This looks even stranger if one is aware that competing equilibrium concepts like Steven Bram's 'Theory of Moves' (Brams, 1994, 2011) are in no respect inferior to Nash's equilibrium view.

[36] If a scientific field is in deep crisis, as is the case in the social sciences now, the theoretical revolution that forms a new consensus usually consists of a set of different lines of attack on the old paradigm. Some of these lines - before entering a coalition with another line to produce what Schumpeter called a *new combination* – are able to refer back in history to an already existing piece of theory that just has to be updated, awakened. This is why the metaphor of the Sleeping Beauty occurs repeatedly in recent literature (see (Kurz, 2011)).

establish maximum welfare. The *neoclassical synthesis*[37], which accompanied the first decades of growth after 1945, was augmented by an element of hyper-rationality: the rational expectations (RE) hypothesis. Macroeconomic dynamics on the basis of RE move even further away from any possible relation to actually observable economic processes. Culminating in Nobel Prize winner Thomas Sergeant's famous textbook on macroeconomic theory (Sargent, 1980) this 'New Classical Macroeconomics' as he prefers to call it, is characterized by the inclusion of a full-fledged internal model-building process that takes place – or better: in equilibrium always has already taken place – in an infinite number of microeconomic units. It clearly is a step towards game theory. The assumptions that (1) all these internal models are identical, are (2) also equivalent to the actual working of the economy, and that (3) every micro agent knows about the first two properties, constitute the RE hypothesis. It evidently is an extremely degenerated case of what John von Neumann had in mind when he talked about modeling stable social constellations. The usual excuse for the 'heroic assumptions'[38] of RE is that less primitive assumptions would lead to insurmountable technical difficulties. This argument barely could hide the fact that popularity and worldwide streamlining of economic theory along the lines of RE was due to its applicability as underpinning for conservative economic policy in the early eighties[39]. In the end, the analytical apparatus had become more cumbersome, attracted (and partly destroyed) more intellectual capacity of potentially creative economists, and was even less in danger to interfere with any actual policy measure[40] – except the permanent unspecified call for more privatization.

It is not surprising that the impetus for a revival of the Neumann-Morgenstern project came from a completely different side, from more engineering inclined areas. The first area was biology, in particular the work done by John Maynard-Smith on evolutionary game theory (Maynard-Smith, 1982, 1988). Experiments in biology had lead researchers to find stable constellations of different behavioral traits within and across species, which allowed for a game-theoretic explanation. It seemed that certain animal populations, like some spiders, as a whole act like a (fictitious) conscious brain of the species would do if it was aiming at maximum reproductive success. Once biologists had jumped on the train of game-theoretic modeling they brought a lot of new ideas on how to extend the narrowed down perspective of standard mathematical treatment.

The second interesting impact on the topics addressed by the Neumann-Morgenstern project came from chemistry and concerned the equilibrium concept. As Ilya Prigogine showed, living systems building-up ordered structures are characterized by processes far away from thermodynamic equilibrium (Prigogine, 1984). Social science as a theory of the particular living system of the human species thus should be based on models that account

---

[37] The proclaimed *synthesis* was always a misnomer. The project of a micro-foundation of a neo-classical macroeconomics had failed dramatically.

[38] Why a theorist shall be considered as a hero if he/she makes assumptions that inhibit their testing remains hidden in what some mathematical circles consider to be their sense of humor.

[39] In 1980 the original idea of RE was already 20 years old and can be found in a paper by John Muth (Muth, 1961).

[40] Policy was left to 'Practical men, who believe themselves to be quite exempt from any intellectual influences .... Madmen in authority, who hear voices in the air ...', as Keynes had put it so aptly in 1936 (Keynes, 1936, p. 383).

for such non-equilibrium processes. The general time-profile of the evolution of living structures therefore rather resembles a sequence of diverging trajectories held together for some time by in-built stabilizers (e.g. institutions), intermitted by substantially shorter periods of revolutionary metamorphosis during which relations and elements vanish and emerge (e.g. by coalitions leading to *new combinations*). This time-profile of living structures, which describes the temporary and spatially limited build-up of neg-entropy, takes place in front of, even opposing, the non-living environment, which is governed by thermodynamic convergence towards maximum entropy. The language of the human species - and game theory can just be a specialized scientific language – has to follow the time-profile of living structures and will not follow the monotone convergence to equilibrium of non-living matter. John von Neumann seemed to have had in mind just the first step of this evolutionary process[41], the production of variety, when he insisted on the importance of his stable set concept and refused the search of a unique equilibrium point. Instead of going for a quest to discover a 'true' solution he turned to the invention of new methods to explore variety, he turned to computer simulation. Today the new discipline of econophysics offers a rich set of knowledge that might help study the evolution of living systems right from analogues to their emergence at bio-chemical roots: Spirals of emergence of variety, selection and extinction of some elements, then emergence of variety at the next level again; all that in parallel and in different (fractal) time scales and spatial dimensions[42]. This is the new methodological background that has emerged, and now waits to challenge those who try to revive and to expand the Neumann-Morgenstern project. It is becoming part of a broader social science, of evolutionary theory[43]. In this broader project the original vision of early game theorists of a new combination of cooperation and conflict will be an important guidance[44].

Finally computer science itself contributes substantially to the new appeal of the Neumann-Morgenstern project. Computer simulation had been crucial for the re-emergence of evolutionary economics in 1982, when Richard Nelson and Sidney Winter published their now famous book on the subject (Nelson and Winter, 1982). Simulation made it possible to 'formalize' heterogeneous micro- and meso-agents in an economy; finally the straight-jacket of mathematical treatability could be disposed of. Experiment by simulation became the correlate to the successful experimental methods in the natural sciences[45], an area not to be

---

[41] Following Herbert Simon an evolutionary process can simply be summarized as consisting of two elements: (1) the generator of variety, and (2) a test selecting the survivors. (Simon, 1969, p. 52)

[42] Fractal analysis in the social sciences promises to improve the understanding of the build-up of self-similar structures (see (Mandelbrot, 1983), (Brown, 2010)).

[43] Evolutionary theories for the social sciences are themselves a heterogeneous variety, often using slightly different names for similar concepts. E.g. Kurt Dopfer and Jason Potts propose a general theory of economic evolution based on 'a process of coordination and change in rules.' (Dopfer and Potts, 2008, p. xii). The ideas of cooperation, conflict, and algorithmic formulations are shining through, but details how to model them are still a matter of controversy among evolutionary economists.

[44] The focus on cooperative aspects had been lost not only by those following John Nash (compare (Strathern, 2002, pp. 293-327)) but also by some biologists, e.g. Richard Dawkins, producing some semi-economic metaphors of 'selfish genes'.

[45] Compare (Hanappi, 1994, p. 171 -175) and (Hanappi, 2011) for a discussion of simulation methods.

confused with the type of experimental economics, which uses observations of reactions of human individuals in test laboratories to produce hypothesis on innate economic traits. But computer simulation methods had an even wider impact on the new Neumann-Morgenstern project. To mention a further area that really exploded due to increased computation capabilities a look at network theory is mandatory (see (Newman, 2010)). 'Games on networks' as well as the evolution of networks in the form of dynamic games are part of a scientific sub-discipline that attracts an ever growing community of researchers. Finally it is remarkable that computer support today is providing an enormous amount of socioeconomic data right to the fingertips of social scientists, a working environment that at the times of Neumann and Morgenstern simply did not exist. The missing centuries of works of a Tycho de Brahe and his colleagues who prepared the ground for theoretical physics (as explained in the introductory chapter of (Neumann and Morgenstern, 1944)) can certainly not be replaced by these technical facilities - but they can be shortened.

To enumerate all the different currents of thought and scientific sub-disciplines, which will in the near future lead to an even more visible renaissance of the Neumann-Morgenstern project goes beyond the aspirations of this author – and surely beyond the scope of this chapter. John von Neumann died in 1957, Oskar Morgenstern in 1977, approximately half a century after they left the active debate of their project, and after many decades of more or less subconscious influences of their masterpiece on the general intellectual climate, the current crisis in the social sciences seems to be prepared for a fulminate comeback of the Neumann-Morgenstern project.

Author details

Hardy Hanappi
University of Technology Vienna, Austria

6. References

Barabási L., 2002, *Linked. The New Science of Networks*, Perseus Publishing, Cambridge MA.

Bellman R., 1969, *On 'Colonel Blotto' and analogous games*, SIAM Review Vol. 11, No. 1.

Borel E., 1921, *La théorie du jeu les équations intégrales à noyau symétrique*. Comptes Rendus del'Académie 173, 1304–1308 (1921); English translation by Savage, L.: The theory of play and integral equations with skew symmetric kernels. Econometrica 21, 97–100 (1953).

Brams S., 1994, *Theory of Moves*, Cambridge University Press.

Brams S., 2011, *Game Theory and the Humanities. Bridging Two Worlds*. MIT Press, London.

Brown C. and Liebovitch L., 2010, *Fractal Analysis*. Series: Quantitative Applications in the Social Sciences, Sage, Los Angeles (CA).

Dopfer and Potts, 2008, *The General Theory of Economic Evolution*, Routledge, London.

Dore, Chakravarty, and Goodwin (eds), 1989, *John von Neumann and Modern Economics*, Clarendon Press, Oxford.

Einstein A., 1905a, *Über einen die Erzeugung und Verwandlung des Lichtes betreffenden heuristischen Gesichtspunkt*. In: Annalen der Physik. 17, pp. 132–148.

Einstein A., 1905b, *Über die von der molekularkinetischen Theorie der Wärme geforderte Bewegung von in ruhenden Flüssigkeiten suspendierten Teilchen*. In: Annalen der Physik. 17, pp. 549-560.

Gödel K., 1931, *Über formal unentscheidbare Sätze der Principia Mathematica und verwandter Systeme, I.* Monatshefte für Mathematik und Physik 38: 173–98.

Hanappi H., 1994, *Evolutionary Economics. The evolutionary revolution in the social sciences.* Avebury Ashgate, Aldershot (UK).

Hanappi H., 2011, *Signs of Reality - Reality of Signs. Explorations of a pending revolution in political economy*. Published as MPRA Paper No. 31570.

Hanappi H. and Hanappi-Egger E., 1999, *Norbert Wiener's Cybernetic Critique of the Information Society - An Update.* International Conference Cybernetics 99, proceedings published by the University of Las Palmas de Gran Canary. Web: http://ftp.econ.tuwien.ac.at/hanappi/Papers/Hanappi_Hanappi-Egger_2001.pdf

Hardy G., 1940, *A Mathematician's Apology*. Cambridge: University Press.

Hodges A., 1992 (1983), *Alan Turing: the Enigma*, Vintage - Random House, London.

Kaufman S., 1993, The *Origin of Order. Self-Organization and Selection in Evolution*, Oxford University Press.

Kovenock D. and Roberson B., 2008, *Coalitional Colonel Blotto Games with Application to the Economics of Alliances*, Discussion papers // WZB, Wissenschaftszentrum Berlin für Sozialforschung, Schwerpunkt Märkte und Politik, Abteilung Marktprozesse und Steuerung, No. SP II 2008-02, http://hdl.handle.net/10419/51118.

Kuhn H. W., 1953, *Extensive Games and the Problem of Information*, in H. W. Kuhn and A. W. Tucker (eds.), Contributions to the Theory of Games, Volume II, Princeton University Press, Princeton.

Kuhn T., 1962, *The Structure of Scientific Revolutions*, Oxford University Press.

Kurz H., 2011, *Who is Going to Kiss Sleeping Beauty? On the 'Classical' Analytical Origins and Perspectives of Input-Output Analysis*, Review of Political Economy, vol. 23(1), pp. 25-47.

Leonard R., 1994, *Reading Cournot, Reading Nash: The Creation and Stabilisation of the Nash Equilibrium*, The Economic Journal, Vol. 104, No. 424 (May, 1994), pp. 492-511.

Leonard R., 2010, *Von Neumann, Morgenstern, and the Creation of Game Theory. From Chess to Social Science*, Cambridge University Press.

Macrae N., 1992, *John von Neumann*, Pantheon Books, New York.

Mandelbrot B., 1983, *The fractal geometry of nature*, W.H. Freeman, New York.

Maynard-Smith J., 1982, *Evolution and the Theory of Games*, Cambridge University Press.

Maynard-Smith J., 1988, *Did Darwin get it Right? Essays on Games, Sex and Evolution*. Penguin Books, London.

McCain R., 2009, *Game Theory and Public Policy*, Edward Elgar, Cheltenham UK.

Mirowski P., 2002, *Machine Dreams: Economics Becomes a Cyborg Science.* Cambridge University Press, Cambridge.

Morgenstern O., 1976, *The Collaboration between Oskar Morgenstern and John von Neumann on the Theory of Games*, Journal of Economic Literature, vol. 14, no. 3, pp. 805-816.

Morgenstern O. and Thompson G. L., 1976, *Mathematical theory of expanding and contracting economies*, Lexington, Mass.: Heath, Lexington Books.

Muth J., 1961, *Rational Expectations and the theory of price movements*, Econometrica, vol. 29.

Nelson R. and Winter S., 1982, *An Evolutionary Theory of Economic Change*, Harvard University Press.

Neumann J., 1928, *Zur Theorie der Gesellschaftsspiele*, Mathematische Annalen, vol. 100, pp. 295-320.

Neumann J., 1996 (1932), *Mathematical Foundations of Quantum Mechanics*, Princeton University Press.

Neumann J., 1937, *Über ein Ökonomisches Gleichungssystem und eine Verallgemeinerung des Brouwerschen Fixpunktsatzes, Ergebnisse eines mathematischen Kolloquiums*, vol. 8, pp. 73-83.

Neumann J. and Morgenstern O., 1944, *Theory of Games and Economic Behavior*, Princeton University Press, Princeton.

Neumann J., 2000 (1958), *The Computer and the Brain*, Yale University Press.

Newman M., 2010, *Networks*, Oxford University Press.

Prigogine I., 1984, *Order out of Chaos*, Bantam, New York.

Punzo L., 1989, *Von Neumann and Karl Menger's Mathematical Colloquium*. In: (Dore, Chakravarty, and Goodwin (eds), 1989, pp. 29-68).

Roberson B., 2006, *The Colonel Blotto Game*, Economic Theory (2006) 29: 1–24.

Samuelson P., 1947, *Foundations of Economic Analysis*, Harvard University Press.

Sargent T., 1980, Macroeconomic Theory, Academic Press Inc., New York.

Schrödinger E., 1944, *What is Life?* Cambridge University Press.

Shackle G., 1967, *The years of high theory: invention and tradition in economic thought 1926-1939*, Cambridge University Press.

Shannon C. And Weaver W., 1949, *The Mathematical Theory of Communication*, University of Illinois Press.

Simon H., 1969, *The Sciences of the Artificial*, MIT Press.

Strathern P., 2002, *Dr. Strangelove's Game*, Penguin Books, London.

Turing A., 1936, *On Computable Numbers, with an Application to the Entscheidungsproblem.* Proceedings of the London Mathematical Society 42(1936): 230-65.

Tukey J., 1949, *A Problem in Strategy*, Symposium on the Theory of Games, Econometrica 17(1), (January 1949), p. 71.

Ulam S., 1958, *John von Neumann 1903-1957*, Bull. Amer. Math. Soc. 64 (1958), Part 2:1-49.

Velupillai K.V., 2011, *Towards an algorithmic revolution in economic theory*, Journal of Economic Surveys.

Weber M., 1904, *Die "Objektivität" sozialwissenschaftlicher und sozialpolitischer Erkenntnis*, Archiv für Sozialwissenschaft und Sozialpolitik. 19. Band.

Whitehead A. and Bertrand Russell, 1910, *Principia Mathematica*, University of Michigan, Ann Arbor, Michigan.

Wiener N., 1948, *Cybernetics: or Control and Communication in the Animal and the Machine*. MIT Press.

Wiener N., 1954, *The Human Use of Human Beings: Cybernetics and Society*. Houghton Mifflin, Boston.

Wittgenstein L., 1921, *Tractatus Logico-Philosophicus*, Wilhelm Ostwalds Annalen der Naturphilosophie.

Zermelo E., 1913, *Über eine Anwendung der Mengenlehre auf die Theorie des Schachspiels*, Proceedings of the Fifth Congress Mathematicians, (Cambridge 1912), Cambridge University Press 1913, 501-504.

Chapter 2

Can Deterrence Lead to Fairness?

Riccardo Alberti and Atulya K. Nagar

Additional information is available at the end of the chapter

http://dx.doi.org/10.5772/54121

1. Introduction

Game theory has been applied, in the last five decades, to the resolution of strategic situations in order to analyse rational behaviour. Following the growth of the technical apparatus, many requests of finding adequate ethical basis to the theory have been made.

Schelling, in [20], reminds what kind of contributions game theory could yield to the study of ethical system, and in a similar fashion, what ethical basis should we include into game theory.

First of all, game theory provides a well defined set of mathematical tools to formalise known ethical problems, thus it allows people with different backgrounds to approach moral issues by means of rigorous procedures. Moreover, procedures themselves, in the sense of behavioural dynamics and solution concepts, could raise methodological insights to the study of ethics.

On the other hand, game theory has its own ethical basis deriving from utilitarian morality, whereby an action cannot be judged by itself, but only its consequences define its moral value. Though Schelling warns the researchers about this simplistic conception of morality, in its most speculative examples, game theory, embraces the definition of utility from the utilitarian tradition. One may argue, as Schelling does, that a moral calculation could be at the foundation of particular allocation of payoff values or could explain why psychological tests strongly diverge from theoretic result; therefore the issue of ethical foundations in game theory is not exhaustively resolved by the utilitarian view. It is nonetheless incontrovertible that game theory has no meaningful applications when the actors are not interested in the consequences of their actions or when the final purpose of a game goes beyond the information comprised in the payoff functions.

The type of ethics that emerges from this view is called "situation" ethics by Fletcher, for it strongly depends on specific circumstances i.e. the consequences of an action, rather than on universal principles. Situation ethics is also called by other authors "retributive" ([13]). If we take for granted that game theory is founded on a situation ethics, then individual rationality finds a complete moral justification as the mean to solve conflicting situations. However, in [20], Schelling refers to Rapoport's consideration that in many cases, some form of conscience, is superior to individual rationality. Can "situation" ethics resolve every ethical issue of game theory?

It is from the uncertainty on the philosophical foundations of game theory that the issues treated in this work derive. If rationality assumes that an agent should pursue his own selfish interest, how could they be deterred to choose inefficient outcomes? How can we define any notion of fairness? Is there any connection between deterrence and fairness from, at least, a game theoretic perspective? Throughout this chapter we try to answer these questions.

In the first paragraph, we introduce the theory of deterrence, from its ethical justifications and philosophical significance ([4, 5, 13, 19, 20, 22]), to some practical applications and results ([8, 19, 20, 22]).

In the second paragraph, we expose three different approaches in the formulation of theories of justice ([10, 14, 16]) in order to have a complete understanding of the implications of justice related to fairness.

The following two sections are dedicated to the analysis of some game theoretic models which were designed to grasp the key features of deterrence and fairness; in particular the third paragraph contains a conspicuous number of examples ([3, 7, 11, 22]) about how to include deterrence within game theory and, in the fourth paragraph, we show how different authors have managed to incorporate fairness into game theoretic forms ([2, 12, 15]).

Next we introduce our personal contribution to the subject. We provide the extension to general n-person games of our theory of temporised equilibria proposed in [1], and we cite some simple applications in computer science.

We then explore the connections between our model and some fundamental results in the social choice theory ([6, 18]).

Eventually, in the last section, we draw the conclusions on our research and we include some future developments.

2. Theory of deterrence

In its simplest form, deterrence is basically an attempt by party A to prevent party B from undertaking a course of action which A regards as undesirable, by threatening to inflict unacceptable costs upon B in the event that the action is taken [22].

Such a definition of deterrence has its origins from the numerous studies made by sociologists, game theorists, psychologists and theorists of nuclear strategy during the years of the Cold War. In that period, understanding the nature of deterrence was of primary importance to design policies capable to prevent the impact of a nuclear holocaust and, at the same time, maintaining own's strategic positions over the opponent party.

Though a great variety of specifications have been proposed, the common sense of the term is widely accepted to indicate the *enforcement over a number of opponents to refrain from specific unwanted course of action in anticipation of some retaliatory response*.

2.1. Ethical justification of deterrence

The doctrine of deterrence is strongly related to the idea of punishment that one can find in psychology and pedagogical tradition. In fact a deterrent threat is realised by a shrewd manipulation of others' beliefs, in regard to some state of the world, using the pressure of a potential punishment. For it to be effective the subject making the threat must have the capability of carrying it out, and, the threat must be credible and stable (it must not prompt the undesired behaviour) [22].

Most people have agreed that the institution of punishment in the sense of attaching penalties to the violation of legal rules, is acceptable. As Rawls underlines [13], the question that rises from trying to ethically justify punishment, rotates around the justification of punishment itself and not around whether or not punishment is justifiable. To better understand this diversity, he suggests to clearly distinguish the justification of a practice from the justification of a particular action falling under a practice; if such a separation is intended then utilitarianism can be used to provide moral judgement.

There exists two criteria to advocate a punishment: the retributive view for which *it is morally fitting that a person who does wrong should suffer in proportion to his wrongdoing* [13] and, the utilitarian view for which a punishment is justifiable only if it is capable of promoting the interests of the whole society. Without the separation proposed by Rawls, it is very difficult to comprehend any act of suppression. It is a common mistake to think that if a practice is justified by the utilitarian view then each action falling under such practice will follow the same view, hence to grasp the moral implications of behaviour it is necessary to formalise the conception of practice and, to include into common knowledge, the specifications of a practice. Within this framework questioning an action that lies in a practice is equivalent to questioning the practice itself.

In [20], the author reports a very significant example, borrowed from Piaget's studies on the moral abilities of children, that summarises the argumentation of this section: *In short, kids find truth socially useful; lying is bad because the children have freely and contractually adopted a rule against it; the purpose of punishment is to deter, and to reaffirm the rule.* Punishment is justified when it is used to deter individuals from trying to prevaricate the rules of an accepted practice. Moreover if we assume that the intentions towards the definition of a practice are driven by the utilitarian view or situation ethics [20], then the position held by Fischer in [5] becomes clear. In the investigations on collective irrationality he writes that it must be considered moral to implement violent acts when they prevent collective irrationality.

2.2. Difficulties in developing a unified theory of deterrence

The theory of deterrence has been extensively developed and refined since the introduction of the notion of deterrence in the analysis of conflicting situations but still, many problems arose when researchers tried to formalise the dynamics of rational behaviour, subject to threats, into a strong axiomatic apparatus. As Downs points out in [4] a strong theory must be capable of producing reliable predictions based on some well defined information set. In this case, a prediction requires the generation of models that specify benefits and costs, the shape of utility functions, the size of the stochastic component associated with assessments of the cost and the probability of winning [4].

The main problems in achieving this goal rely on the extreme difficulty to set up objective experiments i.e. finding an unbiased relevant population from which to draw a sample, defining the duration of the experiment, grasping some general acceptance thresholds, understanding the role of uncertainties. Moreover, deducing high principles from subjective responses might lead to distorted results; in fact all the abstract models of deterrence made the assumption that every individual is perfectly rational and calculates benefits and costs of each alternatives before making a choice. This is virtually impracticable in real life where factors like culture, emotions, personality etc. enter into the decision process.

Another potential problem for the experimental approach should arise in what we should call aggregate threats. The idea comes from the argumentation of Schelling about the issue of total disarmament [19]. In a totally disarmed world, the re-arming of a nation might be a threat for its neighbours. Such a threat is ineffective on its own unless it is corroborated by the commitment to use the reacquired military power; sometimes deterrence threats must be aggregate in order to make sense and, such aggregations are difficult to identify.

Downs, recalling the work of Achen and Snidal (*Rational Deterrence Theory and Comparative Case Studies*), distinguishes two kinds of approach to a rational theory of deterrence: a weak and a strong version. If, as we have already mentioned, a strong theory should be verifiable, self contained and general, a theory of deterrence is defined weak if it is limited to the simple view of the outcome of choices as a function of expected benefits and costs and it derives by induction on specific cases.

The ultimate objective of a coherent and powerful theory of deterrence is to guarantee an effective policy design founded on necessary conditions for a certain type of behaviour to take place; a strong theory of deterrence, if implementable in Downs' view, could fulfil this requirement.

2.3. Models of deterrence

The main production of models of deterrence comes from the '60s and '70s of the 20th century, during the years of the Cold War. Economic analysis became of primary importance in the examination of costs and benefits and it did take advantage from the utilisation of game theory to investigate strategy through deterrence. Within this scenario, Kahn, in the book *On Thermonuclear War*, distinguishes two forms of deterrence: deterring an enemy's first nuclear strike and, using threat of our own first strike to deter lesser aggressions [8].

As Schelling pointed out in many of his works [19, 20], deterrence is effective only in very restrictive circumstances, namely, only when each participant rationally calculates his benefits according to a consistent system of values. Thus each model underlies the assumption of perfect rationality.

Perhaps the simplest example of such models is depicted in [8]. The problem studied by David and Robert Levine is that of a friendly country (f) deterring one enemy's (e) inimical activity. Following the notation used in [8] we have:

- $a \in [0, \bar{a}]$ is the level of inimical activity;
- $u^e(a)$ indicates the non-decreasing level of utility for the enemy;
- $u^f(a)$ represents the non-increasing level of utility for the friendly country;

- $r(a)$ is the capability of retaliatory response of the friendly country to the inimical activity level a;
- $u^e(a) - r(a)$ is the overall enemy utility;
- $u^f(a) - cr(a)$ is the overall friendly country utility, where $c > 0$ is the marginal cost of retaliation.

Assuming perfect information, the optimal solution is for the friendly country to commit to a sufficiently high level of response to any inimical activity to deter all inimical activities.

In order to relax the assumption of perfect information one can use the quantal response model developed by McKelvey and Palfrey as depicted in [8]. By including the probability density for an inimical level activity to take place, the problem of the friendly country is to maximise the utility function (U) expressed by:

$$U^f = \int_0^{\bar{a}} |u^f(a) - cr(a)| \frac{e^{\lambda(u^c(a)-r(a))}}{\int_0^{\bar{a}} e^{\lambda(u^c(a')-r(a'))} da'} da \tag{1}$$

λ is intended to be the measure of "rationality" of the enemy, i.e. $\lambda = 0$ the enemy behaves completely randomly.

From the analysis of this new model, the optimal solutions are $r(a) = 0$ and $\bar{r}$, which mean respectively "do not retaliate" and "retaliate at the maximum possible level". Hence strategies of the type all-or-nothing are indeed optimal. However, as carefully pointed out in [8], this is not always the case in real contexts. In a more complex scenario with more than one enemy, increasing penalties might change the distribution of inimical activities, incrementing the occurrence of more dangerous actions, rather than deter the wrongdoing. By using Zagare's terminology, deterrence might loose its stability.

We report another game theoretic model of deterrence that is focused on describing the logical structure of mutual deterrence. This model has been proposed by Zagare in [22]. He starts with the consideration that in an anarchic world, stability is achieved by a balance of power maintained by relationships of mutual deterrence; hence the equilibrium dynamics depends on the connections between alternative outcome. If mutual deterrence must fulfil the requirements of capability, stability and credibility then the relationship among the possible outcomes of a mutual deterrence game generates the structure of a 2×2 Prisoner's Dilemma.

Zagare reconsiders the descriptivity of the Game of Chicken, that has been the most used model to describe mutual deterrence, in favour of the Prisoner's Dilemma. In the following, we propose the main steps of his argument: we shall indicate with A and B the two players, A plays rows and B plays columns. Table 1 represents the bimatrix of a 2×2 Prisoner's Dilemma.

(a_1, b_1)	(a_1, b_2)
(a_2, b_1)	(a_2, b_2)

Table 1. 2-players Prisoner's Dilemma

We designate (a_1, b_1) as the status quo, that is the outcome from which the unilateral defection of one player is undesirable to its opponent. This means that (a_1, b_2) and (a_2, b_1) are less preferred than the status quo, and hence:

$$For\ A,\ (a_1, b_1) > (a_1, b_2) \tag{2}$$

$$\text{For } B,\ (a_1, b_1) > (a_2, b_1) \tag{3}$$

In addition, there would be no need for mutual deterrence if each player preferred the status quo to the outcome it could induce unilaterally by departing from it. Thus we have the conditions:

$$\text{For } A,\ (a_2, b_1) > (a_1, b_1) \tag{4}$$

$$\text{For } B,\ (a_1, b_2) > (a_1, b_1) \tag{5}$$

and, putting 2, 3, 4 and 5 together we obtain:

$$\text{For } A,\ (a_2, b_1) > (a_1, b_1) > (a_1, b_2) \tag{6}$$

$$\text{For } B,\ (a_1, b_2) > (a_1, b_1) > (a_2, b_1) \tag{7}$$

Zagare identifies an important feature of a deterrent threat, that is capability. A threat is said to be capable if the outcome imposed to the recipient of the threat is less desirable than the one he can obtain by a unilateral deviation from the status quo. This further consideration brings to the relations:

$$\text{For } A,\ (a_2, b_1) > (a_2, b_2) \tag{8}$$

$$\text{For } B,\ (a_1, b_2) > (a_2, b_2) \tag{9}$$

Stability requires the recipient to prefer the original status quo to the outcome associated with the threat, therefore:

$$\text{For both players},\ (a_1, b_1) > (a_2, b_2) \tag{10}$$

And eventually credibility requires:

$$B \text{ must perceive that for } A\ (a_2, b_2) > (a_1, b_2) \tag{11}$$

$$A \text{ must perceive that for } B\ (a_2, b_2) > (a_2, b_1) \tag{12}$$

Combining 6, 7, 8, 9, 10, 11 and 12 we obtain the following relations that indeed form the structure of the classical 2×2 Prisoner's Dilemma:

$$\text{For } A,\ (a_2, b_1) > (a_1, b_1) > (a_2, b_2) > (a_1, b_2) \tag{13}$$

$$\text{For } B,\ (a_1, b_2) > (a_1, b_1) > (a_2, b_2) > (a_2, b_1) \tag{14}$$

Though the simplicity of the model described, there are at least two reasons why mutual deterrence theorists have seldom incorporated the analysis of the Prisoner's Dilemma within their case studies:

- the Game of Chicken represents a worst-case scenario, thus it defines some lower bounds to the problem;
- the Prisoner's Dilemma is a pathological example that shows how individual rationality could bring to social inefficiency. Thus, for players to accept the status quo as the solution of the game will require the introduction of novel equilibrium concepts.

A broader critique on the utilisation of normal form games has also been advanced; in fact a normal form game requires players to make simultaneous decisions and assumes perfect information while in a mutual deterrence scenario actions are more likely to be sequential and conditional.

To overcome the last problem Brams and Wittman in their article *Nonmyopic Equilibria in* 2×2 *Games*, cited in [22], introduce the concept of nonmyopic equilibrium (quoting from [22]).

1. both players simultaneously choose strategies, thereby defining an initial outcome of the game, or alternatively, an initial outcome or status quo is imposed on the players empirical circumstances;
2. once at an initial outcome, either player can unilaterally switch his strategy and change that outcome to a subsequent outcome;
3. the other player can respond by unilaterally switching his strategy, thereby moving the game to yet another outcome;
4. these strictly alternating moves continue until the player with the next move chooses not to switch his strategy. When this happens the game terminates, and the final outcome is reached.

In such a practice, recalling Rawls, the nonmyopic equilibrium strategy for the Prisoner's Dilemma will be both stable and socially rational.

3. Fairness and justice

The fundamental idea in the concept of justice is that of fairness [14]; in order to understand the characteristics of fairness, it is, then, indispensable to have a clear comprehension of justice at least in its moral significance. Many philosophical apparatus have been developed to describe justice, but we identified three authors, the work of whom is considered to be seminal for subsequent refinements. These authors are Rawls [14], Nozick [21] and Otsuka [16]. Though their views strongly disagree on many levels, they all share the same assumption that justice is possible only through fairness; the object of fairness is what differentiates the three approaches.

3.1. Justice as fairness

Rawls in his article *Justice as fairness* [14] links justice to the notion of practice. The idea of practice, as we have seen in the paragraph on deterrence, plays a central role in Rawl's philosophy, for it allows the author to distinguish between two levels of moral reasoning. Thus fairness must be referred to the institutions that provides rights and duties rather than to the individual behaviour. In a similar fashion to the principle of equal consideration of interests, depicted by Singer in *Practical Ethics*, the object of Rawls' fairness is equality of opportunities. In fact, for a practice to ensure justice, the following principle must be met:

- every person who participates in a practice must possess the same amount of freedom;
- inequalities are arbitrary unless they serve to a common good.

In addition, a practice is fair if each participant would benefit from it or when all participants are acting as any other participant would do in similar conditions. The utilitarian influence is quite evident in Rawls' scheme and, as we shall see in the following sections, it has indeed found some interesting implementations in game theory, i.e. fairness equilibrium and Kantian equilibrium.

Since an individual is free when he makes the choice of participating in a practice, he voluntarily bounds his self-interests envisaging some higher common benefit; by accepting the condition of a practice, a participant is requested to behave fairly, in the sense that he cannot expect a principle not to be adopted by others if he is not himself disposed to renounce to it. This notion is called "fair play", and its application will become more evident when placed within a game theoretic framework.

Eventually, Rawls' conception of fairness as justice implies a radical view of a person; people's talents do not belong to them and, hence, one can benefit from his own talents only when every person in a practice can benefit from them. Such assumptions, as we explain in the next section, might represent a failure to treat people as equals, since the disadvantages have partial property rights in other people. If this is so, then the distribution of the products of talents might be unbalanced generating an unfair restriction on individual freedom.

3.2. Justice as individual rights

The apparent violation of people's freedom, resulting from Rawls' conception of a person, is morally unjustified in Nozick's view [10]. Nozick founds his philosophy on the idea of self-ownership and individual natural rights and argues that if a person own himself, then he is the owner of his talents and of the products deriving from the application of them. It is a strong form of libertarianism that provides absolute rights over one's properties. Those rights are indispensable for, by taking them away, one is limited in his options and in the pursuit of a self determined way of life.

Though it seems perfectly reasonable that each individual could have some rights on his properties, Nozick's conception may as well generate inequalities in the distribution of public goods that are unjustifiable on a moral level.

One fundamental aspect of Nozick's theory of justice, that plays and important role in linking deterrence to fairness, is the concept of state. A state is an organisation that threats its citizens to use violence if they do not follow the regulations. A night-watchman state, [21], is then morally legitimate to use force to deter its citizens. In this fashion, it is clear, why Nozick limits centralised justice to the restrictive use of force. In paragraph 6 we shall see how this conception of unbalanced distribution of rights on the use of force could be used to implement fairness from strategic situations.

3.3. Fairness as common ownership

To complete our exposition about the perspective adopted in the definition of a coherent theory of justice, we expose an intermediate position to Rawls' and Nozick's approaches: left-libertarianism. Treated by Otsuka in 1998, it has gained a growing consensus in the last decade [16]. Besides, individuals are self-owned, but they can exercise a property right on some natural resources only with the common consensus of others.

Some difficulties with this egalitarian view may arise when an individual with a scarce productivity would receive the rights to own a bigger part of the world; such a distribution of resources might damage the capability of high efficient producers that would need resources for their work. In this case the assumptions of a fair distribution are violated. One possible way out is to establish a flow of transaction from the capables to the disadvantages, in order to balance the inequalities. Risse, in [16], calls these transaction "solidarity", and utilise the concept as a moral justification for common property.

Eventually, in our view, common ownership and self-ownership together give the most comprehensive definition of fairness.

4. Game theory of deterrence

In the paragraph about deterrence, we have presented two game theoretic models that have been extensively used to study the dynamics of players' behaviour under the influence of some deterrent threat. As Myerson underlines in [11] games are to be considered as simplification of life, thus the understanding of real scenarios through game theoretic models must be accompanied by a process of interpretation. Players are intelligent in the sense that they have a prefect understanding of every aspect of the game and are rational in the sense that they will always choose the action that maximises their individual payoff.

For historical reasons, the strongest argumentations and the most comprehensive models of deterrence were first proposed during the years of the Cold War, hence it is not surprising that the great number of analogies and stories associated to such models are borrowed from peculiar scenarios of those years.

4.1. Strategic form games

Herman Kahn, in his book *On Thermonuclear War*, was the first to link a nuclear crisis to the strategic behaviour of players in the game of Chicken. Table 2 reports the matrix form of the game. Each player is given two choices "cooperate" and "defect". We assume player A plays rows and player B plays columns.

<Draw>	<A wins>
<B wins>	<Disaster>

Table 2. Game of Chicken

For each player i, preferences can be ordered as follows: (1) i wins, (2) Draw, (3) i loses, (4) Disaster. If both cooperate then the game will end up in a draw and if neither player is willing to cooperate than the result is an escalation to nuclear disaster. If only one player behaves cooperatively, it can be exploited by the opponent to his advantage.

As we mentioned in the first paragraph, Zagare criticises the usage of this game form as a model for deterrence and he proofs how the Prisoner's Dilemma might be more suitable. But either models cannot overcome the limits of their theoretic formulation. The critique, is not directly addressed to the game of Chicken or to the Prisoner's Dilemma, but rather to the application of the theory of non-cooperative normal form games to the issue of deterrence. It is the unrealistic assumption that once a strategy has been chosen, players are not given a chance to reconsider and to promote a more compromising attitude.

To overcome the discrepancy with real life, game theorists have developed models based on different type of game forms.

4.2. Extensive form games

The models based on the class of extensive form games allow the presence of sequential move. A common situation studied within this framework provides the presence of two players that we may call *challenger* and *defender*. In the simplest analysis, if the challenger challenges, the defender may resist or submit and if the challenger waits, the status quo continues. Again, a weak defender is one who prefers to submit rather than defend and a strong defender is one who prefers to defend rather than submit. For each type of defender there exists only one solution. When the defender is weak, a challenge is always followed by a submission; when the defender is strong, the challenger waits. Though its sequentiality, this model still has no deep insight into the problem, especially because it makes the assumption of perfect information.

One alternative is to employ a model inspired to the Cuban missile crisis in the 60's. Such a model, called Hawks and Doves, requires the presence of two type of behaviour: a "hawk" behaviour, for both challenger and defender, is assumed to escalate to war with disastrous consequences for both sides, and, a "dove" behaviour from the defender will yield a compromise with a "dove" challenger and a victory to a "hawk" challenger. Eventually a "hawk" defender will win against a "dove" challenger.

This game allow us to characterise the credibility of a threat by using a purely game theoretic argument. In fact, only the subgame perfect equilibria form a credible threat [7]. Figure 1 represents the sequential game of Hawks and Doves, with arbitrary payoff values.

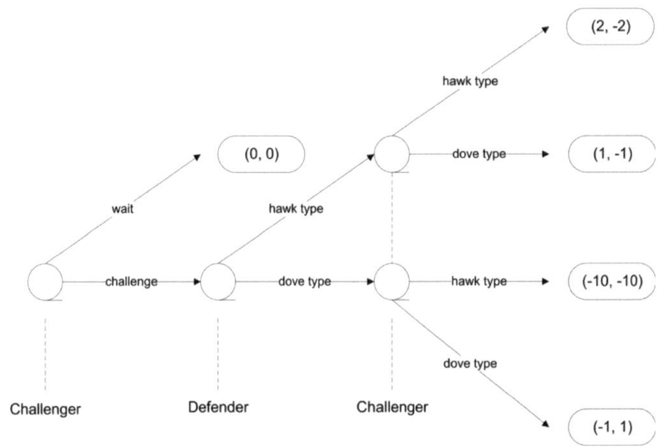

Figure 1. Hawks and Doves

One might complexify the situations represented by this model by adding uncertainty to the knowledge of the challenger about the defender's true preferences. In this way, the challenger never knows which type of defender he is facing, he can only form beliefs about it. The equilibria in the modified example are simply perfect Bayesian since strategies are optimal given beliefs and, beliefs are consistent with strategies on the equilibrium path. We shall say that deterrence has been successful when the challenger is deterred from making demands by the expectation that the defender will always resist so that he will have to backdown with some high probability.

In *Nuclear Deterrence Theory*, Robert Powell proposes another model that grasps the feature of deterrence as being a strategy that leaves something to chances. The idea behind this model is again borrowed from the Cuban missile crisis. In this model, Powell describes a crisis as a ladder of escalation steps that the two sides can in turn climb. Each time a further step is taken, an autonomous risk of nuclear escalation rises. At each step, players are provided with three choices: to surrender, to attack by conducting a full scale nuclear strike and, to take a further step on the escalation ladder. We shall take this example as a strong metaphor for the dynamics of deterrence.

4.3. Repeated Games

Deterrence is a major theme of game theory since, recalling the utilitarian view of rationality, a player's action is measured on the utility it brings rather than on *a-priori* ethical issue related to the action itself. Hence it is not surprising that only a rational decision maker is likely to be moved by deterrent threats.

The theory of repeated games might help us in the analysis of this utilitarian feature of players.

We shall go back to the bimatrix associated to the Prisoner's Dilemma 1 in paragraph 1. As already pointed out in [22] and again underlined in [11], though each player has a deterrent strategy (a_1, b_1) that motivates his opponents to act cooperatively, the issue of credibility is raised from the allocation of payoffs that occurs in this particular example. When B is cooperative (b_1), player A would get a_1 by doing the same as B, but, recalling equation 4, A could get a higher payoff by playing a defective strategy (a_2). Therefore A prefers not to follow his own deterrent strategy when B cooperates, unless A constrains B to follow A's deterrent strategy. But in a pure strategic environment where no communication is allowed, this may not happen thus the credibility of A deterrent threat is compromised. While Zagare bounds his analysis to the explanatory features of the Prisoner's Dilemma, Myerson, in [11], pushes the equilibrium analysis further, considering the implications related to the repeated game.

In such a scenario the concept of reputation becomes even more important. We shall define reputation as the attitude of a player towards a specific strategy, i.e. if A plays the cooperative strategy in a certain number of games, then he gains the reputation of using such strategy.

Let us suppose that A has the reputation of using the strategy "do same as B" against which B plays the cooperative strategy. This will lead to the Pareto efficient outcome. If A changes his strategy and loses his reputation, then the game will eventually end up in the Nash equilibrium which is worse off for both.

The main conclusion one can draw from the repeated game analysis is that the only way to maintain a deterrence threat credible is to keep a sufficiently high level of reputation for this will cause the solution of the game to be the deterrent strategy for both players, at least on the long run.

So far we have introduced classical game theory models and their implications to the theory of deterrence. In the next section we discuss a kind of games, in the class of qualitative games, that are expressly developed for the study of deterrence.

4.4. Games of deterrence

In [3], the authors introduce the concept of games of deterrence as inheriting its key structural characteristics from Isaac's attempt to develop a consistent theory of qualitative games. In fact, instead of dealing with real valued, continuous functions over the product space of individual strategy, each player is provided with a binary valued index that maps a point from the joint strategy space to the set $\{0, 1\}$. An outcome is unacceptable if it is labelled with 0 and it is acceptable if it is labelled with 1. Obviously a rational player will look for an acceptable outcome.

We can distinguish three types of strategies:

1. "no risk": a strategy that guarantees that a player ends up in an acceptable outcome, whether his opponent is rational or not;
2. "limited risk": a strategy that guarantees that a player ends up in an acceptable outcome, as long as his opponent is rational;
3. "high risk": a strategy that give a player an unacceptable outcome, whether his opponent is rational or not.

If a player has no "high risk" strategies, then his strategies are termed positively playable; if he has no positively playable strategies, then his strategies are said playable by default. Eventually, a strategy which is not positively playable nor playable by default is termed non playable. Table 3 represents a simple example of game of deterrence. We assume player E plays rows (e_1, e_2) and player R plays columns (r_1, r_2).

$(1, 0)$	$(1, 1)$
$(1, 1)$	$(1, 1)$

Table 3. Game of Deterrence

Both E's strategy and r_2 for R are "no risk" and playable, and r_1 for R is not playable.

The idea behind the solution concept for this type of games is rather simple: since a rational player will always select an available strategy of, at most, "limited risk", it follows that an equilibrium point is any pair of playable strategies. If we associate to a strategy x a positive playability index $J(x)$ such that, $J(x) = 1$ if x is positively playable and $J(x) = 0$ if not, a solution is, obviously, the set $J(x)$.

Another important feature, of this particular class of games, is the introduction of the *graphs of deterrence*. A graph of deterrence is a bipartite graph such that an arc with origin in x and extremity in y represents the fact that the player who selects x will obtain 0 in the joint strategy (x, y). In reference to 3, the graph of deterrence should be:

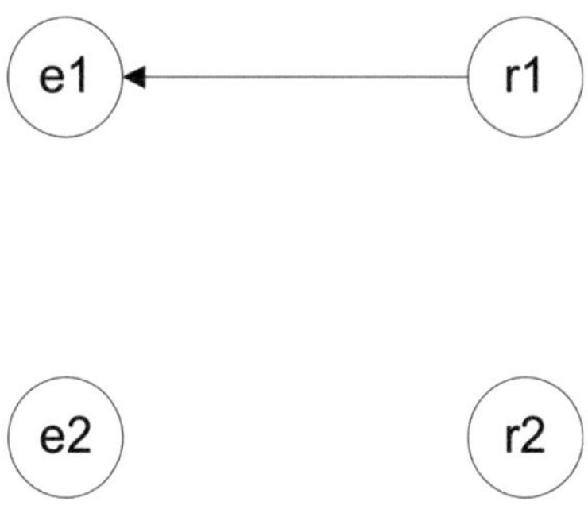

Figure 2. Graph of deterrence of game 3

The games of deterrence have been effectively used to model business processes and network congestion problems.

5. Games of fairness

Most economic models are based on the assumption that rational agents purse their own selfish interests and do not care about any common outcome. Such a behaviour is evident in various examples like the Tragedy of the Commons and the Prisoner's Dilemma.

The Standard View, as Ambrus-Lokatos calls it in [2], considers games as purely means to understand the dynamics of rationality in strategic situations, thus it is not surprising that inefficient solutions can (often) arise. However, as many psychological experiments have remarked, individuals, sometimes act in ways that diverge from the game theoretic expectations; emotions, culture and, in particular, altruism have received some consideration in the realisation of economic and psychological models.

5.1. Fairness as a psychological evidence

One approach to incorporate some sort of moral reasoning into game theory, is closely related to the studies on altruistic behaviour made in psychology. It appears that altruism is more complex than "acting as to benefit the well-being of the others"; in fact, people are inclined to adopt an altruistic behaviour with other who are altruistic to them and tend to hurt those who hurt them. In addition, individuals are more inclined to reduce their well-being in favour of some social goal when they believe that the other will do the same. Such a position implies a form of fairness that the game theoretic rational behaviour cannot incorporate. Rabin, in [12], develops a framework for including this kind of factors within a coherent mathematical model. His work is based on three assumptions that are corroborated by a long series of experiments:

- individuals are inclined to be kind with those who are kind with them;
- individuals are inclined to be aggressive with those who are aggressive with them;
- the smaller the cost of an "emotional behaviour" the greater the effect of such a behaviour on strategic choices.

To have a clearer view let us consider the strategic form of the Ultimatum Game.

A proposer offers a decider to split an amount of money X, if the decider is satisfied, than the money are split according to the offering, otherwise they both get no money. The canonical equilibrium for this game is for the proposer to always propose no more than the smallest unit of X (let us say 1 č) and, for the decider, to always accept it.

This solution is far from being fair, and the behaviour of the proposer is rather aggressive. Rabin's assumptions are capable to address this issue. In fact they do. As resulted from many experiments, [2], the response of the decider to the aggressiveness of the proposer is to not accept the offering and adopt an aggressive behaviour.

Moreover, when the value of the smallest unit of X increases (let us say from 1č to 1000č), then the effect of the "emotional" behaviour on the game outcome becomes almost insignificant. This result, is a *de facto* proof of the third assumption, for which an individual in more inclined to pursue his own interest in proportion to the profitability of doing so.

To formalise the concept of fairness in 2 × 2 games, Rabin's model includes elements from the Geanakoplos, Pearce and Stacchetti framework on psychological games. The utility function depends on players' beliefs as well as on their actions following the relation:

$$U_i(a_i, b_j, c_i) = \pi_i(a_i, b_j) + f_j(b_j, c_i)(1 + f_i(a_i, b_j)) \quad (15)$$

where:

- a_i is the strategy of player i;
- b_j represents his beliefs about what strategy j will play;
- c_i indicates i's beliefs about what strategy j believes that i will play;
- $\pi_i(a_i, b_j)$ is the "material payoff";
- f_l is the kindness function that measures how kind i is to j ($f_i(a_i, b_j)$) and how kind i perceives j to i himself ($f_j(b_j, c_i)$).

A pair (a_1, a_2) is defined a fairness equilibrium if a_i maximise $U(a_i, b_j, c_i)$ for each player and if $a_i = b_i = c_i$.

Though some authors, in particular Ambrus-Lakatos in [2], have pointed out that Rabin's consideration of players' beliefs are quite unrealistic, it might be useful to view how Rabin's model can be applied to the resolution of common games.

Recalling the Prisoner's Dilemma bimatrix from Table 1, and applying the theory of the games of fairness, we shall observe that the common Nash equilibrium (defect, defect) is also a fairness equilibrium. It is easy to show how the Pareto efficient outcome (cooperate, cooperate) is as well a fairness game. From [12], if it is common knowledge that both players are playing the Pareto efficient outcome, then each player is aware that the other is being kind to him for he renounces to a higher payoff he could get by unilaterally deviating from this strategy. Thus, by means of the first assumption, each player is inclined to play cooperatively. It can be shown that, as long as the gains from deviating is not too large, the strategy (cooperate, cooperate) is a fairness equilibrium.

Furthermore this conclusion seems to grasp a reasonable aspect of social interaction, for an individual is more inclined to cooperate when he is confident that others will not defect.

Another interesting aspect is how the behaviour of a player changes when he is responding to constrained choice. Rabin considers a degenerate version of the Prisoner's Dilemma to be as follows:

(a_1, b_1)
(a_2, b_1)

Table 4. Degenerate Prisoner's Dilemma

The outcome in this example are ordered, for player A, in accordance with 4. Player B will always defect, forcing the game to end up to the solution (defect, cooperate) which is the only fairness equilibrium. As pure altruism is excluded from this model, i.e. B does not choose to cooperate but he is forced to do so, the dynamic of A's response changes; A does not have any moral obligation towards B and hence he maximises his own payoff.

In our view, the idea of fairness equilibrium, could be employed as an additional feature of canonical solution concept. Let us now consider the game of Chicken. In table 5, we provide a numerical example of the game represented by 2; player A plays rows and player B plays columns and, each player can choose to dare (first row/columns) or chicken (second row/columns):

(-3X, -3X)	(2X, 0)
(0, 2X)	(X, X)

Table 5. Numerical example of the game of Chicken

This game has 2 pure strategy Nash equilibria (dare, chicken) and (chicken, dare) and it can be shown that both are not fairness equilibria [12].

As one may notice from the above solution, the set of fairness equilibria is not a subset nor a superset of the set of Nash equilibria. Hence we suggest that the concept of fairness equilibrium could be considered as a *ex-post* characterisation of the Nash equilibrium. In fact, a Nash equilibrium point that is also a fairness equilibrium, is the solution in which rational and "emotional" behaviour conciliate.

5.2. Fairness as a philosophical necessity

So far, we have associated the notion of fairness to the framework of game theory by means of psychological arguments on individual's behaviour. In [15] a completely different approach is proposed; indeed Roemer's idea is to describe fairness as a moral necessity rather than an "emotional" calculation. The key concept of this view is borrowed from Kant's categorical imperative: *one should take those actions and only those actions that one would advocate all others take as well.* Specifically, Roemer is interested in understanding how players make their choices when facing proportional alternative joint strategies.

Given these premises, the definition of a solution concept is quite straightforward, in fact, if no player has incentives in realising an outcome in which all players' strategies are altered by the same multiplicative factor, then the game is in a Kantian equilibrium.

For the interests of our argumentation, we shall investigate the Kantian solution when Roemer's concepts are addressed to the study of the Prisoner's Dilemma. In this manner the reader may have an exhaustive picture of different approaches to the same class of problems. For this purpose we shall use a modified version of Table 1.

$(1, 1)$	(a_1, b_2)
(a_2, b_1)	$(0, 0)$

Table 6. Modified Prisoner's Dilemma

with $a_1 = b_1 < 0, a_2 = b_2 > 1$.

Once again, the point of interest is the cooperative solution (1, 1). In [15], the author proves that that strategy is indeed a Kantian equilibrium if and only if the relation $(a_1 + b_2) \leq 2$ holds. Moreover, if this is the case, (cooperate, cooperate) is also the only non-trivial equilibrium.

The Kantian equilibrium framework has been used to model different classic problems in game theory, such as, public-good economies, oligopolist markets and redistributive taxation, however, we are more interested in understanding the motivations behind the concept. The Kantian solution differs, in principle, from the reasons behind the implementation of "pure altruism", i.e taking on the preferences of others; it is a rather individual moral necessity to behave in a way such that it maximise oneself payoff when all others behave in a similar fashion [15].

As we did for fairness equilibrium, we suggest to employ Roemer's equilibrium to calculate the moral efficiency of strategic solutions; in fact when a Nash equilibrium is also a Kantian equilibrium, the outcome has rational and ethical validity.

In the next section, we will propose our contribution to the field. After providing the motivations for our model, we will discuss the mathematical details of the concept of equilibrium.

6. Deterrence as a mean of fairness

In [1], we have presented the concept of *temporised equilibria* in the case of 2×2 games. Within this framework we claimed the possibility to incorporate fairness into game theory by means of deterrence. What we propose in this paragraph is the extension to our model to the general case of n-person games with an arbitrary (finite) number of strategies.

In its simplest form, the element of a temporised game are: a *judge* who provides the *deterrent* information and a *parametric function* from which each player derives his optimal behaviour. We recall the definition of judge, that resembles in many ways the idea proposed by Schelling in [19] of a centralised organisation as a democratic way to monitor nuclear illicit actions, or as envisaged by Nozick in [21] as the only rational sustainable type of state.

The judge is an entity that is external to the game; he is absolutely reliable and trustworthy and he provides the deterrent information to the players [1].

Our model is inspired by Freud's theory of mind. In his works, Freud considers the human mind as divided into three main element: the id, the superego and the ego. While the id and the superego are completely irrational, the ego serves to balance the demands of the id against those of the superego by realistically assessing the limits imposed by the real world. The ego serves an executive function to maximise the benefits to the whole person.

In our view, there are no substantial differences in the justifications that drove Rousseau to develop contractualism, Freud to define his model of mind, Roemer to produce the concept of the Kantian equilibrium and our framework. They all share the idea of the presence of a *meta* concept that does not directly interact with players, but radically changes, at least in some examples, their strategic behaviour. The novelty or our work is that of linking deterrence to fairness in a game theoretic model.

6.1. Mathematical structure

We provide the extension of the theory of temporised equilibrium to general coupled constrained n-person games. We use a mathematical apparatus in which the temporised solution to 2×2 k-game can be derived as a case of the general theory.

Rosen, in [17], shows that when every joint strategy lies in a compact region R in the product space of the individual strategies and that each player's payoff function ϕ_i, $i = 1, \ldots, n$, is concave in his own strategy, then an equilibrium always exists. Besides if the payoff functions satisfy some additional concavity properties then the equilibrium is also unique. These conclusions are valid both for orthogonal constraint sets (R is the direct product of the individual player's strategy spaces S) and for general coupled constrained sets (R is a subset of S).

In our framework, the coupled constrained set is defined by a system of constraints

$$C_j(\phi_j(x)/M_j, \phi_{j+1}(x)/M_{j+1}, k_j) = 0, \tag{16}$$

where:

- $k_j \in \mathbb{R}/0$;
- $M_j \neq 0$ is the maximum value for $\phi_j(x)$;
- $j = 1, \ldots, n-1$.

C_j expresses the weighted parametric difference between consecutive pairs of payoff functions. A game, in which all constraints are regular and have a minimum number of explicit forms that are continuous maps of compact subsets of $[0,1]^l$ into subsets of $[0,1]^{m-l}$, is called k-game.

6.1.1. *Definitions and existence of the equilibrium points*

The n-person k-game to be considered is described in terms of the individual strategy vector for each of the n players. A strategy profile $x_i \in \mathbb{E}^{m_i}$ is defined for each player $i = 1, \ldots, n$ and the vector $x \in \mathbb{E}^m$ denotes the concurrent strategies of all players where $\mathbb{E}^m$ is the product space $\mathbb{E}^m = \mathbb{E}^{m_1} \times \mathbb{E}^{m_2} \times \ldots \times \mathbb{E}^{m_n}$ and $m = \sum_{i=1}^{n} m_i$. The allowed strategies will be limited by the requirements that x must be selected from the graph of a continuous differential map which is the explicit form of the constraints C_j in some of their variables.

If the payoff function of the ith player is continuous in all the variables and is concave in the ith set of variables for fixed values of the other sets of variables, then these conditions should be satisfied in the convex compact set S, which is the product space of the projections of R onto the subspaces containing the variables of each player. It is evident that S contains R so that, in general, continuity and concavity are required also outside R, that is, outside the investigation region of interest. Figure 3 represents a two players coupled constrained strategy set.

The payoff function for the ith player depends on the strategies of all other players as well as on his own strategy: $\phi_i(x) = \phi_i(x_1, \ldots, x_i, \ldots, x_n)$. It will be assumed that for $x \in S$, $\phi_i(x)$ in continuous, differentiable and is quasi-concave in x_i for each fixed value of opponents strategies. With this formulation an equilibrium point of the n-person k-game is given by a point $x^0 \in R$ such that

$$\phi_i(x^0) = \max_{t_i}\{\phi_i(x_1^0, \ldots, t_i, \ldots, x_n^0) | (x_1^0, \ldots, t_i, \ldots, x_n^0) \in R\} \tag{17}$$

$$(i = 1, \ldots, n) \tag{18}$$

At such a point no player can increase his payoff by a unilateral change in his strategy in R.

The result to follow make use of the notation $x = \langle z_1, \ldots z_{m-n+1}, y_1, \ldots y_{n-1} \rangle$ to indicate the concurrent strategy vector. y_i are the elements of x that constitute a $(n-1)$-dimensional

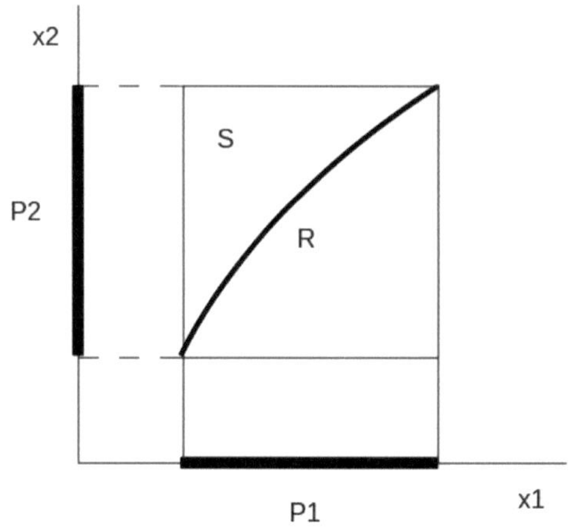

Figure 3. Coupled contrained set with $n = 2$

manifold. Obviously if $z_i \in Z$ and $y_i \in Y$ then $S \supset Z \times Y$. We now prove the equilibrium existence theorem for a n-person k-game.

Theorem 1. *An equilibrium point exists for every n-person k-game.*

Proof. The system of constraints C is an implicit vector-valued function which defines a map from $W \subseteq S \subset \mathbb{R}^{w+n-1} \longrightarrow \mathbb{R}^{n-1}$, following the expression:

$$C(x): \begin{cases} C_1(x) = \phi_1(x)/M_1 - k_1\phi_2(x)/M_2 = 0, \\ \cdots \\ C_l(x) = \phi_l(x)/M_l - k_l\phi_{l+1}(x)/M_{l+1} = 0 \\ \cdots \\ C_{n-1}(x) = \phi_{n-1}(x)/M_{n-1} - k_{n-1}\phi_n(x)/M_n = 0 \end{cases} \tag{19}$$

By definition of k-game, C fulfils the requirements of Dini's theorem; thus $C = 0$, $C = (C_1, \ldots, C_{n-1})$ is solvable in one of its variables.

Formally, if C is differentiable in W and the Jacobian $\partial(C_1, \ldots, C_{n-1})/\partial(y_1, \ldots, y_{n-1})|_{(z^0, y^0)} \neq 0$, then there are neighbourhoods U and V of $z^0 \in \mathbb{R}^w$ and $y^0 \in \mathbb{R}^{n-1}$, respectively, $U \times V = W$ and $w + n - 1 = m$, and a unique mapping $g : U \longrightarrow V$ such that $C(z, g(z)) = 0 \in \mathbb{R}^{n-1}$ for all $z \in U$. Here $g(z^0) = y^0$ and moreover g is differentiable on U. Let exist a compact set $D \subseteq int(U)$ such that $g(d) \in G_d \subseteq int(V), \forall d \in D$ and $D \times G_d \cap S \neq \emptyset$; G_d is compact since it is the image of a continuous function on a compact set. By Tychonoff's theorem $D \times G_d = R^z$ is also a compact set.

The payoff functions are continuous and derivable on R^z, hence by the extreme value theorem they must attain their maximum and minimum value, each at least once.

If $\bar{z} = \arg\max_z \phi_i(z, g(z))$, $i = 1, \ldots, n$ then $(\bar{z}, g(\bar{z})) \in R^z$ is a *temporised* equilibrium point satisfying 17. For suppose that it were not. Then ϕ_l attains its own maximum in $(\bar{z}, g(\bar{z}))$ but $\phi_{l'}$ does not, for some $l = 1, \ldots, n-1$; there would be a point $(\tilde{z}, g(\tilde{z})) \in R^z$ such that $\phi_{l'}(\tilde{z}, g(\tilde{z})) > \phi_{l'}(\bar{z}, g(\bar{z}))$, but this condition violates constraint C_l, therefore $(\tilde{z}, g(\tilde{z})) \notin R^z$. □

We need to focus our attention on the behaviour of g in a limited part of the definition space. Theorem 1 proofs that C defines an implicit function that can be made locally explicit in one of its vector-variables. When such a function exists it is continuous in some open set near a solution point. Let us call SC the set of all C's solution points that fulfil Dini's Theorem requirements; extending the notation from Theorem 1, we indicate with R^s any maximal subset of SC. Therefore for every $s = \langle s_1, \dots s_{m-n+1}, u_1, \dots u_{n-1} \rangle \in SC$ then $(s, g(s)) \in R^s$ for at least one R^s. R^s is called the *local graph* of s. Moreover if $e, t \in SC$ and $(e, g(e)) \in R^e$, $(t, g(t)) \in R^t$ and $e \neq t$ then $R^e \cap R^t = \emptyset$; finally the *global graph* is defined as $R = \bigcup_s R^s$. Following this formalism we can give an alternate definition of a n-person k-game: a n-person game is a k-game if $R \cap S \neq \emptyset$.

6.1.2. Uniqueness of the equilibrium point

In order to discuss the uniqueness of an equilibrium point we must describe the regularity condition of the constraints C more explicitly and discuss the properties of R. Let J^y be the square matrix of the partial derivatives in the y's; in Theorem 1's proof it has been shown that the gradients of the constraints C are linearly independent since $det(J^y) \neq 0$. This is a sufficient condition for the satisfaction of the Kuhn-Tucker constraint qualification. Moreover, if c_l and r_l represent J^y lth's column and row, respectively, from $det(J^y) \neq 0$ it follows that for every $l = 1, \dots, n-1$, if $c_l = \bar{0}$ then $r_l \neq \bar{0}$ and vice-versa. This means that at least one element from each row must be non-zero. We can now state the following:

Theorem 2. *If $(\bar{z}, g(\bar{z}))$ is an equilibrium point satisfying 17 then it is unique in its local graph.*

Proof. Let us suppose that $\frac{1}{M_h} \partial\phi_h(z,y)/\partial y_p - \frac{k_h}{M_{h'}} \partial\phi_{h'}(z,y)/\partial y_p \neq 0$ is the J^y's element at the hth row and pth column. Obviously $\partial\phi_h/\partial y_p$ and $\partial\phi_{h'}/\partial y_p$ cannot be both zero, hence let us assume that $\partial\phi_h/\partial y_p \neq 0$.

It follows that $\nabla\phi_j \neq \bar{0}$ for some y_j and hence at least $(n-1) \mod 2$ payoff functions are strictly monotonic in some R^s for $j = 1, \dots, n-1$. A function which is quasi-concave and monotonic is strictly quasi-concave. Suppose $(w, g(w)), (\bar{z}, g(\bar{z})) \in R^s$ are both maximizers of ϕ_j. Then $w \neq \bar{z}$ implies $\phi_j\left(\frac{1}{2}w + \frac{1}{2}\bar{z}, g\left(\frac{1}{2}w + \frac{1}{2}\bar{z}\right)\right) > min\{\phi_j(w, g(w)), \phi_j(\bar{z}, g(\bar{z}))\} = \phi_j(w, g(w))$ (definition of striclty quasi-concavity) which means that w is not a maximizer of ϕ_j. Consequently each stricly monotonic payoff function attains its own maximum in one unique point $(\bar{z}, g(\bar{z})) \in R^s$, while the other functions are in indifference points. □

We are interested in finding conditions for global uniqueness that translates into uniqueness of the temporised equilibrium point. In order to ensure global uniqueness the following must hold:

Theorem 3. *If $(\bar{z}, g(\bar{z}))$ is an equilibrium point satisfying 17 and SC is a closed set then the equilibrium point is globally unique.*

Proof. Let us consider the local graph R^s at some point $s \in SC$. If SC is closed the maximal closed subset of SC is SC itself, hence there exists a local graph R^s which coincides with SC and $R^s = R$. Applying Theorem 2 to this case, we conclude that there is an equilibrium point $(\bar{z}, g(\bar{z})) \in R$ satisfying 17. The equilibrium point is globally unique for R is the global graph of the game. □

From these conclusions the succeeding holds:

Lemma 1. *The equilibrium point* $(\bar{z}, g(\bar{z}))$ *is an element of* $bd(R^s)$.

Proof. The proof comes directly from the proof of Theorem 2. Considering every smooth and strictly monotonic ϕ_i in the compact set R^s, if there is a maximum $(\bar{z}, g(\bar{z})) \in int(R^s)$ then there exists an arbitrary small quantity ϵ such that $(\bar{z}+\epsilon, g(\bar{z}+\epsilon)) \in int(R^s)$ and $\phi_i(\bar{z}+\epsilon, g(\bar{z}+\epsilon)) > \phi_i(\bar{z}, g(\bar{z}))$.

But this contradicts the hypothesis for which $(\bar{z}, g(\bar{z})) \in int(R^s)$ is a maximum. Hence $(\bar{z}, g(\bar{z})) \in bd(R^s)$. □

Obviously if the hypotheses of Theorem 3 hold then the equilibrium point is an element of $bd(R)$.

6.1.3. 2×2 *games*

In addition to the properties exposed in the previous sections there are some peculiar characteristics that holds for this specific class of games. Figure 7 represent a general 2×2 game.

(a_1, b_1)	(a_2, b_2)
(a_3, b_3)	(a_4, b_4)

Table 7. General 2×2 game

Proposition 1. *The class of* 2×2 *k-games is closed with respect to translation when* $k = \frac{M2}{M1}$.

Proof. C's generic expression for a 2×2 k-game is:

$$zy\left(\frac{a_1 - a_2 - a_3 + a_4}{M_1} - k\frac{b_1 - b_2 - b_3 + b_4}{M_2}\right) + z\left(\frac{a_2 - a_4}{M_1} - k\frac{b_2 - b_4}{M_2}\right) + \\ +y\left(\frac{a_3 - a_4}{M_1} - k\frac{b_3 - b_4}{M_2}\right) + \frac{a_4}{M_1} - k\frac{b_4}{M_2} = 0 \tag{20}$$

For $k = \frac{M2}{M1}$ we can multiply equation 20 by M_1 obtaining:

$$zy\left(a_1 - a_2 - a_3 + a_4 - b_1 - b_2 - b_3 + b_4\right) + z\left(a_2 - a_4 - b_2 + b_4\right) + \\ +y\left(a_3 - a_4 - b_3 + b_4\right) + a_4 - b_4 = 0 \tag{21}$$

It is easy to notice that if we add or subtract a positive quantity T to 21 the equation remains unchanged since in every pair the same quantity is added and subtracted. □

In [1] it has been demonstrated that every symmetric 2×2 game is a k-game and, by definition, $k = 1$. In a symmetric game $M_1 = M_2$ hence *the class of symmetric* 2×2 *games is closed with respect to linear transformation* is a special case of Proposition 1.

We have to make a correction to Theorem 1 in [1] for it is not always true that every 2×2 game has one unique temporised equilibrium.

The following theorem contains the revised version of Theorem 1 in [1].

Theorem 4. *Every* 2×2 *k-game has at most two equilibrium points satisfying 17.*

Proof. The explicit form of the constraint C is in general a homographic function. Let us suppose that C can be made explicit in y, so that $y = g(z,k)$ according to the expression:

$$y = \frac{q(k)z + w(k)}{e(k)z + r(k)} \tag{22}$$

with $q(k) = -\frac{a_2 - a_4}{M_1} + k\frac{b_2 - b_4}{M_2}$, $w(k) = -\frac{a_4}{M_1} + k\frac{b_4}{M_2}$, $e(k) = \frac{a_1 - a_2 - a_3 + a_4}{M_1} - k\frac{b_1 - b_2 - b_3 + b_4}{M_2}$ and $r(k) = \frac{a_3 - a_4}{M_1} - k\frac{b_3 - b_4}{M_2}$. From Theorem 1 there exists at least one value for k such that $y \in [0,1]$ when $z \in [0,1]$. The study of homographic functions reduces to the discussion of three cases:

- $e(k) = 0$: 22 is the equation of a straight line hence $SC = \{z, g(z,k)\}$ is closed and R is the global graph;
- $q(k)r(k) = w(k)e(k)$: 22 is a straight line and it is parallel to the x axis, $SC = \{z, g(z,k)\}$ is closed and R is again the global graph;
- $e(k) \neq 0$ and $q(k)r(k) \neq w(k)e(k)$: 22 is the equation of an equilateral hyperbola with asymptotes parallel to the coordinate axes. In general if the vertical or horizontal asymptote falls in $(0,1)$ then 22 can be discontinuous in one point inside $(0,1)^2$. To avoid such situation it is sufficient to find a value of k such that only one branch of 22 intersects $[0,1]^2$; however this is not always possible (i.e. if $b_1 = b_2 = b_3 = b_4 = 0$ and $a_1 = 5$, $a_2 = -1$, $a_3 = -1$, $a_4 = 1$ then for any value of k both vertices of 22 are in $[0,1]^2$). Obviously each branch of the hyperbola is continuous therefore if only one branch intersects $[0,1]^2$ then $SC = \{z, g(z,k)\}$ is a closed set and R is the global graph, if both branches intersect $[0,1]^2$ then there exist two local graphs R^1 and R^2.

From Theorem 3. when SC is closed there exists one unique point $(\bar{z}, g(\bar{z}))$ satisfying 17, otherwise there are two equilibrium points satisfying 17, one for each local graph. □

6.2. Further considerations

At this point, we study the solution proposed by the equilibrium concept that we have just discussed to the Prisoner's Dilemma. Table 8 represent the numeric version of the game depicted by Table 1. Player A's strategies are indicated with a and player B's strategies are indicated with b.

(4, 4)	(0, 8)
(8, 0)	(1, 1)

Table 8. 2-players Prisoner's Dilemma

By definition, the deterrent in a symmetric game is $k = 1$, hence, the parametric function C takes the form:

$$C(a,b) : \frac{4ab + 8(1-a)b + (1-a)(1-b) - 4ab - 8a(1-b) - (1-a)(1-b)}{10} \tag{23}$$

The two explicit forms of C coincide and are $a = b$; hence when deterred, players' utility functions are:

$$f_1(a) = -3a^2 + 6a + 1, \quad (24)$$

and

$$f_2(b) = -3b^2 + 6b + 1 \quad (25)$$

It is evident from Figure 4, that if A maximises is own utility function on a and B maximises his own utility function on b, then both player will agree to play the cooperative strategy. If this is the case, the solution point is the temporised equilibrium and it is stable and unique.

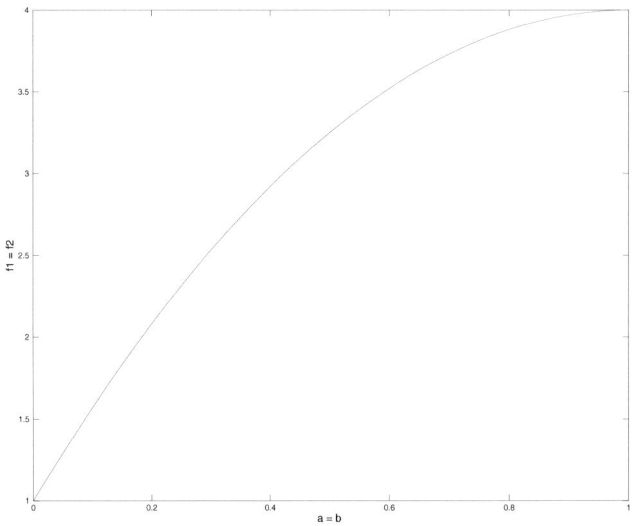

Figure 4. Utility functions in the temporised Prisoner's Dilemma

The key result of our research is that, in most cases, i.e. 8, rationality alone is not sufficient to ensure that each party obtains an optimal outcome - a proper combination of rationality and deterrence can instead implement a fair optimal solution.

6.3. Applications to network analysis

One of the practical field that has mostly benefited from game theory, is the analysis of traffic in telecommunication networks. In particular, our concern is to show the solutions proposed by the temporised framework to classic network problems that have already been thoroughly studied by means of canonical solution concepts. As we noted in [1], the power of the temporised equilibrium lies is that it does not require to include any incentive mechanism to obtain efficient outcomes, for such a feature is already part of the notion of equilibrium.

Thus, for the sake of completeness, we cite four examples of 2×2 games, taken from [1] and for each of them we compare the Nash solution with our solution.

6.3.1. Forwarder's dilemma

Suppose we have a simple network made of four nodes, two senders (S_1, S_2) and two receivers (R_1, R_2). Each sender has an assigned receiver and for a packet to go from a sender to its receiver it has to pass through the other sender which then will forward it to its destination.

Again, each sender, when operates as a forwarder, has two options: to forward the packet or to drop it. The cost of forwarding a packet is expressed by the value $0 << c << 1$, the reward for having a packet correctly sent to destination is 1 and the gain for a lost packet is 0. Such a configuration can be formalised as a Prisoner's Dilemma game as depicted by Table 9; S_1 plays rows and S_2 plays columns.

(1-c, 1-c)	(-c, 1)
(1, -c)	(0, 0)

Table 9. Forwarder's Dilemma

It is well known that the Nash equilibrium for this game prescribes each node to drop other's packet.

As we have shown before in this section the temporised solution is to play cooperatively that, eventually, leads to the outcome $(1 - c, 1 - c)$, which is Pareto efficient.

6.3.2. *Joint packet forwarder game*

In this case, we have a network made of one sender, one receiver and two step nodes S_1, S_2. The sender needs to send a packet to the receiver but, due to the network topology, the packet must pass through S_1 and S_2. Each step node has two choices: to forward the packet or to drop it. Again the cost of forwarding a packet is $0 << c << 1$ for each step node. If both S_1 and S_2 successfully forward the packet, then they gain 1. Table 10 represents the bimatrix associated with the game: again S_1 plays rows and S_2 plays columns.

(1-c, 1-c)	(-c, 0)
(0, 0)	(0, 0)

Table 10. Joint packet forwarder

There exist two Nash equilibria that provide the outcomes $(1 - c, 1 - c)$ and $(0, 0)$.

Let us suppose that the value of the deterrent information is $k = 1$; in this case the temporised solution would be as advantageous as one of the Nash equilibria, prescribing the outcome $(1 - c, 1 - c)$.

6.3.3. *Multiple access game*

We have two senders S_1 and S_2 that want to access a shared communication channel to deliver some packets to their receivers. Each sender has to option: to access the channel to send the packet or to wait. Let us assume that senders and receivers are in the same power range, thus their transmissions create mutual interference. The cost of accessing the channel and send the packet is $0 << c << 1$ for each sender. A packet is successfully delivered, and it is rewarded 1, when no collisions occur.

Table 11 is the bimatrix associated with the game: as usual S_1 plays rows and S_2 plays columns.

One may notice that this is the game of Chicken. This game has three Nash equilibria which give the outcomes: two in pure strategies $(0, 1 - c)$, $(1 - c, 0)$, and one in mixed strategies $(1/4 - 1/2c, 1/4 - 1/2c)$. Since the game of Chicken is symmetric, by definition $k = 1$, thus

(0, 0)	(0, 1-c)
(1-c, 0)	(-c, -c)

Table 11. Multiple access

the temporised equilibrium provides the payoff $1/4c^2 - 1/2c + 1/4$. Such payoff is clearly more efficient than the Nash mixed strategy equilibrium as predicted by Corollary 1 in [1].

6.3.4. *Jamming game*

In the last example we have one sender S_1 and one receiver. Let us assume that the wireless medium is split into two channels and that S_1 is able to select which channel to use at each time step of transmission. Suppose there exist another sender S_2 that is also able to send packet via one of the channels and that S_2's aim is to jam S_1's transmission. Eventually, S_1 receives 1 if its signal is not jammed and -1 otherwise; S_2 receives 1 if it jams S_1's signal and -1 otherwise.

The bimatrix of the game is represented by Table 12. S_1 plays rows and S_2 plays columns.

(-1, 1)	(1, -1)
(1, -1)	(-1, 1)

Table 12. Jamming game

In this last example, the Nash and the temporised solutions coincide.

7. Social choice theory

As many economists have stressed, there exist some basic requirements that a social choice function should possess in order to provide a sustainable structure for its participants. The social choice function must be Pareto optimal in the sense that if every individual prefers any alternative x to another alternative y, then society must prefer x to y. It should also guarantee a mild form of individual liberty, which means that if there are at least two individuals, then each individual has at least a pair of alternatives over which he is decisive. And, every set of individual orderings must be included in its domain set. The Paretian liberal paradox, formulated by Sen in [18], establishes that there is no social choice function that can simultaneously be Pareto optimal and liberal. Such a result has been thoroughly studied by many authors, with different solutions in both economics and game theory.

One fundamental assumption made by Sen is that the Pareto optimality and the liberal conditions must be simultaneously combined to restrict the domain of the social choice function. If this is the case, there is no way out of the paradox, as explained in [18] and [6], for, with at least two individuals, if the domain set of the social choice function contains at least two alternatives, then some form of circular ordering of preferences may always arise.

In our view, Nozick's way out of the paradox with the inclusion of individual rights into states selection, is the most conclusive. In fact when an individual exercises his own rights over possible states of the world, he is actually putting constraints on the set of alternatives open to the social choice. A right is the possibility for a group of individuals to restrain the set of social states to a subset of the original set of states.

Rights do not establish any ordering over the states of the world, but divide them into classes [6]. Once a class is defined, the exercise of a right might exclude it from any further consideration in terms of collective choice. Following the Kantian argument, for which rights must precedes welfare, Nozick depicts a two step procedure for implementing an effective social choice:

1. each individual or group of individuals exercises some right, thus restraining the set of available alternatives;
2. some selection mechanism over the set of remaining alternatives is employed to determine the final choice.

Given this procedure, it is possible to redefine the Pareto optimality condition, given in [18]: if every individual prefers any alternative x to another alternative y, *among the alternatives that are still available after the rights have been exercised*, then society must prefer x to y.

In [6], Gardenfors pushes the analysis of Nozick's ideas even further by creating a rights system formal model. However, to the scope of our argumentation, we underline one aspect of the rights-based approach: (a) *the introduction of a procedure to bound the set of relevant social alternatives to a subset of non-conflicting states.*

A problem, that is strictly related to Sen's paradox, has been proposed by Schelling, as reported by Fischer in [5]. He notices how irrational collective outcomes, in the sense of an inefficient allocation of resources, might arise from individual rational behaviour. His analysis is centred on what structural features a liberal society should possess in order to prevent irrational social outcomes. If societies must guarantee, at least, a mild form of freedom and an efficient distribution of resources, then the relationship between Schelling's problem and Sen's paradox becomes evident.

From the examination of how indeed society are structured, Schelling observes that, the most part of a society consists of institutional arrangements, or practices using Rawls' terminology. The purpose of these arrangements is to overcome individual irrationalities in the interest of a common goal. Such arrangements are not only institutions in the economic sense, but they include forms of popular wisdom, some traditions taken to the level of moral principles or the shrewd and undisputed use of violence. As remarked by Nozick in [21], violence has profound connections with deterrence and fairness when the authority which uses it is confined to a night-watchman state. Finally, in the interest of our discussion, we would like to point out an aspect that emerges from these ideas: (b) *an effective allocation of resources can be reached when rational individuals are forced to follow certain behavioural patterns.*

One might notice that condition (b) might include condition (a); in fact, when some individuals' behaviour is constrained, the set of alternative social state they can reach could be restrained. This is the key concept of temporised equilibria.

When the judge provides the deterrent information, that is by definition symmetric and incomplete [1], it coerces the players to calculate the explicit forms of Equation 19 in order to have a clear understanding of the deterrent threat. A temporised equilibrium is then a point in the coupled constrained set of strategy (a), where individual rationality coincides with collective rationality, for each player is forced, by the judge reputation, to not deviate from the designated pattern (b).

This analogy is also supported by the examples proposed in the preceding section, where it is evident how the effective restraining of the set of alternatives leads to a more convenient overall outcome.

From the analysis in this paragraph, we shall, eventually, draw the conclusions that: the temporised solution realises a social outcome that is Pareto efficient and liberal, and, Nozick's system of rights is somehow equivalent to the application of an accurate utilisation of deterrence as a mean of fairness.

8. Conclusions

In this chapter we have discussed some of the fundamental aspects which link game theory to the broader issues of justice, fairness and social choice. Our argumentation moved from the philosophical foundation of deterrence and justice to the operative definition of game models inspired by such theoretic frameworks. Our own contribution is then the mathematical synthesis of the work from many authors on the subject of deterrence and fairness.

Moreover, we have drawn a correspondence between the temporised equilibrium theory and Nozick's proposal to overcome Sen's paradox in the context of the social choice theory. From such a correspondence we have been able to utilise a game theoretic model to analyse the inefficiencies of rational behaviour when measured on a social scale.

Though most of the significance of the temporised solution is theoretical, we have identified two fields where practical applications could be prolific: artificial intelligence and opportunistic network analysis.

In our view, the analogy with Freud's theory of mind, that inspired the concept of temporised equilibria, could be further exploited: in fact, the mathematics involved could be implemented into an artificial system which can mimic the dynamics of the mind when it struggles to balance the requests from its internal elements. Together with an automated and effective method to assign preferences over a set of states of the world, i.e. hedonistic calculus, and an intelligent system of sensors to collect the stimuli from the environment, it could be possible to realise an artificial conscience, in which a conscious decision is represented by the temporised equilibrium. Such a solution will be optimal, in the sense that it is the best possible allocation taking into consideration the preferences of the building blocks of the mind.

Another interesting area of application is the analysis of opportunistic network [9]. Opportunistic networks (Oppnets) differ from traditional networks in which the nodes are all deployed together and where the size of the network and locations of all its nodes pre-designed (at least the initial locations for mobile networks). In oppnets, we first deploy a seed oppnet, which may be viewed as a pretty typical ad hoc network. The seed then self-configures itself, and then works to detect "foreign" devices or systems using all kinds of communication media-including Bluetooth, wired Internet, WiFi, ham radio, RFID, satellite, etc. Detected systems are identified and evaluated for their usefulness and dependability as candidate helpers for joining the oppnet. Best candidates are invited into the expanded oppnet. A candidate can accept or reject the invitation (but in life-or-death situation it might be ordered to join). Upon accepting the invitation, a helper is admitted into the oppnet. How to select the efficient helper nodes is a vital research field in Opportunistic Network. The resources of the admitted helper are integrated with the oppnet, and tasks can be offloaded to or distributed amongst this and all other helpers. A decentralised command centre presides over the operations of the oppnet throughout its life.

One may notice how the architecture of a typical opportunistic network resembles the theoretic model of temporised equilibria. Though in this field of application some work has already been done by the authors, i.e. the definition of the helper selection protocol and the formalisation of the set of rules to manage the assignment of shared resources, there is still a long way to go to design an efficient and self-sufficient system for finding helper nodes in a more articulated scenario.

Author details

Riccardo Alberti and Atulya K. Nagar
Centre for Applicable Mathematics and System Science (CAMSS), Department of Mathematics and Computer Science, Liverpool Hope University, United Kingdom

9. References

[1] Alberti, R. (2010). Temporised equilibria: a rational concept of fairness into game theory. *International Journal of Computing Science and Mathematics*, Vol. 3, No. 3, (December 2010)
[2] Ambrus-Lakatos, L. (2002). On Preferences for Fairness in Non-Cooperative Game Theory. (June 2002)
[3] Bestougeff, H., Rudnianski, M. (1998). Games of Deterrence and Satsficing Models Applied to Business Process Modelling. (1998)
[4] Downs, G. W. (1989). The Rational Deterrence Debate. *World Politics*, Vol. 41, No. 2, (January, 1989) page numbers (225-237)
[5] Fischer, C. S. (1981). Review: Solving Collective Irrationality. *American Journal of Sociology*, Vol. 87, No. 2, (September, 1981) page numbers (438-444)
[6] Gardenfors, P. (1981). Rights, Games and Social Choice, *Noûs*, Vol. 15, No. 3 (September 1981), page numbers (341-356)
[7] Langlois, J. P. (2002). Applicable Game Theory, (2002) Chapter 3
[8] Levine D. K., Levine R. A. (2005). Deterrence in the Cold War and the "War on terror". (September, 2005)
[9] Lilien L.; Bhuse V. & Gupta A., (2006). Opportunistic Networks: The Concept and Research Challenges in Privacy and Security. *International Workshop on Research Challenges in Security and Privacy for Mobile and Wireless Networks*, (March 2006)
[10] Lukasova, A. (1995). Nozick's Libertarianism: A Qualified Defence, (1995), 2pp
[11] Myerson, R. B. (2006). Force and restraint in strategic deterrence: a game-theorist's perspective. *based on a talk presented at the Chicago Humanities Festival on Peace and War*, (November, 2006)
[12] Rabin, M. (1993). Incorporating fairness into game theory and economics. *The American Economic Review*, Vol. 83, (1993) page numbers (1281-1302)
[13] Rawls, J. (1955). Two Concepts of Rules. *The Philosophical Review*, Vol. 64, (1955) page numbers (3-32)
[14] Rawls, J. (1957). Justice as fairness. *The Journal of Philosophy*, Vol. 54, No. 22, (October, 1957) page numbers (653-662)
[15] Roemer, J., E. (2010). Kantian Equilirium. *Scandinavian Journal of Economics*, Vol. 112, No. 1, (2010) page numbers (1-24)

[16] Risse, M. (2004). Does Left-Libertarianism Have Coherent Foundations?. (April, 2004)
[17] Rosen, J. B. (1965). Existence and Uniqueness of Equilibrium Points for Concave N-Person Games, *Econometrica*, Vol. 33, No. 3 (1965), page numbers (520-534)
[18] Sen, A. (1970). The Impossibility of a Paretian Liberal, *Journal of Political Economy*, Vol. 78, (1979), page numbers (152-157)
[19] Schelling, T. C. (1962). The Role of Deterrence in Total Disarmament. *Foreign Affairs*, Vol. 40, No. 3, (April, 1962) page numbers (392-406)
[20] Schelling, T. C. (1968). Game theory and the study of ethical systems. *The Journal of Conflict Resolution*, Vol. 12, No. 1, (March, 1968) page numbers (34-44)
[21] Vallentyne, P. (2006). Robert Nozick, Anarchy, State and Utopia. *Central Works of Philosophy*, Vol. 5, (2006) page numbers (86-103)
[22] Zagare, F. C. (1985). Toward a Reformulation of the Theory of Mutual Deterrence. *International Studies Quarterly*, Vol. 29, (1985) page numbers (June 155-169)

Chapter 3

Inductive Game Theory: A Simulation Study of Learning a Social Situation*

Eizo Akiyama, Ryuichiro Ishikawa, Mamoru Kaneko and J. Jude Kline

Additional information is available at the end of the chapter

http://dx.doi.org/10.5772/54181

1. Introduction

Inductive game theory (IGT) aims to explore sources of beliefs of a person in his individual experiences from behaving in a social situation. It has various steps, each of which already involves a lot of different aspects. A scenario for IGT was spelled out in Kaneko-Kline [15]. So far, IGT has been studied chiefly in theoretical manners, while some other papers targeted applications and conducted an experimental study. In this chapter, we undertake a simulation study of a player's learning about some details of a social situation. First, we give a brief overview of IGT, and its differences from the extant game theories. Then, we explain several points pertinent to our simulation model.

1.1. Developments of inductive game theory

The scenario for IGT given in [15] consists of three main stages:

(1) Experimental Stage: making trials-errors and accumulating experiences;

(2) Inductive Derivation Stage: construction of an individual view from accumulated experiences;

(3) Analysis/Use Stage: uses of the derived view for behavioral revision.

Fig.1 describes the relationships among the three stages[1]. The process starts with the experimental stage, where a player makes trials-errors, and accumulates memories from experiences. In the second stage, each player constructs an individual view from accumulated experiences, which is based on induction; this is the reason for the title "inductive" game

*The authors are partially supported by Grant-in-Aids for Scientific Research No.21243016 and No.2312002, Ministry of Education, Science and Culture.

[1] We may regard (1) and (2)-(3), respectively, as corresponding to the *experiencing self* and the *remembering self* in Kahneman [11, p.381]. Kahneman talks about various examples and aspects relevant to this distinction.

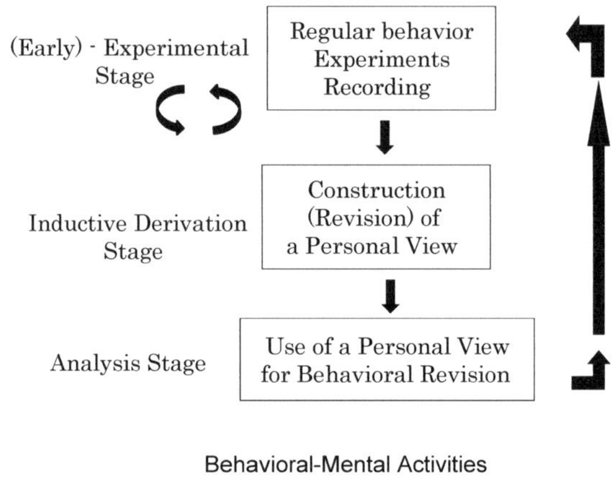

Figure 1. Three Stages of IGT

theory. In the third stage, once a player has built his view, he uses it for his decision making or behavioral revision. After the third stage, the process goes to the first stage, and those stages may cycle.

Each stage already includes a lot of new problems. To study those problems, we borrow concepts from the extant game theories, but often we need to think about whether some can or cannot be used for IGT and whether to modify them for IGT, since they often rely upon the presumptions of the extant game theories.

In Kaneko-Matsui [19] and Kaneko-Kline [15], [16], [17], we have focused on the second and third stages. The first stage of making trials-errors and accumulating memories was discussed, but described in the form of informal postulates. Taking the resulting sets of accumulated memories from trials and errors as given, the second and third stages are formulated in a theoretical manner. However, the first stage is of a very different nature from the other two, and each player's bounded cognitive ability is crucial. For this, we may take two approaches: experimental and simulation. Takeuchi *et al.* [22] conducted an experimental study, and here, we take a simulation method.

It would be helpful to discuss, before giving a description of our simulation study of IGT, how IGT differs from two main stream approaches in the recent game theory literature: the classical *ex ante* decision approach and the evolutionary/learning approach. The contrasts between them will motivate our use of a simulation study.

The focus of the classical *ex ante* decision approach is on the relationship between beliefs/knowledge and decision making (cf., Harsanyi [8] for the incomplete information game and Kaneko [13] for the epistemic logic approach to decision making in a game). In this approach, the beliefs/knowledge is given *a priori* without asking their sources. Thus, IGT is relevant for exploring sources of beliefs and knowledge in experiences.

Contrary to this, the evolutionary/learning approach (cf., Weibull [24], Fudenberg-Levine [6], and Kalai-Lehrer [12]) targets "learning". However, this approach does not ask the question of the emergence of beliefs/knowledge; instead, their concern is typically the convergence of the distribution of actions to some equilibrium. The term "evolutionary/learning" means that some effects from past experiences remain in the distribution of genes/actions. It is not about an individual's learning of the structure or details of the game; typically it is not specified who the learner is and what is learned. When we work on an individual's learning, we should make these questions explicit.

If the learner is an ordinary person, the convergence of behavior in the limit is not very relevant to his learning. Finiteness of life and learning must be crucial. Here, "finite" is "shallowly finite", rather than the negation of infinity in mathematics. Consequently, we conduct simulations over finite spans of time corresponding to the learning span of a single human player. Our simulation indicates various specific components affecting one's finite learning, while they are not relevant in the limiting behavior.

1.2. Simulation study of a social situation

Now, we discuss several important points of our simulation model.

(1): *An ordinary person and an every-day situation in a social world:* We target the learning of an ordinary human person in a repeated every-day situation, which we regard only as a small part of the social world for that person. We choose a simple and casual example called "Mike's Bike Commuting". In this example, the learner is Mike, and he learns the various routes to his work. Using this example, the time span and the number of reasonable repetitions for the experiment become explicit.

We study a one-person problem, but it should not be regarded as isolated from society. It is a small part of Mike's social world.

(2): *Ignorance of the situation:* At the beginning, Mike has no prior beliefs/knowledge about the town. His colleague gave a coarse map of possible alternative routes without precise details, and suggested one specific route from his apartment to the office. Mike can learn the details of these routes only if he experiences them. We question how many routes Mike is expected to learn after specific lengths of time.

(3): *Regular route and occasional deviations:* Mike usually follows the suggested route, which we call the regular route. Occasionally, when the mood hits him, he takes a different route. This is based on the basic assumption that his energy/time to explore other routes is scarce. Commuting is only a small part of his social world, and he cannot spend his energy/time exclusively for exploring those routes.

(4): *Short-term and long-term memories:* We distinguish two types of memories for Mike: short-term and long-term. Short-term memories form a finite time series consisting of past experiences, and they will be kept only for some finite length of time, perhaps a few days or weeks; after then they will vanish. However, when an experience occurs with a certain frequency, it becomes a long-term memory. Long-term memories are lasting.

In our theory, the transition from a short-term to a long-term memory requires some repetition of the same experience within a given period of time. This is based on the general idea that memory is reinforced by repetition. Our formulation can be regarded as a simplified version of Ebbinghous' [5] retention function.

(5): *Finiteness and complexity:* Our learning process is formulated as a stochastic process. Unlike other learning models, we are not interested in the convergence or limiting argument. As stated above, the time structure and span are finite and short. In our example, we discuss how many times Mike has experienced a particular route after a half year, one year, or ten years. We will find many details, which are highly complex even in this simple example. We analyze those details and find the lasting features in Mike's mind.

(6): *Marking salient choices as important:* Although the situation is extremely simple, it is difficult for Mike to fully learn the details of the entire town even after several years. We consider the positive effect on learning by "marking", introduced in Kaneko-Kline [14]. If Mike marks some "salient" choice as "important", and restricts his trial-deviations to the marked choices, then we find that his learning is drastically improved. Imperfections in a player's memory make marking important for learning. Without marking, experiences are infrequent and lapse with time. Consequently, his view obtained from his long-term experiences could be poor and small. By marking, he focuses his attention on fewer choices, and successfully retains more as long-term memories.

Up to here, we study how many times Mike needs to commute in order to learn some routes. Precise objects Mike possibly learns are not targeted. There are two directions of departure from this study. One possibility is to study Mike's learning of internal components of routes, and the other is about relationships between routes. Of course, to study both in an interactive way is possible. In this paper, however, we consider a problem of the latter sort, namely, Mike's learning of his own preferences from experiences.

(7): *Learning preferences:* Here, we face new conceptual problems. We should make a distinction between having preferences and knowing them. We assume that Mike has well-defined complete preferences, but his knowledge is constrained to only some part by his experiences. Also, it is important to notice that learning one's preferences differs from keeping a piece of information. Since the feeling of satisfaction is relative and likely to be more transient than the perception of a piece of information, we hypothesize that learning one's preferences needs comparisons of outcomes close in time. Consequently, marking alternatives becomes even more important for obtaining a better understanding of his own preferences.

In our simulation study up to Section 4, we will get some understanding of relevant "shallowly finite" time spans for ordinary life learning. Our study on learning preferences in Section 5 is more substantive than the studies up to Section 4. However, we will not go to the direction to a study of learning of internal structures of routes. This will be briefly discussed in Section 7.

The chapter is organized as follows: In Section 2, we specify our model and simulation frame. In Section 3, we give simulation results and discuss them to see how much Mike can learn for given time spans. In Section 4, we introduce the concept of "marking", and observe its positive effects on learning. In Section 5, we consider the problem of learning his preferences. In Section 6, we carry out a sensitivity analysis of changing various parameters describing Mike's learning and memory characteristics. Section 7 is devoted to a discussion our results and their implications for IGT as well as suggesting some future directions for simulations studies.

2. Mike's bike commuting

Mike moves to a new town and starts commuting to his office everyday by a bike. At the beginning, his colleague gives him a simple map depicted as Fig.2 and indicates one route shown by the dotted line. Mike starts commuting every morning and evening, five days a week, that is, 10 times a week. From the beginning, he wants to know the details of those

Figure 2. A Map of the Town

routes, but the map is simple and coarse. He decides to explore some alternative routes when the mood hits him, but typically he is too busy or tired and resorts to the *regular route* suggested by the colleague[2].

The town has a lattice structure: His apartment and office are located at the south-west and north-east corners. To have a route of the shortest distance from his apartment to the office, he should choose "North" or "East" at each lattice point; such a route is called a *direct* route. There are 35 direct routes. He enumerates these routes as $a_0, a_1, ..., a_{34}$, where a_0 denotes the regular route.

In our simulation, we assume that Mike follows a_0 with probability $4/5 = 1 - p$ and he makes a deviation to some other route with $p = 1/5$. This probability p is called the *deviation probability*. When he makes a deviation, he chooses one route from the remaining 34 routes with the same probability $1/34$. His behavior each morning or evening can be depicted by the tree in Fig.3. He himself may not be conscious of these probabilities or of this tree. In sum, on average, he makes a deviation twice a week to any of the other routes with equal probability.

After following route a_l, he gets some impressions and understanding of a_l. In this paper we do not study the details of a_l that he learns; instead, we study conditions for an experience to remain in his mind as a long term memory.

As mentioned in Section 1, he has two types of memories: *short-term* and *long-term*. A short-term memory is a time series of experiences of the past m trips. An experience disappears after m trips of commuting. If the same experience, say a_l, occurs at least k times in m trips, experience a_l becomes a long-term memory. Long-term memories form a set of experiences without time-structure or frequency[3].

[2] We may start with only the assumption that he is given the regular route, without having a map. This case is more faithful to IGT given in Kaneko-Kline [15]. However, this makes our simulation study much more complicated. We will keep our study as simple as possible.

[3] This lack of time structure and frequency is motivated by bounded rationality of the player. Limitations on his memory and computation abilities lead him to ignore some aspects like the time structure and frequency of long term memories.

In our simulation, we specify the parameters (m, k) as $(10, 2)$, meaning that Mike's short-term memory has length 10, and if a specific experience occurs at least two times in his short-term memory, it becomes a long-term memory. This situation is depicted in Fig.4, where at time $t-1$, the routes a_0, a_2 are already long-term memories, and at time t, route a_1 becomes a new long-term memory.

We consider another parameter T, denoting the total number of trips (time span). For example:

$$\begin{aligned} \text{after a half year, } T &= 2 \times 5 \text{ (days)} \times 25 \text{ (weeks)} = 250; \\ \text{after 1 year, } T &= 2 \times 5 \text{ (days)} \times 50 \text{ (weeks)} = 500; \\ \text{after 10 years, } T &= 2 \times 5 \text{ (days)} \times 500 \text{ (weeks)} = 5000. \end{aligned}$$

Our simulation will be done by focusing on the half year and 10 year time spans. In Mike's Bike Commuting, the number of available routes is 35, but later, this will also be changed, and the number of routes will be denoted as a parameter s. Listing all the parameters, we have our *simulation frame* F:

$$F = [s, p; (m, k)]. \tag{1}$$

We always assume that in the case of a deviation, a route other than a_0 is chosen with equal probability $1/(s-1)$.

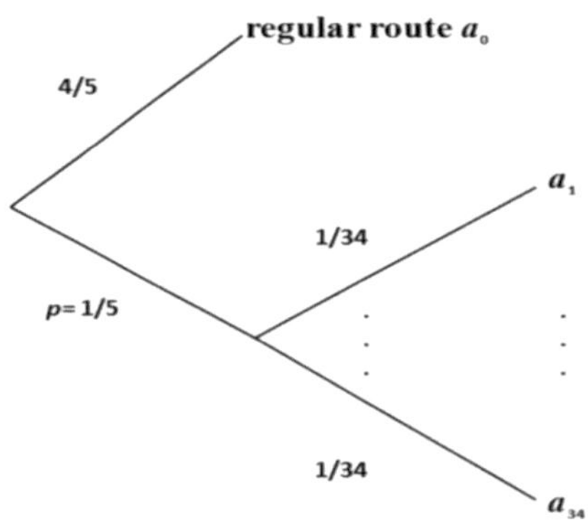

Figure 3. Decision Tree of each Trip of Commuting

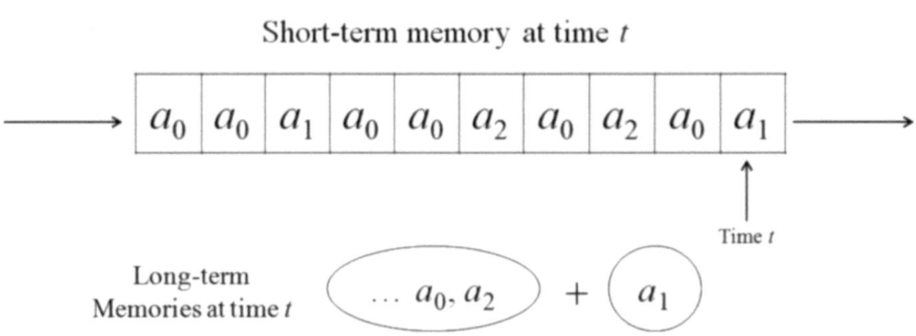

Figure 4. Short-Term and Long-Term Memories

The stochastic process is determined by the simulation frame F and a given T, which consists of T component stochastic trees depicted in Fig.3. This process is denoted by $F[T] = [s, p; (m, k) : T]$. Our concern is the probability of some event of long-term memories at time T. For example, what is the probability of the event that a particular route a_l is a long-term memory at T? Or, what is the probability that all routes are long-term memories? We calculate those probabilities by simulation. In Section 3, we give our simulation results for $F = [s, p; (m, k)] = [35, 1/5; (10, 2)]$ and $T = 250, 5000$.

Before going to these results, we mention one analytic result: For the stochastic process $F[T] = [35, 1/5; (10, 2) : T]$,

$$\text{the probability that all routes become long-term memories tends to 1 as } T \text{ tends to infinity.} \tag{2}$$

This can be proved easily because the same experience occurs twice in a short-term memory at some point of time almost surely if T is unbounded. This result does not depend on the specification of parameters of F. Our interest, however, is in finite learning. Our findings by simulation for the finite learning periods of $T = 250$ and $T = 5000$ differ significantly from the convergence result. This suggests that focusing on convergence results does not inform us about finite learning.

3. Preliminary simulations and the method of simulations

We start in Section 3.1 by giving simulation results for the case of $s = 35$. The results show that it would be difficult for Mike to learn all the routes after a half year. After ten years, he learns more routes, but we cannot say much about which specific routes he learns other than the regular one. In Section 3.2, we give a brief explanation of our simulation method and the meaning of "probability".

3.1. Simulation results for $s = 35$

Consider the stochastic process determined by $F = [s, p : (m, k)] = [35, 1/5; (10, 2)]$ for up to $T = 250$ (a half year) and $T = 5000$ (10 years). Table 1 provides the probabilities of the event that a specific route a_l is a long-term memory at $T = 250, 5000$, and also at a large T.

The row for a_0 shows that the probability of the regular route a_0 being a long-term memory is already 1 at $T = 250$ (a half year). This "1" is still an approximation result meaning it is very close to 1.

The row for a_l $(l \neq 0)$ is more interesting. The probability that a specific a_l is a long-term memory at $T = 250$ and 5000 is 0.069 and 0.765, respectively. Our main concern is to evaluate these probabilities from the viewpoint of Mike's learning.

T	250	5000	28252 (> 56 years)
a_0	1	1	1
a_l $(l \neq 0)$	0.069	0.765	0.99

Table 1

r	1	2	3	4	5	$\cdots$
	0.089	0.223	0.272	0.213	0.121	$\cdots$

Table 2

r	$\cdots$	25	26	27	28	29	$\cdots$
	$\cdots$	0.109	0.159	0.153	0.153	0.124	$\cdots$

Table 3

Some reader may have expected that the probability for $T = 250$ would be much smaller than 0.069, because in each trip, the probability of route a_l $(l \neq 0)$ being chosen is only $1/5 \times 1/34 = 1/170 = 0.00588$. However, it is enough for a_l to occur in a consecutive sequence of length 10 (short-term memory) at some $t \leq 250$, and there are 240 such consecutive sequences. Hence, the probability turns out not to be negligible[4]. The accuracy of this calculation will be discussed in Section 3.2.

The rightmost column is prepared for a purpose of reference. The number of trips 28252 ($>$ 56 years) is obtained from asking the time span needed to obtain the probability 0.99 of a_l $(l \neq 0)$ being a long-term memory. The length of 56 years would typically exceed an individual career[5], and thus we regard the limiting convergence result (2) as only a reference.

The cases of $T = 250$ and 5000 are relevant to our analysis. Nevertheless, a single probability 0.069 or 0.765 tells us little about what Mike might be expected to learn in those time spans. We next look more closely at the distribution of routes he learns for each of those time spans.

For $T = 250$, we give Table 2, which describes the probability of exactly r routes (the regular route and $r - 1$ alternative routes) being long-term memories in 35 routes:

After $r = 5$ routes, the probability is diminishing quickly, so we exclude those numbers from the table. According to our results, Mike typically learns a few routes (the average is about 3.33) after half a year. For $r = 3$, one route must be regular, but the other two are arbitrary. We have $\binom{34}{2} = 561$ cases, so the probability of a particular 3 routes being long-term memories is only $0.272/561 = 0.000485$ which is very small. This means that although Mike learns about 2 alternative routes, it is hard to predict with much accuracy which pair would be learned.

At $T = 5000$, i.e., ten years later, Mike's learning is described by Table 3.

Again, we show only the values of r having high probabilities. The average of the number of routes as long-term memories is about 27. Because most of the distribution lies between 25 and 29 routes, we find that there are many more cases to consider than after half a year. For example, consider 0.109 for $r = 25$, which is the probability that exactly 25 routes are learned. This probability can be obtained from the probability 0.765 in Table 1 by the equation:

$$\binom{34}{24} \times (0.765)^{24} \times (1 - 0.765)^{9} \doteqdot 0.109.$$

[4] A famous example called the *birthday attack* may be indicative for this fact: In a class consisting 50 students, what is the probability of finding at least one pair of students having the same birthday? Since each student has the probability 1/365 of an arbitrary given day of a year being his birthday, it might be expected not to have a pair of students of the same birthday. However, the exact calculation tells that the probability is about 0.97.

[5] Our model without decay of long-term memories is likely to be inappropriate for 56 years.

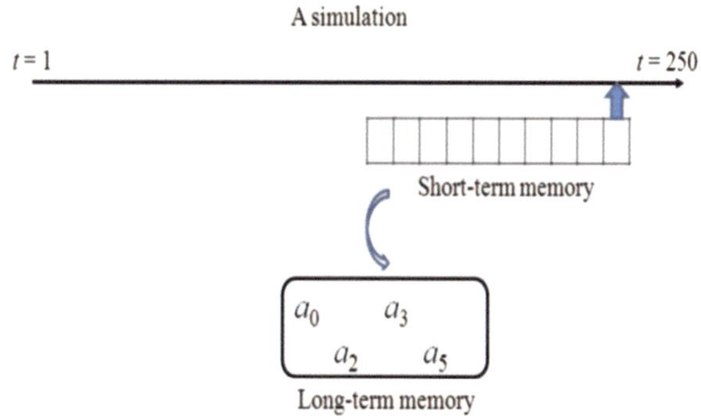

Figure 5. A Simulation up to $T = 250$

Looking at this equation, we obtain the probability that a specific set of 25 routes are long-term memories is only $0.109/\binom{34}{24} = 8.31 \times 10^{-10}$. In sum, Mike learns about 27 alternative routes after 10 years. However, the number of combinations of 24 routes from 34 is enormous at about 1.3×10^8 and much larger than the $\binom{34}{2} = 561$ cases we need to consider after only half a year.

Finally, we report the average time for Mike to learn all the 35 routes as long-term memories, which is 28.4 years (14, 224.3 trips). If he is very lucky, he will learn all routes in a short length of time, say, 10 years, which is an unlikely event of probability 9×10^{-5}. The probability of having learned all routes in 35 years is much higher at 0.806.

After all, the above calculations indicate that "finiteness" involved in our ordinary life is far from "large finiteness" appearing in the convergence argument in mathematics. In this sense, we are facing shallowly finite problems, which was emphasized in Section 1. In Sections 4 and 5, we will discuss related problems to this issue from different perspectives.

3.2. Simulation method

We now explain the concept of "probability" we are using, and discuss the accuracy of this concept. First we mention why this is not calculated in an analytic manner. The analytic computation is feasible up to about $T = 30$, but beyond $T = 40$, it is practically impossible in the sense that for $T = 50$, it takes decades to calculate with current (year 2007) computers using our analytical method. This is caused by the limited length of short-term memory and multiple occurrences needed for a long-term memory.

We take the relative frequency of a given event over many simulation runs instead of computing probabilities analytically. We use the Monte Carlo method to simulate the stochastic process up to a specific T for the simulation frame $F = [s, p : (m, k)] = [35, 1/5 : (10, 2)]$. The frame has only two random mechanisms depicted in Fig.3, but they are reduced into one random mechanism. This mechanism is simulated by a random number generator. Then, we simulate the stochastic process determined by F up to $T = 250$ or $T = 5000$ or some other time span. A simulation is depicted in Fig.5. One simulation run gives a set of long-term memories: In Fig.5, routes a_0, a_2, a_3, a_5 are long-term memories at some time before $T = 250$.

We run this simulation 100,000 times. The "probability" of a_l is calculated as the relative frequency:

$$\frac{\#\{\text{simulation runs with } a_l \text{ as a long-term memory}\}}{100,000} \tag{3}$$

In the case of $T = 250$, this frequency is about 0.069 for $l \neq 0$, and it is already 1 for $l = 0$ in our simulation study.

We compare some results from simulation with the results obtained by the analytical method. For $T = 20$ and $s = 35$, the probability of a_l being a long-term memory can be calculated in an analytic manner using a computer. The result coincides with the frequency obtained using simulation to an accuracy of 10^{-4}.

The robustness of the frequency (probability) 0.069 in Table 1 is evaluated further by looking at 1,000,000,000 simulation runs. In these runs, we have 68,594,265 runs where a_1 is a long-term memory. Counting also simulation runs where a_l ($= a_2, ..., a_{34}$) is a long-term memory, we find that the smallest (and largest) number of runs where a_l is a long-term memory is 68,569,941 (respectively, 68,596,187), both of which translate to the frequency 0.069 when rounding off to three decimal places.

In sum, we calculate the "probability" of an event as the relative frequency over numerous simulation runs since the analytic calculation is difficult for the large finite time spans and simulation frames under consideration.

4. Learning with marking: Simulation for $s = 5$

We now show how "marking", introduced in Kaneko-Kline [14], can improve Mike's learning. By concentrating his efforts on a few "marked" routes, he is able to learn and retain more experiences. This is because the likelihood of repeating an experience rises by reducing the number of alternative routes. In Section 4.1, we consider the case where Mike marks only four alternative routes in addition to the regular one. We see a dramatic increase in his learning of alternative routes. In Section 4.2, we show how a more planned approach can improve the effect of "marking" on his learning.

4.1. Marking five salient routes and simulation results

Suppose that Mike decides to mark some routes from his map for his exploration. He uses two criteria:

(i) He chooses routes having a scenic hill or flowers;

(ii) He avoids construction sites.

Then, he marks only four alternative routes, which are depicted in Fig.6. Adding the regular route a_0, we denote the five marked routes by a_0, a_1, a_2, a_3, a_4.

The above situation is described by changing the simulation frame to $F = [s, p : (m, k)] = [5, 1/5 : (10, 2)]$ for $T = 250$ or 5000. The probability of a_l ($l \neq 0$) being a long-term memory is calculated by our simulation method and is given in Table 4:

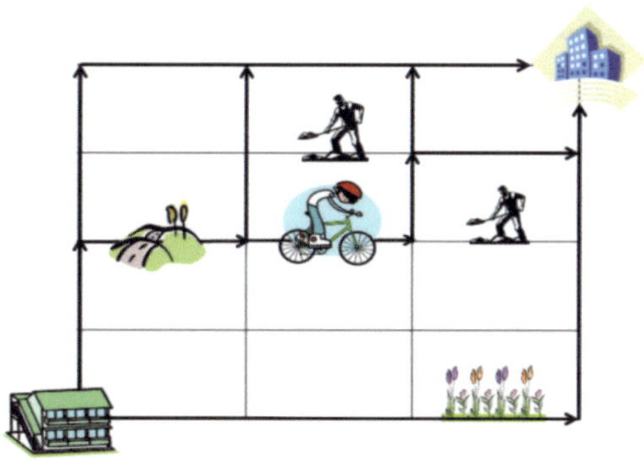

Figure 6. Five Marked Routes

T	250	5000
$s = 5$	0.970	1.00
$s = 35$	0.069	0.765

Table 4

Table 5 lists the length of time needed to obtain the probability 0.99 that an alternative route a_l $(l \neq 0)$ is a long-term memory. With marking he needs only 425 trips (10.2 months), as opposed to the 28,253 trips (more than 56 years) without marking.

T	425	28253
$s = 5$	0.990	1.000
$s = 35$	0.114	0.990

Table 5

We also have calculated, and presented in Table 6, the probability that exactly r $(= 1, 2, 3, 4, 5)$ routes are long-term memories at $T = 250$. The average number of routes learned is 4.9. Table 7 states that the average time for Mike to learn all 35 routes is about 100 times the average time to learn 5 routes by marking. This suggests that Mike might be able to use marking in a more sophisticated manner to learn all 35 routes in a shorter period of time than the 28.4 years required without marking. We will look more closely at this idea in Section 4.2.

r	1	2	3	4	5
	8.00×10^{-7}	1.04×10^{-4}	5.05×10^{-3}	0.109	0.886

Table 6

	$s = 5$	$s = 35$
the average number of trips to learn all	151.8 3.6 months	14,224.3 28.4 years

Table 7

4.2. Learning by marking and filtering

Suppose that Mike has learned all four marked alternative routes in addition to the regular route after a half year. He may then want to explore some other routes. He might plan to explore the other 30 routes by dividing them into 6 bundles of 5 routes, trying to learn each bundle one by one. We suppose that he explores one bundle for a half year, and he moves to the next bundle storing any long-term memories in the process. Thus, Mike has discovered a method of filtering to improve his learning.

According to the result of Section 4.1, Mike most likely learns all five routes within a half year. By his filtering he reduces the expected time to learn all 35 routes from 28.4 years to only $250 \times 7 = 1750$ (3.5 years).

The probability of that he finishes his entire exploration in 3.5 years is $(0.886)^7 \doteqdot 0.427$, and with the remaining probability 0.573, at least one route is not learned after 3.5 years. If some routes still remain unlearned, then we assume that he rebundles the remaining routes into bundles of 5. However, we expect a rather small number of unlearned routes to remain; the event of 3 remaining is rare event occurring with only probability 0.03. With high probability, Mike's learning finishes within 4 years.

If we treat the above filtering method alone, forgetting the original constraint such as the energy-scarcity mentioned in Section 1.2, the extreme case would be that he chooses and fixes one route for two trips and goes to another route. In this way, he could learn all routes with certainty in precisely 35 days. However, this type of short-sighted optimal programming goes against our original intention of exploration being rather rare and unplanned. Commuting is one of many everyday activities for Mike, and he cannot spend his energy/time exclusively on planning and undertaking this activities. Though our example is very simplified, we should not forget that many unwritten constraints lie behind it, which are still significant to Mike's learning.

5. Learning preferences

Here, we consider Mike's learning of his own preferences. Mike finds his own preferences based on comparisons between experienced routes. First, we specify the bases for our analysis, and then we formulate the process by which Mike learns his own preferences. We simulate this learning process in Section 5.1, and show that learning of his preferences is typically much slower than learning routes. Consequently, notions like "marking" become even more important. In Section 5.2, we consider the change of the process when he adopts a more satisfying route based on his past experiences.

5.1. Preferences

Since Mike has no idea of details along each route at the beginning, one might wonder if he has well-defined preferences over the routes or what form they would take. By recalling the original meaning of "preferences", however, we can connect them with experiences. Since an experience of each route gives some level of satisfaction, comparisons between satisfaction levels can be regarded as his preferences. Here, preferences are assumed to be inherent, but they are only revealed to Mike himself when he experiences and compares different outcomes. In this way, Mike may come to know some of his own preferences.

trips	Prob. of comparison a_0 vs. a_l	Prob. of comparison a_l vs. $a_{l'}$
250 (a half year)	0.981	0.053
5000 (10 years)	1.000	0.671
10000 (20 years)	1.000	0.892

Table 8

We assume that Mike's inherent preference relation over the routes is complete and transitive. A preference between two routes is experienced only by comparing the two satisfaction levels from those routes[6] [7]. A feeling of satisfaction typically emerges in the mind (brain) without tangible pieces of information. Such a feeling may often be transient and only remain after being expressed by some language such as "this wine is better than yesterday's". We assume, firstly, that satisfaction is of a transient nature, and secondly, that the satisfaction from one route can be compared with that of another only if these have happened closely in time.

We formulate a preference comparison between two routes as an experience. This experience has a quite different nature from a sole experience of a route. The former needs the comparison of two experienced satisfaction levels. To distinguish between these different types of experiences, we call a sole experience of a route a *first-order experience*, while a pairwise comparison of two routes is a *second-order experience*. Our present target is second-order experiences.

Consider Mike's learning of such second-order experiences in the simulation frame $F = [s, p : (m, s)] = [5, 1/5 : (10, 2)]$ with $T = 250$ or 5000. A short-term memory is now treated as a sequence of length 10. Consecutive routes can be compared to form preferences over pairs. For example, in Fig.7, the short-term memory is the sequence of 10 pairs $\langle a_1, a_0 \rangle, \langle a_0, a_0 \rangle, ..., \langle a_3, a_0 \rangle$. We treat them as unordered pairs, e.g., the pairs $\langle a_1, a_0 \rangle$ and $\langle a_0, a_1 \rangle$ in $t-9$ and $t-5$ are treated as the same. These second-order experiences may become long-term memories.

For a second-order experience to become a long-term memory, however, it must occur at least twice in a short-term memory. In Fig.7, $\langle a_0, a_1 \rangle$ occurred twice, and hence it becomes a long-term memory. We require these consecutive unordered pairs be disjoint; for example, (a_0, a_3) and (a_3, a_0) occurred twice having the intersection a_3, so these occurrences are not counted as two.

a_1	$^{a_1}\ a_0$	$^{a_0}\ a_0$	$^{a_0}\ a_0$	$^{a_0}\ a_0$	$^{a_0}\ a_1$	$^{a_1}\ a_2$	$^{a_2}\ a_0$	$^{a_0}\ a_0$	$^{a_0}\ a_3$	$^{a_3}\ a_0$
$\to$	$t-9$	$t-8$	$t-7$	$t-6$	$t-5$	$t-4$	$t-3$	$t-2$	$t-1$	t $\to$

Figure 7

[6] This should be distinguished from the notion of "revealed preferences" (cf. Malinvoud [20]) where a preference is defined by a (revealed) choice from hypothetically given two alternatives. It is our point that this hypothetical choice is highly problematic from the experiential point of view.

[7] Our problem is how a person learns his own preferences from experiences, but not how his preferences emerge. In this sense, our problem is not "endogenous preferences". Nevertheless, our problem includes partial and/or false understanding of one's own preferences; thus, it is potentially related to the field of endogenous preferences. See Bowles [2] and Ostrom [21] for the literature on endogenous preferences, and see also Kahneman [11] for other aspects related to this literature as well as our problem.

r	4	5	6	7	8	9	10
10 years later	1.07×10^{-3}	0.0155	0.079	0.215	0.329	0.269	0.0913
20 years later	1.59×10^{-15}	7.86×10^{-5}	0.0016	0.0179	0.111	0.366	0.504

Table 9. Probabilities of preference learning after 10 and 20 years

The computation result is given in Table 8 with $l, l' = 1, 2, 3, 4$ and $l \neq l'$. In the column of a_0 vs. a_l, the probability of the preference between a_0 and a_l being a long-term memory is given as 0.981 for $T = 250$. After only about 2 years, the probability is already 1[8].

We find in the right column of Table 8 that Mike's learning is very slow. After a half year, Mike hardly learns any of his preferences between alternative routes. An experience of comparison between a_l vs. $a_{l'}$ happens with such a small probability, because both deviations a_l and $a_{l'}$ from the regular route a_0 are required consecutively and also twice disjointedly. This means that his learned preferences are very incomplete even after quite some time.

For example, suppose that Mike's original preference relation is the strict order, a_3, a_4, a_0, a_1, a_2 with a_3 at the top, which is depicted as the left diagram of Fig.8. After half a year, he likely learns his preferences between a_0 (regular) and each alternative $a_l, l = 1, 2, 3, 4$, which is illustrated in the middle diagram of Fig.8. It is unlikely that he learns which of a_3 or a_4 (or, a_1 or a_2) is better. Even if he believes *transitivity* in his preferences, he would only infer from his learned preferences that both a_3 and a_4 are better than a_1 and a_2.

Ten years later, Mike's knowledge will be much improved. By this time, with probability 1, he will have learned his preferences between a_0 and each alternative $a_l, l = 1, 2, 3, 4$. He will also likely have learned his preferences between some of the alternatives. Table 9 lists the probabilities that exactly r of his preferences are learned. Recall that there are $\binom{5}{2} = 10$ comparisons. Even after 10 years, Mike is still learning his own preferences over alternative routes. After 20 years, however, he learns much more about his preferences. As it happens, by the time Mike is able to get to taste the rough with the smooth, he is already old.

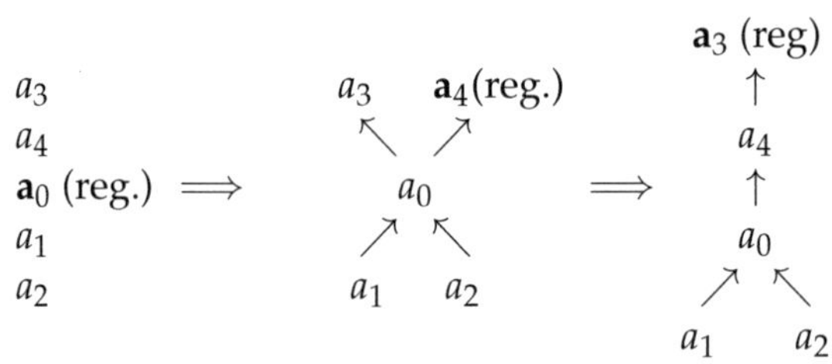

Figure 8

[8] One might wonder why the value of 0.981 for a comparison between a_0 and a_l is higher 0.970 for just learning a route a_l in Table 4. This can be explained by the counting of pairs at the boundary. For example, the comparison between a_0 and a_1 appearing in Table 8 becomes a long-term memory from the short-term memory at time t. However, in our previous treatment of memory of routes, a_1 would not be a long-term memory.

5.2. Maximizing preferences

The results of the previous subsection tell us that it is difficult for Mike to learn his complete preferences. However, completeness should not be his concern. For him, it would be important to find a better route than the regular one, and to change his regular behavior to the best route he knows. This idea is formulated as follows:

(1) He continues to learn his preferences until he can compare each marked alternative to the regular one;

(2) If he finds a better route a_l than a_0 in those comparisons, then he chooses a_l (arbitrarily, if there are multiple) as the new regular route;

(3) He stores a_0 and the alternative routes less preferred than a_0;

(4) He makes an exploration of his preferences over the remaining marked alternatives with the new regular route a_l;

(5) He repeats the process determined similarly by (1) – (4) until he does not find a better route than the regular one.

The final result of this process gives a highest preference. Our concern is the length of time for this process to finish, and his knowledge about his preferences upon finishing.

Suppose that Mike's original (hidden) preferences are described by the left column of Fig.8; he has a strict preference ordering $a_3 \succ a_4 \succ a_0 \succ a_1 \succ a_2$, where a_0 is the regular route. After some time, he learns his preferences described in the middle diagram. In this case, it is very likely that only his preferences between a_0 vs. a_l $(l \neq 0)$ are learned. The arrow $\rightarrow$ indicates the learned preferences.

Here, let us see the average time to finish his learning for preference maximization, under the *assumption* that as soon as he finishes his learning of the preferences between the regular route and alternative ones, he moves to learning the unlearned part. The transition from the left column to the middle one in Fig.8 needs the average time 136.2 (3.3 months). When he reaches the middle diagram, he stores the preferences over a_0, a_1 and a_2.

In the middle diagram of Fig.8, he starts comparing between a_3 and a_4. Here, a_4 is taken as the new regular route. Once he obtains the preference between a_3 and a_4, he goes to the right diagram and he plays the most preferred route a_3. The average time for this second transition is 11.0 trips (1.1 week). Hence, the transition from the left diagram of knowing no preferences, to the rightmost diagram takes the average time of $136.2 + 11.0 = 147.2$ trips (3.5 months).

We have $5! = 120$ possible preference orderings over a_0, a_1, a_2, a_3, a_4 and a_5. We classify them into 5 classes by the position of a_0. Here we consider only the other two cases: a_0 is the top or the bottom. When a_0 is the top, only one round of comparing a_0 to other a_l is enough to learn that a_0 is his most preferred route. This takes the average time 136.2 (3.3 months), which is the same as the time for the transition to the middle of Fig.8. In the case with the top a_0, however, Mike learns no other preferences.

Consider the case where a_0 is the bottom. There are several cases depending upon his choice of new regular routes. But now there are four possibilities for the choice of the next regular route. Depending upon this choice, he may finish quickly or needs more rounds. The more quickly he finishes, the more incomplete are his preferences. Alternatively, the slowest case

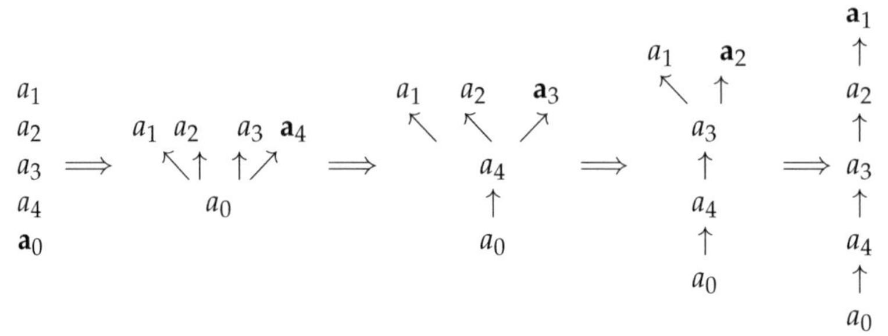

Figure 9. Transitions with learning preferences

for finding the top needs 4 transitions. Fig.9 depicts the slowest case: The total average time is $136.2 + 78.0 + 36.4 + 11.0 = 261.6$ (6.3 months); the bold letter means the regular route. By this process he finds his complete preferences, still, with the help of transitivity.

In Sum, if Mike learns the top quickly, he learns virtually nothing about his preferences between the other alternatives. On the other hand, if he finds the top slowly, he would have a much richer knowledge of his own preferences[9].

6. Sensitivities with parameter changes

We have seen the effects of changes of s and T on Mike's learning determined by the simulation frame $F = [s, p; (m, k)]$. In this section, we briefly consider the sensitivity of the simulation results to the other parameters p (deviation probability), m (length of a short-term memory), k (threshold number).

The deviation probability p and the other two parameters (m, k) are of a different nature. First, we keep in mind that our intention is to capture casual everyday learning. While p is regarded as externally given, it may be controlled by Mike in an effort to learn more about alternative routes. The parameters m and k may also be within Mike's control, but because they describe his memory ability, changing them may require greater effort on his part than increasing p. Whether or not these are in Mike's control, it is still interesting to find out how sensitive his learning is to these parameters.

We start with a sensitivity analysis of learning to changes in m and k. Let $p = 1/5$ and $s = 5$. Table 10 gives the probability of a specific route a_l $(l \neq 0)$ being a long-term memory for the cases of $k = 1, 2, 3$ with $m = 10$. Focusing on $T = 250$, the drop in probability from 0.970 for $k = 2$ to 0.488 for $k = 3$ suggests that Mike's learning is quite sensitive to changes in k.

On the other hand, Table 11 suggests that his learning is less sensitive to the change in the length m of each short-term memory.

When m and k change simultaneously for $s = 5, 35$, we have the results listed in Tables 12 and 13.

[9] Some reader may wonder what implications this argument has on the discounted sum of future utilities. Even under the stationarity assumption that preferences are time independent, this problem of time preferences requires the 3rd-order experiences, i.e., a preference between a present outcome and a next outcome should be compared with another preference. Without the stationary assumption, experiences of any orders are required. In this sense, from the experiential point view, the discounted sum of future utilities are out of the scope.

	$T = 250$	$T = 5000$
$k = 1$	1.000	1.000
$k = 2$	0.970	1.000
$k = 3$	0.488	1.000

Table 10. $s = 5$ and $m = 10$

	$T = 250$	$T = 5000$
$m = 7$	0.930	1.000
$m = 10$	0.970	1.000
$m = 20$	0.995	1.000

Table 11. $s = 5$ and $k = 2$

(m, k)	$T = 250$	$T = 5000$
$(10, 2)$	0.970	1.000
$(20, 3)$	0.840	1.000

Table 12. $s = 5$

(m, k)	$T = 250$	$T = 5000$
$(10, 2)$	0.069	0.765
$(20, 3)$	0.007	0.140

Table 13. $s = 35$

$p \setminus T$	250	5000	Av.no
0.05	0.259	0.998	1720
0.1	0.655	1.000	488.6
0.2	0.970	1.000	151.7
0.3	0.999	1.000	80.24

Table 14. $s = 5$

Table 13 shows that increasing both k and m implies that Mike's learning can also be affected a lot. In the case of $s = 35$, his learning of a single alternative becomes much worse. However, from Table 12, we find the implication that "marking" still helps Mike a lot.

Finally, we consider how sensitive Mike's learning is with respect to the probability of deviations p. We look at how his learning changes when p changes from $1/5$ to 0.05, 0.1 and 0.3. We focus on the probability that a specific a_l $(l \neq 0)$ becomes a long-term memory for the cases of $s = 5, 35$ and $T = 250, 5000$. The results are given in Tables 14 and 15:

We find that the probability of a_l $(l \neq 0)$ being a long-term memory is quite sensitive to a change in p. In the case of $s = 5$, when $p = 0.1 = 1/10$ or $0.05 = 1/20$, the probability of an alternative route becoming a long-term memory after a half year is much smaller than at $p = 1/5$. In the case of $s = 35$, the decrease in this probability is even more dramatic. On the other hand, increasing p to 0.3 has quite a large effect of raising the probability to almost 1 even for half a year. The rightmost columns of Tables 14 and 15 also list the average number

$p \setminus T$	250	5000	Av.no
0.05	0.005	0.091	215707
0.1	0.018	0.312	54893
0.2	0.069	0.765	14223
0.3	0.143	0.957	6548.5

Table 15. $s = 35$

of trips needed to have all routes being long-term memories. These numbers are seen to also be highly sensitive to changes in p.

The changes of deviation probability p should be interpreted while taking (1) of Section 1.2 into account. That is, if commuting is a small part of his entire social world, then p should be a relatively small value such as 0.2 or 0.05. If Mike is not busy with other work, and he keeps enough energy and curiosity about details of the routes, it may be as high as 0.3. On the other hand, 0.3 means that he uses his energy three times in a week, and his behavior may be interpreted as shirking by his boss.

7. Concluding discussions

The example of Mike's bike commuting is a small everyday situation and provides insights to our everyday behavior. It is designed to capture several aspects of a human behavior in a social world. One important aspect is that the life span of a human being has a definite upper bound. Mike's bike commuting is used to compute what learning is possible within his life span. Also, our target situation is partial relative one person's entire social world. In this respect, the regular behavior is a consequence of time/energy saving and infrequent deviations are exploration behavior. We conducted various simulations to see effects of those aspects.

Consider some implications of our simulation study to related literatures. Our original motivation was, from the viewpoint of IGT, to study the origin/emergence of beliefs/knowledge of the structure of the game. Long-term memories are the source for such beliefs/knowledge. Our results have the implication that it would be difficult for a person to learn the full structure of a game, unless it is very simple. Even with marking, the learning will typically be limited. A focus on limiting cases is no longer appropriate. This leads us to deviate entirely from the literature of evolutionary/learning approach mentioned in Section 1.1.

Our research is more related to everyday memory in the psychology literature (Linton [9], [10] and Cohen [3]). Yet, there is a large distance between our study and experimental psychology. To build a bridge between those fields, we need to develop our theory as well as experimental and simulation studies. Kaneko-Kline [14] had a theoretical study in this direction by introducing a measure of the size of an inductively derived view and considering the effects of marking. This is one direction among many other possible extensions.

In the following, we mention several other possible extensions.

Aspect 1: Long-term memories and decaying: We assume that once an experience becomes a long-term memory, it will last forever. However, it would be more natural to assume that even long-term memories are subject to decay unless they are kept experienced once in a while. In

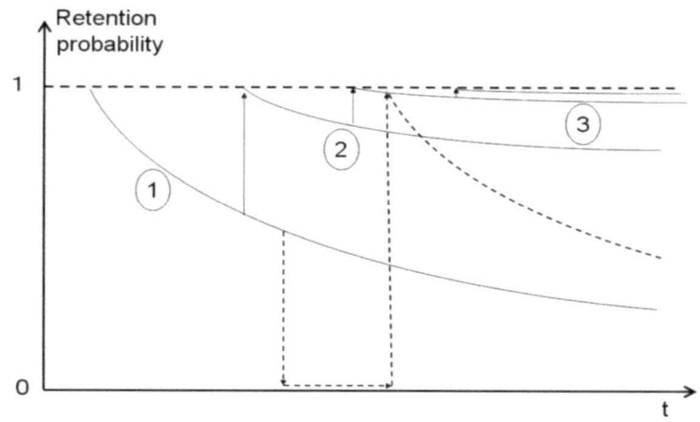

Figure 10. Ebbinghous' Retention Function

particular, when the regular behavior changes as in Section 5, decay or forgetfulness about past regular behavior might become important. This is relevant to the problem of Section 4.2.

The above problem is related to Ebbinghous' [5] retention function which was used to describe experimental results of memory of a list of meaningless syllables. There, no distinction is made between a short-term memory and a long-term memory. The retention function is typically considered as taking the shape of a curved line depicted in Fig.10, where the height denotes the probability of retaining a memory and it is diminishing with time[10].

It is more relevant to our research that repetitive learning makes the probability of retention diminish more slowly. In Fig.10, the second solid curve is obtained when the second experience occurs while the first experience still remains as a memory. On the other hand, the dotted curve is obtained if the first experience disappeared from his memory before the second experience. Thus, the shape of the dotted curve is the same as the first solid one. The second solid curve is flatter than the first one because of repetitive reinforcement. If the third experience occurs soon enough, we move to the third solid curve which is even flatter.

Our treatment of memory can be expressed similarly. For this, consider $(m, k) = (10, 2)$. Once the subject has an experience at t_1, he keeps it as a memory for 10 periods. In Fig.11, the second experience does not come to him within 10 periods, but it comes later at t_2. Then the third experience comes within 10 periods after t_2, and the memory remains forever.

In Ebbinghous' case, the retention function becomes flatter with more experiences, meaning that the memory has a longer expected life. A longer lived memory is more likely to be repetitively reinforced, and so the memory may persist. Our treatment can be seen as a simplification of Ebbinghous' retention function, where we distinguish between a short-term and a long-term memory without decay.

This direction may become even more fruitful with an experimental study.

Aspect 2: Intensities of experiences and preferences: We also ignored intensities of stimuli from

[10] His experiments are interpreted as implying that the retention function may be expressed as an exponential function. By careful evaluations of Ebbinghous' data, Anderson-Schooler [1] reached the conclusion that the retention function can be better approximated as a power function, i.e., the probability of retaining a memory after time t is expressed as $P = At^{-b}$.

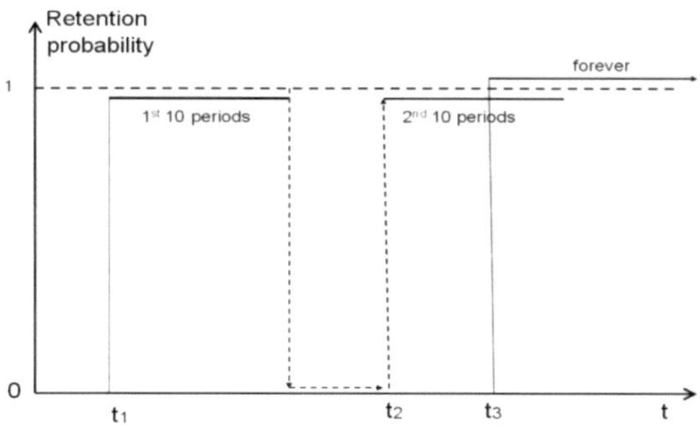

Figure 11. Our Retention Function

experiences. This aspect could be important in the treatment of preferences in Section 5. For example, only preference intensities that are beyond some threshold remain in short-term memories. The use of thresholds is similar to the need for repetition. The concept of "marking" (saliency) is closely related to this problem. It is a topic for future work.

Aspect 3: Two or more learners: We have concentrated our focus on the example of Mike's bike commuting. Our original interests are in learning in game situations with two or more learners (persons)[11]. This has other new features: For example, how does his learning affect the other's learning? In particular, when we consider the other person's understanding, possibly by switching social roles, it affects the persons' behaviors drastically, e.g. emergence of cooperation may be observed. These possibilities are studied in Kaneko-Kline [18]. In that setting, the domain of experiences plays essential roles, for which a simulation study must be informative[12].

These extensions may generate a lot of implications for IGT. We can even introduce more probabilistic factors related to decaying of long-term as well as short-term memories. However, more essential extensions are related to the consideration of internal structures of routes and inductive derivations of individual views from experiences.

Aspect 4: Internal Structures and subattributes: We ignored the internal structure and subattributes of each route in the town by treating it as one entity. Nevertheless, IGT is about the formation of a person's beliefs about the structure of a game situation. The internal structure and subattributes are relevant to this type of analysis. In fact, the introduction of such internal structures will be a key for essential developments of our simulation study as well as IGT itself.

When this is taken into account, an inductive derivation may be regarded as drawing a picture by connecting one subattribute with another. This is originally motivated in Kaneko-Kline

[11] Hanaki *et al.* [7] studied the convergence of behaviors in a 2-person game, where each player's learning of payoffs is formulated in the way of the present paper but his behavior is formulated as a mechanical statistical process following the learning literature. Then, they studied behavior of outcomes in life spans of middle range. Their approach did not take purely the viewpoint of IGT in that a player consciously makes a behavior revision once he has a better understanding of a game situation. Nevertheless, it would give some hint to our further research on IGT.

[12] These aspects are considered in an experimental context in Takeuchi, *et al.* [22], but are not connected to a simulation study.

[15]. Such a process is partially discussed in a theoretical manner in Kaneko-Kline [17]. However, a simulation study will give more detailed information. One immediate question from this to Mike's bike commuting: When Mike is told only one route from the colleague without a map of the town, what kind of a map can Mike construct? After given periods of time, how correct and complete is it?

In sum, simulation studies of those new aspects provide implications for IGT and a lot of new directions for research.

Author details

Eizo Akiyama, Ryuichiro Ishikawa, Mamoru Kaneko
Faculty of Engineering, Information, and Systems, University of Tsukuba, Japan

J. Jude Kline
School of Economics, University of Queensland, Australia

8. References

[1] Anderson J. R. and L. J. Schooler, Reflections of the environment in memory, *American Psychological Society* 2, 396-408.
[2] Bowles, S. (1998), Endogenous Preferences: The Cultural Consequences of Markets and Other Economic Institutions, *Journal of Economic Literature*, 36(1), 75-111.
[3] Cohen, G., (1989), *Memory in the Real World*, Lawrence Erlbaum Associates Ltd., Toronto.
[4] Deutsch, D. and J. A. Deutsch, eds. (1975), *Short-term Memory*, Academic Press, New York.
[5] Ebbinghous, H., (1964, 1885), *Memory: A contribution to experimental psychology*, Dover Publications, New York.
[6] Fudenberg, D., and D.K. Levine, (1998), *The Theory of Learning in Games*, MIT Press, Cambridge.
[7] Hanaki, N., R. Ishikawa, and E. Akiyama, (2009), Learning Games, *Journal Economic Dynamics & Control* 33, 1739-1756.
[8] Harsanyi, J. C., (1967/68), Games with Incomplete Information Played by 'Bayesian' Players, Parts I,II, and III, *Management Sciences* 14, 159-182, 320-334, and 486-502.
[9] Linton, M., (1975), Memory for Real-World Events, in D. A. Norman & D. E. Rumelhart, eds., *Exploration in cognition*, Freeman Publisher, San Francisco.
[10] Linton, M., (1982), Transformations of memory in everyday life. In U, Neisser ed. *Memory Observed: Remembering in natural contexts*, Freeman Publisher, San Francisco.
[11] Kahneman, D. (2011), *Thinking, Fast and Slow*, Penguin, London.
[12] Kalai, E., and E. Lehrer, (1993), Subjective Equilibrium in Repeated Games, *Econometrica* 61, 1231-1240.
[13] Kaneko, M., (2002), Epistemic Logics and their Game Theoretical Applications: Introduction. *Economic Theory* 19, 7-62.
[14] Kaneko, M., and J. J. Kline, (2007), Small and Partial Views derived from Limited Experiences, SSM. DP. No.1166, University of Tsukuba.
http://www.sk.tsukuba.ac.jp/SSM/libraries/pdf1151/1166.pdf

[15] Kaneko, M., and J. J. Kline, (2008a), Inductive Game Theory: A Basic Scenario, *Journal of Mathematical Economics* 44, 1332-1363. Corrigendum: the same journal, 46, pp.620-622, 2010.

[16] Kaneko, M., and J. J. Kline, (2008b), Information protocols and extensive games in inductive game theory, *Game Theory and Applications* 13, 57-83.

[17] Kaneko, M., and J. J. Kline, (2009a), Partial Memories, Inductively Derived Views, and their Interactions with Behavior, to appear in *Economic Theory*. DOI: 10.1007/s00199-010-0519-0

[18] Kaneko, M., and J. J. Kline, (2009b), Transpersonal Understanding through Social Roles, and Emergence of Cooperation. DP. No.1228, University of Tsukuba. http://www.sk.tsukuba.ac.jp/SSM/libraries/pdf1226/1228.pdf

[19] Kaneko, M., and A. Matsui, (1999), Inductive Game Theory: Discrimination and Prejudices, *Journal of Public Economic Theory* 1, 101-137. Errata: the same journal, 3, p.347, 2001.

[20] Malinvaud, E., (1972), *Lectures on Microeconomic Theory*, North-Hollond. Amsterdam.

[21] Ostrom, E. (2000), Collective Action and the Evolution of Social Norms, *Journal of Economic Perspective*, 14(3), 137-158.

[22] Takeuchi, A., Y. Funaki, M. Kaneko, and J. J. Kline, (2011), An Experimental Study of Behavior and Cognition from the Perspective of Inductive Game Theory, DP. No.1267, University of Tsukuba. http://www.sk.tsukuba.ac.jp/SSM/libraries/pdf1251/1267.pdf

[23] Tulving, E., (1983), *Elements of Episodic Memory*, Oxford University Press, London.

[24] Weibull, J. W., (1995), *Evolutionary Game Theory*, MIT Press. London.

Chapter 4

A Tale of Two Ports: Extending the Bertrand Model Along the Needs of a Case Study

Naima Saeed and Odd I. Larsen

Additional information is available at the end of the chapter

http://dx.doi.org/10.5772/54425

1. Introduction

Competition between container terminals may occur if they serve the same hinterland or handle transshipment for container flows with the same origin and/or destination. In this study we focus on the first case. Competition may take place both between terminals located within the same port and those located in different ports. Disregarding terminal charges for container handling and storage, different container terminals will rarely be perfect substitutes from a user perspective. They may differ with respect to transport cost for the inland leg, efficiency, level of service in terms of vessel calls, freight rates charged by container lines, etc.

In a competitive situation with few players and an inhomogeneous product or service, the outcome in terms of market shares and prices can often be treated as the result of a game where each player maximizes profit, but with due consideration to the expected reaction of its competitors. When the competitor's actions are confined to setting the prices of their own product (service), the outcome can be modeled as in the Bertrand equilibrium [1]. Bertrand model is named after Joseph Louis François Bertrand (1822-1900) and was formulated in 1883 by [2] in a review of [3] book in which Cournot had put forward the Cournot model. The model examines the pricing behaviour of interdependent firms in a product market with few rival firms. The idea was developed into a mathematical model by [4].

Our case study deals with the 4 container terminals serving the Pakistani market and is somewhat more complicated than a simple Bertrand situation.

The questions we pose and try to answer are the following:

1. Can the present situation with respect to market shares and container handling fees be interpreted as the outcome of a Bertrand game when we apply our best 'guesstimates' of the parameters of the problem?

2. What are the impacts of the policy pursued by the landlord port in a competitive setting like this? In particular, we are interested in the trade-off between annual rent paid by a terminal and fee on containers handled by the terminal.
3. What can the terminals in the port of Karachi (and Karachi port) in total gain by cooperating (i.e. by forming a coalition)? This potential gain can be the source of a possible cooperative game between the Port of Karachi and the two private terminals.

The rest of the chapter is organized as follows: The following section presents a detailed description of the case study and the information and data available on the terminals. In next section we present a number of research works related to the application of game theory to the port sector. This is followed by the game's solution for the present situation, the rent/unit fee trade-off, the coalition aspect and finally the conclusion.

2. Case studies

Pakistan has the following three major seaports:

- *Karachi Port*, the premier port of Pakistan, is located between the towns of Kiamari and Saddar, close to the heart of old Karachi. It handles about 75% of the entire national trade.
- *Port Muhammad Bin Qasim* is Pakistan's first industrial and multi-purpose deep sea port. Located in the Indus delta region at a distance of 50 km southeast of Karachi, it is well connected to the whole country through modern modes of transportation.
- *Gawadar Port* has just been constructed as the third port of Pakistan. Situated on the Baluchistan coast, it is about 460 km from Karachi and 120 km from the Iranian border (see Figure 1). This port will not be considered in this research as it has just started its operation.

Figure 1. Location of three ports in Pakistan

2.1. Karachi port

Pakistan has about 1062 km of coastline on the Arabian Sea, spreading from the Indian border to the Persian Gulf. At the time of partition, Pakistan (then West Pakistan) had only one functional deep water port at Karachi which not only catered for the entire seaborne cargo of northern India but also provided a transit trade facility to landlocked Afghanistan.

Cargo handled at the Karachi port since the commencement their operation is shown in the figure 2:

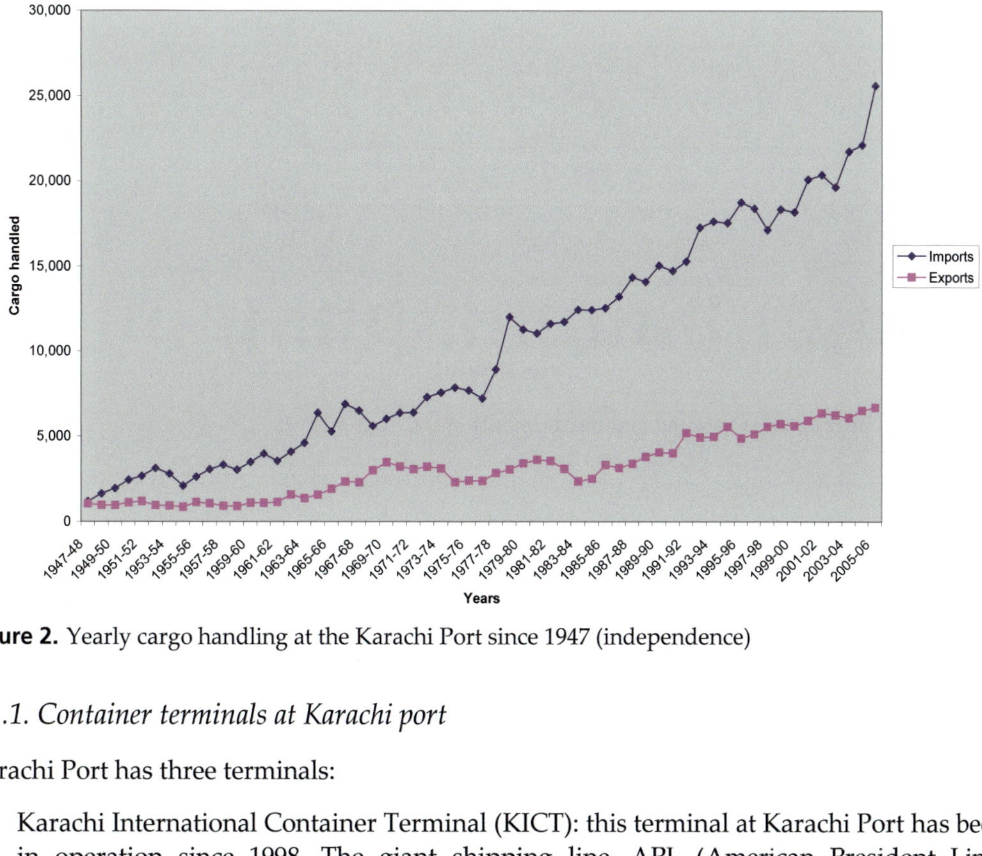

Figure 2. Yearly cargo handling at the Karachi Port since 1947 (independence)

2.1.1. Container terminals at Karachi port

Karachi Port has three terminals:

- Karachi International Container Terminal (KICT): this terminal at Karachi Port has been in operation since 1998. The giant shipping line, APL (American President Line) invested in KICT on a build-operate-transfer (BOT) basis. BOT is the classic case of concessions in which the public sector does not lose ownership of the port infrastructure, and new facilities built by private firms are transferred to the public sector after a specified period of time. Now the terminal has been bought and is managed by Hutchison, Hong Kong.
- Pakistan International Container Terminal (PICT): this terminal at Karachi Port was privatised in August 2002. It was also developed on a BOT basis, specifically build, operate and transfer after 21 years. It is the only container terminal in Pakistan which is sponsored and owned by Pakistanis[1].

[1] See http://www.pictcntrtrack.com/. Accessed 5th April 2007.

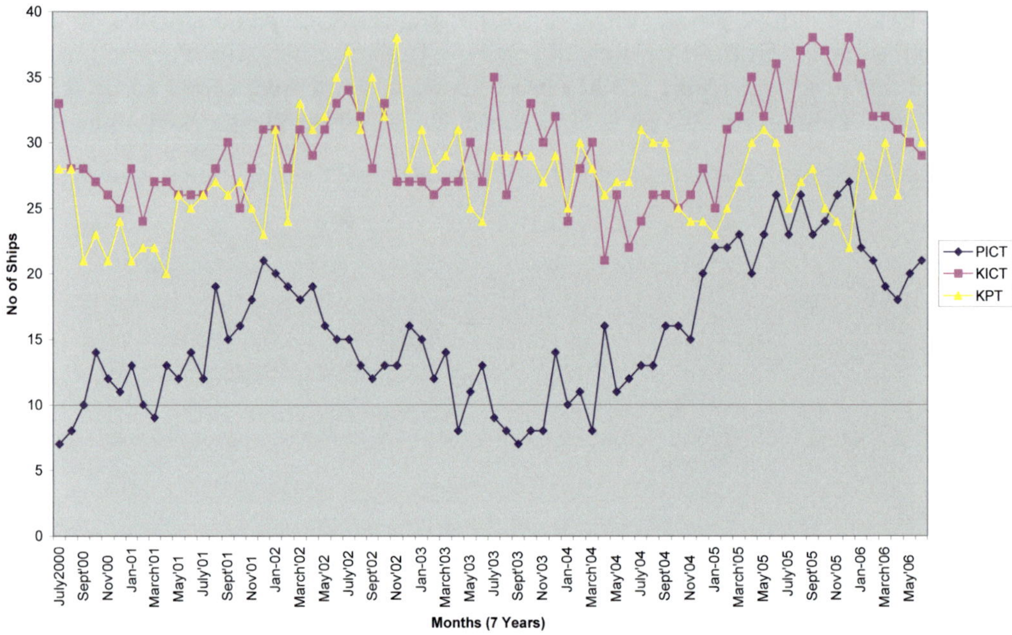

Figure 3. Number of ships handled at three terminals in the Karachi Port

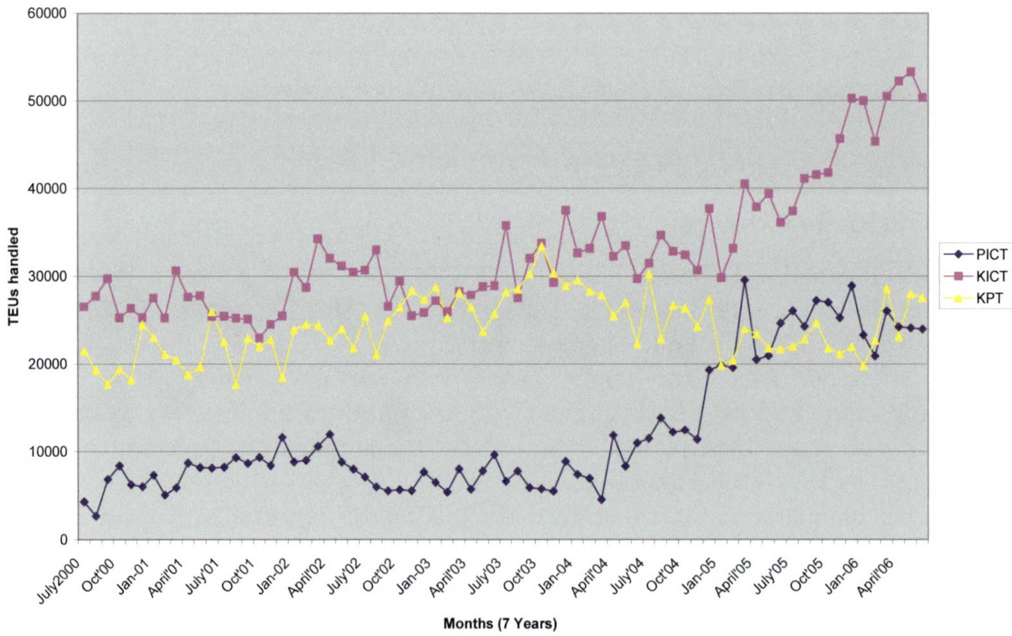

Figure 4. Total TEUs handled at three terminals in the Karachi Port

- KPT [2]: this is a conventional terminal with priority berthing for geared vessels, mainly feeders. Unlike KICT and PICT, this terminal does not have modern equipment like gantry cranes to handle containers. Regular container service started at this terminal in 1973.

The total number of ships and total TEUs (Twenty Foot Equivalent Units) handled at the three terminals of Karachi port, for the period July 2000 to June 2006, are shown in the Figures 3 and 4.

2.2. Port Muhammad Bin Qasim

In the past half century of Pakistan's existence, its seaborne cargo handling has increased tremendously; hence the need for another port was felt. This need became a necessity when the Pakistan Steel Mill project was conceived in the 1970s.

Construction of port Qasim, the second sea port of Pakistan, was started in the mid 1970s and completed and opened to shipping in 1990; it has been in operation since then.

Cargo handled at Port Muhammad Bin Qasim, since the commencement its operations, is shown in the Figure 5[3]:

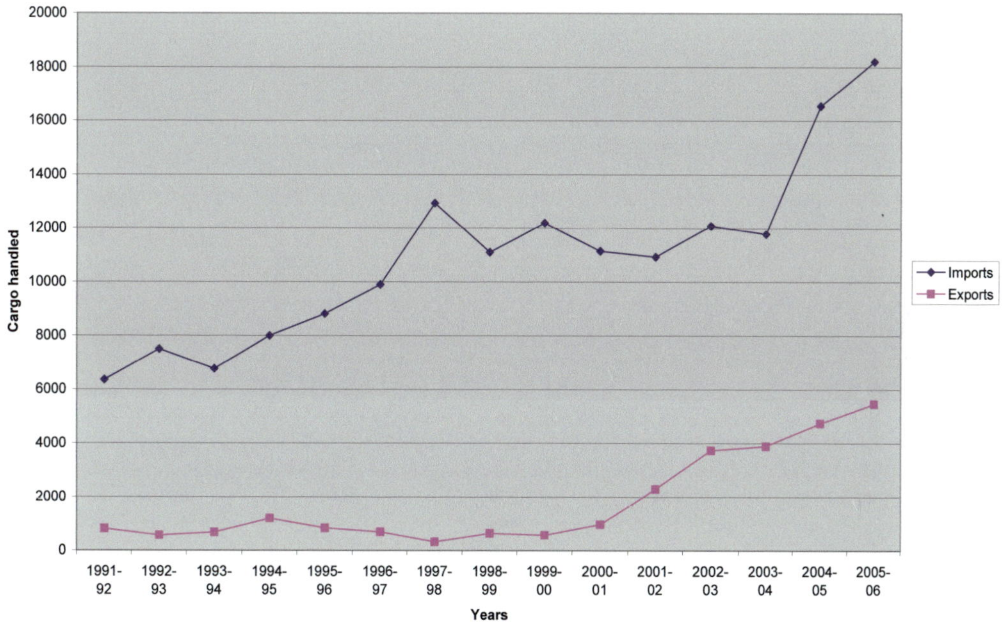

Figure 5. Yearly cargo handling at Port Qasim since 1991

[2] This public terminal does not have any specific name like KICT or PICT. In the official records of Karachi port authority, it is simply written "Geared Vessels", probably because only geared vessels call at this terminal. For simplicity it is mentioned as KPT in this paper. Moreover, as this terminal is owned by the port authority, both the port and KPT are referred to as KPT.

[3] The data used in the graph for Karachi Port was provided by Karachi Port Authority, while the data for Port Qasim is available online at http://www.statpak.gov.pk/depts/fbs/publications/yearbook2007/transport/20.3.pdf http://www.paksearch.com/Government/STATISTICS/bulletin00/Transport/TC1.html Accessed 15 July, 2008.

2.2.1. Terminal at Port Muhammad Bin Qasim

Port Muhammad Bin Qasim has only one container terminal known as Qaim International Container Terminal (QICT). It was built at Port Muhammad Bin Qasim on a build-own-operate (BOO) basis. In the case of BOO, parts of the seaport are transferred to the private operators for development. Initially, Maersk invested in QICT, which has now been bought by Dubai port investors. This terminal was incorporated in 1996 and is located approximately 45 km from Karachi.

Figure 6 shows the performance, in terms of handling container lines, of all four terminals of the two ports. Since 2003, QICT has been the biggest terminal in terms of throughput of ships and containers. Although KPT lacks modern equipment it handles a considerable number of ships and containers. The reason is that feeder and geared vessels tend to prefer this terminal due to its low handling charges. Moreover, only KPT has sheds to store goods de-stuffed from containers.

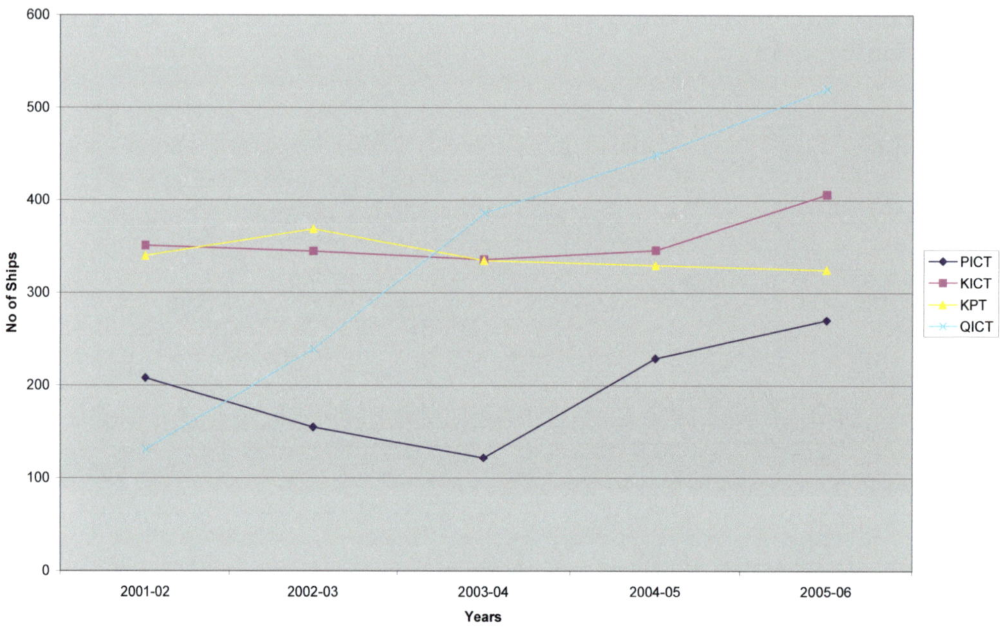

Figure 6. Total traffic handled at two ports in Pakistan

2.3. Cost structure of the two ports

The port of Karachi faces high fixed costs as compared to Port Muhammad Bin Qasim. The reason for this is that Karachi Dock Labor Board (KDLB), which was established to provide regular work and income for dock workers, was previously employed on a casual basis. Up to 2006, there were 3673 dock workers, but the government closed down KDLB in December 2006. KDLB adds significantly to the cost. Firstly, the cargo handling companies are obliged to employ unnecessary KDLB staff and in most cases, pay unnecessary incentive payments.

Secondly, the cargo handling companies have to pay a 'cess' (i.e., a levy) to provide minimum salaries for KDLB staff when they are not 'working' and medical, as well as other, benefits. Moreover, the KDLB impairs competition between the KPT and Port Qasim Authority as the latter does not have a Dock Labor Board.[4] In short, KDLB has contributed to the increase in the fixed cost of Karachi port as compared to Port Muhammad Bin Qasim.

The costs of container operations at Qasim are lower than they are at Karachi, because they do not have to pay for a dock labor board and they pass only 60 percent of the wharfage charges back to the PQA. These savings reduce costs by about US$29 per TEU (US$17 for the dock labor board and US$12 for the lower wharfage) compared with KPT. On the other hand, QICT has the additional costs of inland transport to/from Karachi and the cities to the north: the weighted average additional land transport cost of using Qasim rather than the KPT is about US$31 per TEU. The net result is that Qasim suffers from a cost disadvantage of about US$2 per TEU, which is trivial in relation to total costs and the importance of service quality.

2.4. Performance indicators of the two ports

Pakistan's ports now rate quite highly on the two most important performance indicators: handling speeds are generally up to international standards and tariffs are only slightly higher by international standards.

2.4.1. Handling speeds

The ship–shore handling speeds for containers at Karachi is 25 moves per crane/berth hour, while at Port Qasim it is 22–24 moves per crane/berth hour. However, at the conventional terminal at Karachi port, productivity is lower at about 17 moves per crane hour. But it has the advantage over private terminals in terms of services, which are reported to be about US$30 per TEU less expensive than those at the KICT and QICT.

2.4.2. Tariffs

Total container handling charges at Port Karachi's specialized terminals are rather on the high side by international standards. They are estimated at US$113 per twenty foot (not mentioned for individual terminals). There are two main reasons for the relatively high charges. Firstly, the shipping lines impose several additional charges including, in some cases, a shipping surcharge whose justification is now no longer clear. Secondly, a 'terminal handling charges' (THC) is effectively charged twice: (i) by the shipping line and (ii) by the container terminal. If the THC were charged only once, the cost would be around US$88 per TEU, which is more in line with international charges, including those in Indian ports. Handling charges for Port Qasim are US$105.

---

[4] See http://www.brecorder.com/index.php?id=483970&currPageNo=1&query=&search=&term=&supDate Accessed 20 April, 2007.

Port entry charges are high at both Karachi and Qasim. The combined KPT charges on ships for port entry, tugs, pilotage and berth hire amount to about US$0.82 per GRT (gross registered tons). This would be equivalent to US$26 per TEU on the assumption of a 35,000 GRT ship handling 1000 TEU. Although these charges are well above international benchmarks, they are not significantly higher than at the main Indian ports. However, they are over five times higher than those at the dominant container transshipment ports of the region of Colombo, Dubai and Shalala. The KPT is aware that their charges are high and has reduced the average port entry cost from over US$1 per GRT to US$0.82 in the last year.

The PQA's port entry charges are slightly lower than those of the KPT. The combined charges on ships for port entry, tugs, pilotage and berth hire at PQA amount to about US$0.72 per GRT. These tariffs contribute to large financial surpluses for the KPT and the PQA. They deter lines from calling with large ships, as these are the main reason for the Pakistan Port surcharge. The PQA reduced port charges by 15 percent in May 2005.

2.5. Ports' finances

Both ports make excessive profits. The KPT's budget revenues of US$140 million for 2004–05 are more than twice as high as their budgeted costs. A surplus at this level is unusual in the port industry, as is the very high level of income from investments (43 percent). The PQA also makes a large profit, with operating revenue of US$32.1 million compared with operating costs of US$21.2 million in the fiscal year 2003. This shows that about 40 percent of the net surplus comes from income from investment, property and storage while about 60 percent comes from operations[5].

2.5.1. Revenue from private container terminals to port authorities

According to the agreement with KPT, KICT pays $6.03 per move to KPT, while PICT pays $12.54 per move[6]. Similarly, according to the lease agreement (30-year renewable lease) in 1994, QICT pays a flat rent of Rs 48 million per annum, a wharfage of Rs 400 per TEU, a 5 per cent royalty of the load on/load off revenue on cargo exceeding 150,000 TEUs to PQA per annum[7].

The average market share (in terms of TEUs handled) for all terminals, calculated for the years 2001–2006[8],and handling charges, obtained with the help of questionnaire[9] are given in Table 2-1.

---

[5] See Transport Competitiveness in Pakistan: Analytical Underpinning for National Trade Corridor Improvement Program, 2006. *World Bank report available online at* http://www-wds.worldbank.org/external/default/WDSContentServer/WDSP/IB/2006/08/14/000160016_20060814091138/Rendered/PDF/36523.pdf Accessed 25 April, 2008.

[6] Source of information is personal interview with authority at Karachi Port Trust.

[7]See http://www.pakistaneconomist.com/database2/cover/c98-21.asp Accessed 25 April, 2008.

[8] Data about TEUs handled by all three terminals at Karachi port was obtained from port authority. While data for QICT is collected from official website of terminal www.qict.net Accessed 20 April, 2007.

	QICT	*KICT*	*PICT*	*KPT*
Market Share	31%	32%	13%	23%
Handling Charges	76$	94$	69$	54$

Table 1. Market share & handling charges

3. Game theory, the model

3.1. Game theory

Game theory was created by Von Neumann and Morgenstern in their classic book *The Theory of Games and Economic Behavior,* published in 1944. Game theory provides a framework for the study of the interactions of decision-makers whose interests are related, though distinctly different, and whose actions jointly determine all outcomes. There are several different ways of characterizing a game, generally not equivalent to one another; however, they all have certain elements in common. The common elements are the set of decision-makers, called *players,* the rules and regulations concerning the possible decisions that each decision maker can choose, sometimes called the *set of strategies,* and rules and regulations governing the way that the players' decisions are related to the reward or payoffs they receive [5].

According to [6], the theory of games can be divided into two distinct approaches: the *strategic* or *noncooperative* approach and *coalitional* or *cooperative* approach. In noncooperative game theory players treat each others as competitors and each individual's payoff is affected by the strategies chosen by the other players. When choosing a strategy, a player therefore picks one that yields his most preferred outcome given the strategies chosen by the others. This behaviour implies rationality. Nash equilibrium is the result of rational play: each player's strategy is an optimal response to what he believes the other players will do, and the belief is correct, while cooperative game theory is concerned with those situations in which players can negotiate before the game is played about what to do in the game. Moreover, it is assumed that these negotiations can be concluded by the signing of a *binding* agreement. Under these conditions, the precise strategies available in the game will not matter very much. What matters is the *preference* structure of the game, since it is this that determines what contracts are feasible [7].

In Oligopoly models, one side of the market typically consists of either price or quantity setters, with price takers on the other side. With homogenous products, the Cournot model is most often chosen to describe market interaction, and with differentiated products, the Bertrand model is usually applied. For each of these static, simultaneous-decision models of oligopoly, there is a sequential-decision counterpart. These sequential-decision models are the progeny of von Stackelberg's strategic analysis of quantitiy setting [8]. Stackelberg models rely on leadership by one of the rivals. Originally, von Stackelberg extended the

[9] Questionnaires filled by shipping agents working for foreign principals at Karachi city of Pakistan [See 14] . We have not been able to verify that the average charge reported in the questionnaires corresponds to the actual average charge per container for the respective terminals.

Cournot model to include leadership behavior [9]. That is, the Stackelberg leader's output choice influences the output choices of its rivals, and the Stackelberg leader chooses output in full recognition of its followers' reactions. Von Stackelberg's insight can readily be adapted to the Bertrand pricing model [8]. In this paper, our focus is on Bertrand price model application to a service industry that consists of four container terminals.

3.2. Related literature

Research related to the application of game theory to ports is very limited. [10] developed a game-theoretic best response framework for understanding how competing ports will respond to development at a focus port, and whether the focus port will be able to capture or defend market share by building additional capacity. They applied this model to the investment and competition currently occurring between the ports of Busan and Shanghai. Unlike the analyses on which port expansion plans are typically based, the authors explicitly account for the incentives and opportunities for fellow competitors to respond to investments (or the threat thereof) or to defend or appropriate market share through a game-theoretic response framework. However, the authors did not apply a two-stage Bertrand competition model. They instead, in order to develop a game-based analysis of Busan and Shanghai port development policies, abstracted from the pricing game, focusing instead on strategy in the development game given the observed or projected prices.

[11] applies the Hotelling location model to inter-port competition. They use this model to develop a framework for linking the strategic interdependence between ports' and potential terminal providers' investment desirability. Moreover, considering the importance of the role of the users' cost in selection of a port, the authors explicitly include it in the model in order to examine how changes in the users' cost affect potential operators' decisions. However, the users' cost consists here of port dues charged by the port authority and the service fee of the service providers.

[12] considers both the quantity of competition and price of competition between ports and examines the interaction between hinterland access conditions and port competition. Competition between ports is treated as competition between alternate intermodal transportation chains, while the hinterland access conditions are represented by both the corridor facilities and the inland roads. When ports compete in quantities, an increase in corridor capacity will increase the port's own output, reduce the rival port's output and increase the port's own profit. On the other hand, an increase in inland road capacity may or may not increase the port's own output and profit, owing to various offsetting effects. Finally, inland road pricing may or may not increase the port's own output and profit.

[13] analyses the interaction between the pricing behaviour of the ports and optimal investment policies in port and hinterland capacity. They use the framework of a two-stage game in capacities and prices. The main focus here is on a governance structure where capacity decisions are public, but pricing decisions are private. The game is analysed by backwards induction. The authors obtained the following results. First, profit-maximizing ports internalise hinterland congestion in so far as it affects their customers. Second,

investment in port capacity reduces prices and congestion at both ports, but increases hinterland congestion in the region where the port investment is made. Investment in a port's hinterland is likely to lead to more port congestion and higher prices for port use, and to less congestion and a lower price at the competing port. Third, the induced increase in hinterland congestion strongly reduces the direct benefits of extra port activities. Finally, imposing congestion tolls on the hinterland road network raises both port and hinterland capacity investments.

[14] applies a two-stage game that involves three container terminals located in Karachi Port in Pakistan. In the first stage, the three terminals have to decide on whether to act as a singleton or to enter into a coalition with one or both of the other terminals. The decision at this stage should presumably be based on the predicted outcome for the second stage. The second stage is here modelled as a Bertrand game with one outside competitor, the coalition and the terminal in Karachi Port (if any) that has not joined the coalition. Furthermore, three partial and one grand coalition among the three terminals at Karachi Port are investigated. The concepts of "characteristic function" and "core" are used to analyse the stability of these coalitions. And results revealed that one combination does not satisfy the superadditivity property of the characteristic function and can therefore be ruled out. The resulting payoffs (profits) of these coalitions are analysed on the basis of "core". The best payoff for all players is in the case of a "grand coalition". However, the real winner is the outsider (the terminal at the second port) which earns a better payoff without joining the coalition, and hence will play the role of the "orthogonal free-rider".

[15] analyses the effect of the type of concession contracts on port user surplus and on profits of terminal operators (or port authorities) with the help of game theory. Authors have selected three ports in Pakistan to perform this analysis. These ports function as 'landlords' and have signed concession contracts with private container terminal operators. However, the features of the contracts at present are different for each terminal. Four cases are discussed in this article. The first case is the present situation in which authors treat competition between terminals as a Bertrand game in which each terminal non-cooperatively determines charges for container handling and pays fees to port authorities according to the contract. Furthermore, in the second and third cases, a cost benefit analysis is conducted by solving the Bertrand model. The results reveal that in the long run it is profitable for the Karachi Port to establish a same fixed fee contract with its private terminals. However, users are better off in a situation where a percentage fee concession contract would be adopted instead.

[16] introduces a game theory model to study the scale of container terminals in combination with the market size in order to examine how they affect terminal competition. The starting point is the landlord port management system with long-term concessions agreements shaping the formal relationships between the port authority (who owns the land) and the private terminal operators (who use the land for terminal activities). The model was applied to Karachi and Qasim ports in Pakistan. Results show that the perspectives of port authorities and terminal operators on the balance between economies of scale and intra-port competition are different. Port authorities have a preference for a

number of small terminals inside their ports in view of stimulating intra-port competition. Terminal operators prefer to operate in ports with the smallest number of large terminals (one terminal if possible - monopoly setting).

3.3. The model

3.3.1. The demand for container terminal services

The present model treats the competition among terminals as a Bertrand game and also uses the outcome of the Bertrand game to investigate the payoff (profit) for the concerned entities. The Bertrand game is a natural choice in this setting. In the container terminal industry, competitors offer similar but, from the perspective of individual customers, not quite homogeneous services. To detail the structure of the Bertrand game, the demand function of each service provider must be made explicit.

The term "terminal users" is applied to the agents who pay the cost of container freight and handling, and make the choice of which terminal to use. Different ports and terminals can rarely be considered as perfect substitutes from a user perspective. In addition to the terminals' charges for handling and storing containers, the user will have additional costs, or other user costs (OUC). The components of OUC include the following:

- Inland transport (such as rail and truck) costs for transporting containers to and from terminals within Pakistan.
- Freight rates charged by container lines, in particular any surcharges related to port and terminal efficiency.
- Costs related to transport time, including the cost of container lease or rental. Container lease cost is included in this component because, with the increase in transport time, the lease period will also increase, which will result in increased costs.

The difference between the first and third components is that, in the first case, cost refers to what users pay for inland transportation of containers, while in the third case, cost refers to the costs they have to bear because of the time spent in transporting containers.

Even if terminal charges are equal, differences in OUC may lead to different market shares for competing terminals. On the other hand, differences in OUC may result in persistent differences in terminal charges and market shares even in a competitive setting. The present model assumes that OUC is composed of two components, one that is independent of the volume of containers handled by each terminal, and one that is an increasing function of the volume handled (and decreasing in rated capacity). The rationale for a variable component of OUC is two-fold:

1. The spatial aspect: Marginal customers will, on average, have longer transport distances and higher transport costs to the terminal than the average customer.
2. When the volume of containers approaches or exceeds the rated capacity, different types of delays are likely to increase. Some delays may affect the ship turnaround time and subsequently the freight rates due to congestion surcharges by shipping lines, while other types of delays may affect the dwell time of containers in port.

A counteracting force may be that the level of service for vessel calls will improve with the volume of containers handled. For surface and air transport, this aspect is generally referred to as the Mohring effect [17]. [18] uses the throughput share of the port to capture the Mohring effect. However, in this case with constant capacity, it can be expected that both 1 and 2 will have a stronger negative impact on OUC than the positive Mohring effect. In general, the user cost function for terminal "*i*," OUC (*i*), has the following form:

$$OUC(i) = C0_i + f(\frac{X_i}{CAP_i}) \quad (1)$$

where $C0_i$ is the fixed component, X_i is the volume handled by terminal "*i* and CAP_i is the rated capacity of terminal "*i*." *f* is an increasing function of the ratio.

In the numerical implementation of the Bertrand model, the market share of each terminal is determined by an aggregate multinomial logit model, and the demand for all terminals combined is a function of the logsum from the logit model.

The use of a logit model presupposes that a "utility function" can be assigned to each terminal. The utility functions in an aggregate logit model can be interpreted as a measure of the attractiveness of a terminal as perceived by the "average" user.

The utility functions of terminals are given as follows:

$$U_i = a_i + b\{\ p_i + OUC_i\ \} \quad (2)$$

Where U_i is the "utility" of terminal i $\quad$ *i* = KICT, PICT, KPT, and QICT

p_i is price charged per unit by terminal i $\quad$ *i* = KICT, PICT, KPT, and QICT

OUC_i = other user cost at each terminal i $\quad$ *i* = KICT, PICT, KPT, and QICT

b is the co-efficient of price charged by terminals and a_i is the alternative specific constant for terminal i;

a_{PICT} and $a_{KPT} = 0$, while a_{KICT} and $a_{QICT} > 0$.

The alternative specific constant is included in the utility functions for KICT and QICT, to capture the attributes that enable these terminals to obtain high market shares compared to their competitors. As is apparent from Table 1, the average market share of these two terminals is high compared to PICT and KPT.

As KPT is owned by Port Authority, the port and KPT are treated as one economic entity in the model, and no distinction is made between handling charges and fees. What matters for KPT is the combined revenue from fees paid by the private terminals, and profits from their own terminal's operation.

The market share of terminal "*i*" is given by the following logit expression:

$$Q_i = \frac{e^{U_i}}{\sum_j e^{U_j}} \qquad i = \; KICT,\ PICT,\ KPT \ and \ QICT \tag{3}$$

The logsum is defined by

$$LS = \ln(\sum_j e^{U_j}) \tag{4}$$

Total aggregate demand (in TEUs) for all the players is thus given by

$$X = Ae^{\theta LS} \tag{5}$$

where A and θ are constants and $0 < \theta < 1$,

Individual demand for player "i" is given by the following equation:

$$X_i = X.Q_i \qquad i = \; KICT,\ PICT,\ KPT,\ and \ QICT \tag{6}$$

Therefore, the demand faced by a terminal will depend on handling charges (including unit fee) and OUC for all terminals. The private terminals will keep the handling charge, but the revenue from fees is transferred to the Port Authority. Individual demand is elastic because change in price and other attributes of one terminal will shift the traffic between that terminal and other terminals. There will also be a slight effect on the total demand via the logsum.

3.3.2. Revenue/profit for terminals

The operating surplus of the terminal "i" is the following:

$$\Pi_i = (p_i - w_i - c_i) \cdot X_i \tag{7}$$

where p_i is the handling charge per TEU paid by the users, w_i is the fee paid by private terminals per TEU handled, and c_i is the marginal cost per TEU.

If the contract implies that unit fee is a percentage of the handling charge, the surplus is alternatively given by

$$\Pi_i = \left(p_i.(1 - \delta_i) - c_i\right) \cdot X_i \tag{8}$$

where δ_i is the fee and $p_i.(1 - \delta_i)$ is the share of the handling charges retained by the terminal.

The profit for KPT (including Port Authority) is taken as

$$\Pi_3 = (p_3 - c_3)X_3 + w_1 X_1 + w_2 X_2 \tag{9}$$

where 1= KICT, 2= PICT and 3= KPT.

For any contract between the Port Authority and a private terminal operator that will be viable in the long run there must be;

$$\Pi_i = (p_i - w_i - c_i) \cdot X_i \geq annual\, rent(i) \tag{10}$$

That is, the operating surplus must be greater than the annual rent paid to the Port Authority. This constraint is set in general terms; however, it is not incorporated in the model to get numerical solutions because it never becomes binding in this model.

Insofar as w_i (or δ_i) will influence the outcome of a game between competing terminals, that is, the total revenue ($p_i \cdot X_i$), a contract that specifies the magnitudes of δ_i and annual rent constitute an important strategic decision for a Port Authority that attempts to maximize total revenue.

In a competitive situation with few players and an inhomogeneous product, the outcome in terms of market shares and prices can often be treated as the result of a game where each player maximizes profit, but with due consideration of the expected reaction of its competitors. When the competitor's actions are confined to setting the prices of their own product (service), the outcome can be modeled as Bertrand equilibrium [1].

Whatever price other terminals are charging, terminal i's profit is maximized when the incremental profit from a very small increase in its own price is zero. Therefore, in order to find the best reply for player i, it is necessary to differentiate its profit function with respect to p_i and set the derivative equal to zero. The Bertrand Nash equilibrium is characterized by the first-order conditions:

$$\frac{\partial \Pi_i}{\partial p_i} = 0, \qquad i = KICT,\ PICT,\ KPT \text{ and } QICT \tag{11}$$

The profit function, say, for terminal 1 is given by:

$$\Pi_1 = (p_1 - c_1) \cdot X_1 \tag{12}$$

since

$$X_1 = Ae^{\theta LS} Q_1 \tag{13}$$

By substituting the value of X_1 in equation (12) one gets

$$\Pi_1 = (p_1 - c_1) \cdot Ae^{\theta LS} Q_1 \tag{14}$$

$$\Pi_1 = p_1 . Ae^{\theta LS} Q_1 - c_1 . Ae^{\theta LS} Q_1 \tag{15}$$

By taking the derivative of equation (15) and setting it equal to zero, we get

$$\frac{\partial \Pi_1}{\partial p_1} = Ae^{\theta LS} Q_1 + p_1 \frac{\partial (Ae^{\theta LS} Q_1)}{\partial p_1} - c_1 \frac{\partial (Ae^{\theta LS} Q_1)}{\partial p_1} = 0 \tag{16}$$

$$\frac{\partial \Pi_1}{\partial p_1} = Ae^{\theta LS}Q_1 + (p_1 - c_1)\frac{\partial (Ae^{\theta LS}Q_1)}{\partial p_1} = 0 \tag{17}$$

By taking the log of the equation (13) we get

$$\ln(X_1) = \ln(A) + \theta LS + U_1 - LS \tag{18}$$

$$\frac{\partial 1n(X_1)}{\partial p_1} = \frac{\partial X_1}{\partial p_1} \cdot \frac{1}{X_1} \tag{19}$$

or

$$\frac{\partial X_1}{\partial p_1} = \frac{\partial 1n(X_1)}{\partial p_1} \cdot X_1 \tag{20}$$

By taking the derivative of equation (18) with respect to P_1 we get:

$$\frac{\partial 1n(X_1)}{\partial p_1} = \frac{\theta}{\sum_j e^{U_j}} \cdot \frac{\partial (\sum_j e^{U_j})}{\partial p_1} + \frac{\partial U_1}{\partial p_1} - \frac{1}{\sum_j e^{U_j}} \frac{\partial (\sum_j e^{U_j}}{\partial p_1} \tag{21}$$

$$= \frac{\theta}{\sum_j e^{U_j}} \cdot e^{U_1}(b) + b - \frac{1}{\sum_j e^{U_j}} \cdot e^{U_1}(b) \tag{22}$$

$$= \theta b \frac{e^{U_1}}{\sum_j e^{U_j}} + b - \frac{e^{U_1}}{\sum_j e^{U_j}} \cdot b \tag{23}$$

since $Q_1 = \frac{e^{U_1}}{\sum_j e^{U_j}}$

By substituting the value of Q_1 in above equation we get:

$$\frac{\partial 1n(X_1)}{\partial p_1} = \theta b Q_1 + b - bQ_1 \tag{24}$$

$$= b(\theta Q_1 + 1 - Q_1) \tag{25}$$

By substituting equations (13) and (25) in equation (20) we get:

$$\frac{\partial X_1}{\partial p_1} = Ae^{\theta LS}Q_1\left[b(\theta Q_1 + 1 - Q_1\right] \tag{26}$$

By substituting equation (26) into equation (17) we get:

$$\frac{\partial \Pi_1}{\partial p_1} = Ae^{\theta LS}Q_1 + (p_1 - c_1)\left[b(\theta Q_1 + 1 - Q_1)\right]Ae^{\theta LS}Q_1 = 0 \tag{27}$$

$$Ae^{\theta LS}Q_1 + (p_1 - c_1)\left[b(\theta Q_1 + 1 - Q_1)\right]Ae^{\theta LS}Q_1 = 0 \tag{28}$$

$$Ae^{\theta LS}Q_1\left\{1 + (p_1 - c_1)\left[b(\theta Q_1 + 1 - Q_1)\right]\right\} = 0 \tag{29}$$

By solving the above equation for p_1 we get:

$$p_1 = c_1 - \frac{1}{b(\theta Q_1 + 1 - Q_1)} \tag{30}$$

This is the implicit reaction curve (pricing rule) for player 1(i.e. KICT). The reaction function cannot be given on a closed form in this model. The prices of the other players enter via Q_1, as can be seen in (2) and (3). Similarly, reaction curves for the other three terminals can be derived. Solving these reaction functions yields the Nash equilibrium in prices.

3.3.3. *Cooperative Game with external competitors*

In the case of cooperative game, three terminals within Karachi port can establish different combinations of coalition. In this case, the profit function for each terminal will be different from equation (10). For instance, if all the terminals at Karachi port decided to work under one decision unit, then the profit function of the coalition, for instance, for KICT will be as follows:

$$\Pi_1 = \left[X_1(p_1 - c_1) + X_2(p_2 - c_2) + X_3(p_3 - c_3)\right] \tag{31}$$

This will give 3 conditions, one for each price.

Again Bertrand Nash equilibrium is characterized by the first-order conditions, thus by taking the derivative of equation (31) and setting it equal to zero we get the condition:

$$\begin{aligned}\frac{\partial \Pi_1}{\partial p_1} = &\frac{\partial\left(Ae^{\theta LS}Q_1\right)}{\partial p_1}(p_1 - c_1) + Ae^{\theta LS}Q_1 + \frac{\partial\left(Ae^{\theta LS}Q_2\right)}{\partial p_1}(p_2 - c_2)\\ &+ \frac{\partial\left(Ae^{\theta LS}Q_3\right)}{\partial p_1}(p_3 - c_3) = 0\end{aligned} \tag{32}$$

From Equation (26) we have:

$$\frac{\partial(Ae^{\theta LS}Q_1)}{\partial p_1} = Ae^{\theta LS}Q_1\left[b(\theta Q_1 + 1 - Q_1)\right]$$

The third (and fourth) term is the cross derivatives.

$$\frac{\partial\left(Ae^{\theta LS}Q_2\right)}{\partial p_1}(p_2-c_2)=Ae^{\theta LS}Q_2[b(\theta Q_1-Q_1)](p_2-c_2)$$

This should give us:

$$\frac{\partial \Pi_1}{\partial p_1}=Ae^{\theta LS}Q_1\left[b(\theta Q_1+1-Q_1)\right](p_1-c_1)+Ae^{\theta LS}Q_1+$$
$$Ae^{\theta LS}Q_2[b(\theta Q_1-Q_1)](p_2-c_2)+Ae^{\theta LS}Q_3[b(\theta Q_1-Q_1)](p_3-c_3)=0$$

Now:

$Ae^{\theta LS}Q_1$ Cancels out, leaving

$$\left[b(\theta Q_1+1-Q_1)\right](p_1-c_1)+1+Q_2[b(\theta-1)](p_2-c_2)+Q_3[b(\theta-1)](p_3-c_3)=0 \quad (33)$$

This is the reaction curve for KICT when all 3 terminals have formed a coalition within Karachi Port. Similarly, reaction curves for other two terminals can be derived. Moreover, in this case we have not considered the fee paid by private terminals to Karachi port in the profit function because this is a matter of internal transfers within the coalition. Similarly in other combinations of coalition, fee of that terminal will not be included in the profit function which will become the partner with KPT.

3.3.4. Assumptions about the parameters of the model

3.3.4.1. Assumed value for b

b is the coefficient of price at ports or cost for customers (shipping lines). In other words, this is the coefficient of price of the choices faced by decision makers. There is only one research, by [19] in which this value has been estimated, by discrete choice methodology, taking any port as a case study. [19], estimated a logit model for container terminal selection by shipping companies on a dataset for the 4 terminals treated here and obtained a statistically significant parameter of -0.0624 for the container handling charge (in US$). Therefore, based on this value we assume that the value for price parameter, in our model, is -0.050.

3.3.4.2. Assumed value for a

In general terms, Equation (2) can be written as by dividing utility into two additive parts. For instance, for two alternatives, A and B, the utilities can be written as follows:

$$U_{An}=V_{An}+\varepsilon_{An} \quad and \quad U_{Bn}=V_{Bn}+\varepsilon_{Bn} \quad where \quad n=1,\ldots\ldots..N. \quad (34)$$

where N is the number of decision makers (or users); $V_{An} and V_{Bn}$ are the systematic (or representative) components of the utility of A and B; and ε_{An} & ε_{Bn} are the random parts and are called the disturbance (or random components). a is the alternative-specific constant and reflects the mean of $\varepsilon_{Bn} - \varepsilon_{An}$; that is, the difference between the utilities of alternatives A and B when "all else is equal". The values of alternative specific constant for QICT, KICT and Gwadar terminal are arbitrary chosen.

3.3.4.3. Assumed value for cost

The basis of all port tariffs should be short-run marginal cost, which measures the resources used up by supplying a unit of port service. However, strictly setting a price equal to the marginal cost is best only in a perfectly competitive free economy or in an efficient socialist economy [20].

For the marginal costs of terminals, the average cost of PICT for 2005 is calculated. Figures are obtained from the annual report of the terminal[10]. The terminal's operating cost is Rs 695,915,000 divided by total containers handled in 2005, i.e. 206,764 TEUs, which gives an average cost of US$ 57. On the basis of this figure, the marginal costs for the three private terminals are assumed.

3.3.4.4. Assumed value for θ

Demand for port calls, port trans-shipment and supplementary service is derived from demand for the goods involved and is thus a function of economic growth, industrial production and industrial trade [21]. Thus, a change in price and other attributes of one terminal will shift the traffic to that terminal from other terminals. It will not much affect the total demand, but will affect the market share of all four terminals. That is why the value for θ is quite low.

3.3.4.5. Input Parameters

Tables 2 and 3 provide information about the input parameters used in the model. The values of the log sum parameter and price parameter are assumed on the basis of the literature review. The values of user cost constants are also assumed; a high value is set for KPT because this is a conventional terminal and does not have modern equipment like gantry cranes to handle containers. Moreover, the user cost for QICT is set at US$ 7 because it suffers from a cost disadvantage of about US$ 2 per TEU as compared to KICT and PICT. The values for marginal costs for private terminals are explained in previous section. However, the values for marginal cost for KPT and Gwadar Port are assumed and arbitrary chosen. Data about capacity is collected from the official website of each terminal.[11]

After completion of the Makran **Coastal Highway,** Gwadar Port will be connected to Karachi; however, it is still located far from the industrial area, which is why the user cost is set at $ 9. Moreover, although the terms of a 40 year concession agreement between **Gwadar Port Authority and PSA Gwadar Ltd.**, are not publicly available, on the basis of available information, for instance, provision of 40 years tax holiday from the government of Pakistan, we will assume that it pays 3% as a royalty for cargo exceeding 200,000 TEUs.

---

[10] Available online at http://www.pictcntrtrack.com/docs/annualreport.pdf. Accessed 25 April, 2008.

[11] See www.qict.net, http://www.kpt.gov.pk/, http://www.kictl.com/, http://www.pictcntrtrack.com/. Accessed 20 April, 2007.

KPT charges Rs 445 per square meter from KICT. The total area for KICT is 218,300 square meters. Thus, the annual rent is: 218,300 × 445 = Rs 97,000,000 or $ 1616 in thousands. Similarly, KPT charges Rs 473 per square meter from PICT. The total area for PICT is 210,000 square meters. Thus, the annual rent is: 210,000 × 473 = Rs 99,000,000 or $ 1659 in thousands. QICT pays a flat rent of Rs 48,000,000, or $ 800 in thousands, per annum to Port Qasim authority.

Level of Demand (A)	*Logsum parameter (θ)*	*Price parameter (λ)*
1550,000	0.010	-0.050

Table 2. General parameters of demand

	QICT	*KICT*	*PICT*	*KPT*	*Gwadar*
Alt.spec. constant (α_i)	0.1	0.5	0	0	0.3
User cost constants in $ ($C0_i$)	7	5	5	40	9
Marginal cost in $ (c_i)	50	55	55	27	48
Capacity (CAP_i)	600,000	525000	400,000	300,000	675,000
Terminal fee[12] in $ (w_i, δ_i)	5%of price	6.03	12.54		3% of price
Annual rent (In 000 US$)	800	1616	1650		

Table 3. Terminal specific parameters

The chosen form of the user cost function is shown below:

$$OUC(i) = C0_i + 0.5\left(\frac{X_i}{CAP_i * 0.8}\right)^4 \tag{35}$$

This function implies that the user cost starts to rise sharply when throughput exceeds 80% of rated capacity.

With these parameters, the Bertrand equilibrium is defined by the system of non-linear equations that can be solved numerically by an equation solver to give equilibrium rates for container handling and the market shares.

4. Bertrand solution

With the available information, a model consisting of equations 1, 2, 3, 12, 30, 33 (for each terminal) and 5 is solved using an equation solver. In other words by solving the equilibrium of the Bertrand game we can get the pricing rule set by the players, which will yield the Nash equilibrium.

---

[12] Source of information for unit fee and annual rent is personal interview with authority at Karachi port and for port Qasim see http://www.pakistaneconomist.com/database2/cover/c98-21.asp Accessed 25 April 2008.

4.1. Independent terminals with fee (present situation)

In this case, we take the present contracts between port authorities and private terminals as fixed and assume that each terminal sets the handling fee so as to maximize operating surplus. Each terminal operator has full information and knows the reaction of the other operators to its own actions.

Results obtained for the present situation, which is when private terminals are paying fees to the port authorities are depicted in Tables 4 and 5. If we compare market shares presented in Table 4, with actual market share (See Table 1), we found that they are close to the actual figures.

	QICT	*KICT*	*PICT*	*KPT*
Equilibrium Price US$/TEU	81.60	90.10	91.90	53.20
User Cost US$/TEU	7.40	5.80	5.30	41.40
Market share	0.30	0.31	0.18	0.21
Profit (In Mill US$)	12.4	13.8	6.6	14.3[13]

Table 4. Bertrand equilibrium (λ=-0.05)

Total Demand in 1000s TEUs	*1502*
Combined profit Karachi Port (Terminal 2-4) In Mill US$	34.7

Table 5. Total demand for two ports & combined profit for the Karachi port

In order to analyse the effect of fees on the overall profit of the port of Karachi, we assume that port authorities do not charge fees from private terminals. Results presented in Tables 6 and 7 show that this results in low prices for all players and consequently low profit for QICT and KPT. This is similar to what has been described in international trade: the suggestion that government intervention can raise national welfare by shifting oligopoly rents from foreign to domestic firms. The crucial point is that the home firm can increase its profits by persuading foreign firm to charge a *higher* price than the Nash equilibrium. To do this, it must commit to a higher price than would be optimal. To achieve this, government must impose an export tax [22]. Similarly, in this situation, 'fee' charged by port authorities, plays the same role as played by 'government tax' in international trade.

	QICT	*KICT*	*PICT*	*KPT*
Equilibrium Price US$/TEU	77.60	84.20	80.90	51.20
User Cost US$/TEU	7.30	5.90	5.70	40.70
Market share	0.28	0.32	0.23	0.18
Profit (In Mill US$)	11.5	13.9	9.0	6.4

Table 6. Bertrand equilibrium without fee

[13] Including fees paid by KICT and PICT

Total Demand in 1000s TEUs	*1506*
Combined profit Karachi Port (Terminal 2-4) In Mill US$	29.3

Table 7. Total demand for two ports & combined profit for the Karachi port

4.2. Independent terminals with fees and less elastic price

A crucial assumption in the numerical model is the price parameter in the logit model. In order to test for the sensitivity of this assumption, we change the value of this parameter from –0.05 to –0.03. Tables 8 and 9, show the results with less price sensitive demand.

		QICT	*KICT*	*PICT*	*KPT*
Equilibrium Price	US$/TEU	99.40	109.50	109.00	70.40
User Cost	US$/TEU	7.30	5.90	5.40	41.60
Market share		0.28	0.32	0.20	0.21
Profit (In Mill US$)		18.6	23.3	12.4	20.6

Table 8. Bertrand equilibrium (λ=-0.03)

Total Demand In 1000s TEUs	*1522*
Combined profit Karachi Port (Terminal 2-4) In Mill US$	56.4

Table 9. Total demand for two ports & combined profit for the Karachi port

Not unexpectedly, it turns out that less sensitive demand results in higher handling charges and higher profits for the terminals, but moderate changes in market shares. However, less price sensitive demand, in general, will outweigh the higher handling charges, which increase the value of the logsum and leads to higher total demand.

4.3. Rent/fee trade-off

As can be seen from Table 2, the fee per TEU paid by PICT to Karachi port authority is about twice the fee paid by KICT. The reason for this is not clear, but it probably reflects the fact that the PICT contract has more recently been negotiated. In the long run – when contracts have to be renewed – it is reasonable to assume that Karachi port authority will charge the same fee from both private terminals.

Annual rent paid by private terminals to the landlord port has to come out of operating profit. Now profit for the Karachi port will has 3 sources, the profit from KPT as a terminal, the transfer from container handling at KICT and PICT and the annual rent. According to Figure 7, Karachi port can maximize its profit by increasing the fee up to the level $75. However, it can maximize the combined profit of all terminals by charging the fee up to $40. So it might actually be profitable to set the fee so as to maximize combined profit and extract some of the profit from KICT and PICT as annual rent. However, the port might still do better by maximizing the combined profit, taking away all competition within the Port of Karachi, i.e. a duopoly.

4.4. Coalition/Duopoly

In this section, a hypothetical situation is created in which all three terminals at Karachi Port have formed a coalition. Hence they work under one single decision unit. The reason for forming a coalition may be two-fold. Firstly, as a coalition, the terminals in Karachi Port will increase their market power and the game is transformed to a game of duopoly. Secondly, by forming a coalition they may increase the combined capacity which will result in reduction in average waiting time. The reason is that before a coalition, there are three queues at three independent terminals (servers), but after formation of a coalition, there will be one queue and three servers. This will cause a decrease in average waiting time for existing customers. Moreover, it will also increase their efficiency and reduce total cost.

Results presented in the Tables 10 and 11 show that this coalition results in high profit for all cooperating units as well as for their competitor.

		QICT	*KICT*	*PICT*	*KPT*
Equilibrium Price	US$/TEU	83.00	102.00	102.00	74.00
User Cost	US$/TEU	8.50	5.30	5.10	40.10
Market share		0.42	0.28	0.17	0.12
Profit (In Mill US$)		18.1	20	12.2	8.6[14]

Table 10. Bertrand equilibrium – Duopoly

Total Demand in 1000s TEUs	*1495*
Combined profit Karachi Port (Terminal 2-4) In Mill US$	41

Table 11. Total demand for two ports & combined profit for Karachi port

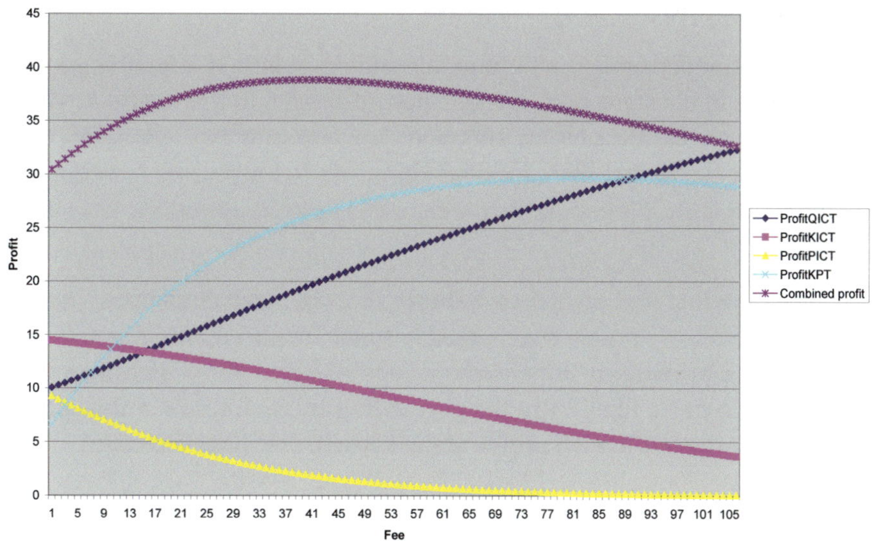

Figure 7. Relation between fee and profit of terminals

[14] Excluding fee from KICT and PICT

4.4.1. Addition of Gwadar port as a player

Gwadar port, due to political conflicts, at present handles very small volume of cargo. However, in the long run when this port will be fully functional, it is expected that due to geo-political importance Gwadar port will tend to capture transit traffic to/from Iran, Afghanistan and China. In addition, Gwadar port will also compete, for Pakistani trade that presently goes through Karachi and Port Qasim.

Thus we have included Gwadar port as a player to analyze how the additional player may influence the formation of coalition or duopoly.

	QICT	*KICT*	*PICT*	*KPT*	*GWD*
Equilibrium Price US$/TEU	76.50	90.90	90.90	62.90	77.70
User Cost US$/TEU	7.30	5.10	5.10	40.10	9.20
Market share	0.27	0.22	0.13	0.09	0.28
Profit (In Mill US$)	9.2	11.8	7.2	5.1	11.7

Table 12. Bertrand equilibrium - Duopoly

Total Demand in 1000s TEUs	*1507*
Combined profit Karachi Port (Terminal 2-4) In Mill US$	24.1

Table 13. Total demand for two ports & combined profit for the Karachi port

In this situation, according to results presented in Tables 5-9 and 5-10, formation of coalition will not benefit that much to the Karachi port as did in first case.

5. Discussion and policy implications

Whether working independently or forming a coalition, which is a feasible proposition for terminals operating in the same port, is a question addressed in this analysis. We presented the Bertrand solution of the present situation prevailing at two ports, as well as two hypothetical situations. Market shares and handling fees in the present case, obtained by the Bertrand model, are quite close to the actual figures. Hence, they confirm the validity of the proposed model.

A comparison of the first and second case shows that the Nash equilibrium results in high prices for all terminals, when they are charged with fee by port authorities and are working independently. This increase in price is more for PICT and KICT as compared to QICT, because they have to pay a high fee to the port authority. As a result of this, profit of PICT has decreased. Nevertheless overall profits of the Karachi port have increased.

This situation is similar to what has been described in international trade where the suggestion is that government intervention can raise national welfare by shifting oligopoly rents from foreign to domestic firms. The starting point of this debate was several papers by [23, 24], who showed that the government policies can serve the 'strategic' purpose of altering the subsequent incentives of firms, acting as a deterrent to foreign competitors.

Further, [25] have extended the application of strategic trade policy to the Bertrand competition, with firms (home and foreign) taking each others' prices as given.

Each firm's best responses describe a reaction function that is upward sloping. The Nash equilibrium is at the point where two curves intersect. The crucial point is that the home firm can increase its profit by persuading foreign firm to charge a *higher* price than at the Nash equilibrium. To do this, it must commit to a higher price than would be optimal. To achieve this, government must impose an export tax.

A situation similar to the international trade exists in this case. Reaction functions for the two ports are drawn into the price space (see fig. 8). Nash equilibrium is at point N where two curves intersect with each other. Now Karachi can increase its profit only by moving northeast along the Qasim reaction function. This can be achieved when the port authority imposes fees on the private terminals, forcing them as well as their competitor to raise their prices and to earn greater profits.

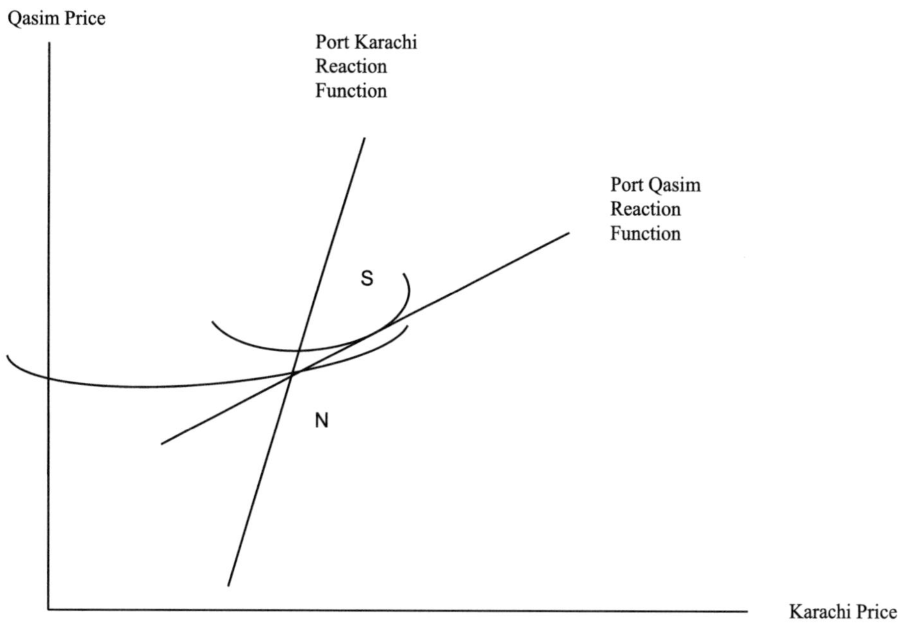

Figure 8. Reaction functions for two ports

We also analyzed the situation with less sensitive price. All the terminals can earn excess profit by charging high prices in this situation even when they are working independently. However, the situation can become more favourable if three terminals at Karachi port work under one single decision unit, creating the situation of duopoly. However, considering revenue only as unit of analysis for port performance is not appropriate. Other factors should also be considered which contribute to the overall efficiency of both ports. For instance, as mentioned earlier as a result of forming a coalition, the combined capacity at Karachi port will be greater as compared to the capacity of individual terminals. This will result in the reduction in the average waiting time of existing customers.

Author details

Naima Saeed and Odd I. Larsen
Department of Economics, Informatics and Social Science, Molde University College, Specialized University in Logistics, Norway

6. References

[1] Pindyck, R.S. and Rubinfeld, D.L. (2001). *Micoreconomics.* New Jersey: Prentice Hall International.
[2] Bertrand, J. (1883). Book review of theorie mathematique de la richesse sociale and of recherches sur les principles mathematiques de la theorie des richesses, Journal de Savants, 67, 499–508
[3] Cournot, A. 1838. Researches into the Mathematical Principles of the Theory of Wealth. Trans. N.T. Bacon, New York: Macmillan, 1929.
[4] Edgeworth, Francis (1889). The pure theory of monopoly, reprinted in Collected Papers relating to Political Economy 1925, vol.1, Macmillan.
[5] Friedman, W.J. (1977). *Oligopoly and the Theory of Games.* Amsterdam: North-Holland Publishing Company.
[6] Neumann, V.J. and Morgenstern, O. (1967). *Theory of Games and Economic Behaviour.* New York: John Wiley & Sons.
[7] Binmore, K. (1992). *Fun and Games: A Text on Game Theory.* Lexington, Mass: D.C. Heath and Company.
[8] Higgins, R.S. (1996). An economic theory of leader choice in Stackelberg models, *Journal of Economic studies*, 23, 79-95.
[9] Von Stackelberg, H. 1952. *The Theory of the Market Economy*, William Hodge and Company Ltd, London.
[10] Anderson, C.M., Park, Y.A., Chang, Y.T., Yang, C.H., Lee, T.W., Luo, M. (2008). A game-theoratic analysis of competition among container port hubs: the case of Busan and Shanghai. *Maritime Policy and Management*, 35, 5-26.

[11] Kaselimi, E.N. and Reeven, P.V. (2008). The impact of new port terminal operating schemes on inter-port competition. *Proceedings of the International Association of Maritime Economist (IAME) Conference*, 4-6 April, Dalian, China.

[12] Zhang, A. (2008). The Impact of Hinterland Access Conditions on Rivalry between Ports. Sauder School of Business, University of British Columbia, Canada. Retrieved 1st February, 2009, from http://www.internationaltransportforum.org/jtrc/DiscussionPapers/DP200808.pdf

[13] De Borger, B., Proost, S. and Van Dender, K. (2008). Private Port Pricing and Public Investment in Port and Hinterland Capacity. *Journal of Transport Economics and Policy*, 42, 527-561.

[14] Saeed, N. and Larsen, O.I. (2010a) An application of cooperative game among container terminals of one port. *European Journal of Operational Research*, 203, 393-403.

[15] Saeed, N. and Larsen, O.I. (2010b) Container terminal concessions: a game theory application to the case of the ports of Pakistan. *Maritime Economics and Logistics*, 12(3), 237-262.

[16] Kaselimi, E. N., Notteboom, T. E., & Saeed, N. October 2011. "A Game Theoretical Approach to the Inter-relation between Terminal Scale and Port Competition." *Proceedings of the International Association of Maritime Economists (IAME) Annual Conference.* Santiago, Chile.

[17] UNITE (2003) *Unification of accounts and marginal costs for Transport Efficiency.* Project funded by the EC under the Fifth Framework Transport RTD.

[18] Veldman, S.J.H. and Rachman, A. (2008). A Model of Transhipment Port Competition: A test with cross-section and time-series data for the Mediterranean. *Proceedings of the International Association of Maritime Economist (IAME) Conference,* Dalian, China.

[19] Saeed, N., (2009). Competition and Cooperation among Container Terminals in Pakistan:with Emphasis on Game Theoretical Analysis, PhD theses in Logistics, Molde University College, Norway.

[20] Bennathan, E. and Walters, A.A. (1979). *Port Pricing and Investment Policy for Developing Countries*. World Bank Publications, Washington, D.C.: Oxford University Press.

[21] Meersman, H., Van de Voorde, E. and Vanelslander, T. (2003). Port Pricing. Considerations on Economic Principles and Marginal Costs. *European Journal of Transport and Infrastructure Research,* 3(4), 371-386.

[22] Krugman, P.R., 1989, Industrial Organization and International Trade. In: *HandBook of Industrial Organization,* edited by R.Schmalense, R.D.Williy, Vol. 11 (Elsevier Science Publishers), pp. 1181-1223.

[23] Brander, J.A. and Spencer, B.J. (1983). International R&D Rivalry and Industrial strategy. In: Schmalense, R. and Williy, R.D. (Eds.), (1989). *HandBook of Industrial Organization* (Vol. 11., pp. 1181-1223). Elsevier Science Publishers.

[24] Brander, J.A. and Spencer, B.J. (1985). Export subsidies and international market share rivalry. In: Schmalense, R. and Williy, R.D. (Eds.) (1989). *HandBook of Industrial Organization* (Vol. 11., pp. 1181-1223). Elsevier Science Publishers.

[25] Eaton J and Grossman GM (1986). Optimal trade and industrial policy under oligopoly. In: Schmalense R and Williy RD (Eds.) (1989). *HandBook of Industrial Organization* (Vol.11., pp. 1181-1223). Elsevier Science Publishers.

Section 2

Game Theory in Engineering

A Game Theoretic Approach Based Adaptive Control Design for Sequentially Interconnected SISO Linear Systems

Sheng Zeng and Emmanuel Fernandez

Additional information is available at the end of the chapter

http://dx.doi.org/10.5772/50488

1. Introduction

Adaptive control has attracted a lot of research attention in control theory for many decades. In the certainty equivalence based adaptive controller design [4, 5], the unknown parameters of the uncertainty system are substituted by their online estimates, which are generated through a variety of identifiers, as long as the estimates satisfy certain properties independent of the controller. This approach leads to structurally simple adaptive controllers and has been demonstrated its effectiveness for linear systems with or without stochastic disturbance inputs [10] when long term asymptotic performance is considered. Yet, the certainty equivalence approach is unsuccessful to generalize to systems with severe nonlinearities. Also, early designs based on this approach were shown to be nonrobust [13] when the system is subject to exogenous disturbance inputs and unmodeled dynamics. Then, the stability and the performance of the closed-loop system becomes an important issue. This has motivated the study of robust adaptive control in the 1980s and 1990s, and the study of nonlinear adaptive control in the 1990s.

The topic of adaptive control design for nonlinear systems was studied intensely in the last decade after the celebrated characterization of feedback linearizable or partially feedback linearizable systems [7]. A breakthrough is achieved when the integrator backstepping methodology [8] was introduced to design adaptive controllers for parametric strict-feedback and parametric pure-feedback nonlinear systems systematically. Since then, a lot of important contributions were motivated by this approach, and a complete list of references can be found in the book [9]. Moreover, this nonlinear design approach has been applied to linear systems to compare performance with the certainty equivalence approach. However, simple designs using this approach without taking into consideration the effect of exogenous disturbance inputs have also been shown to be nonrobust when the system is subject to exogenous disturbance inputs.

The robustness of closed-loop adaptive systems has been an important research topic in late 1980s and early 1990s. Various adaptive controllers were modified to render the closed-loop systems robust [6]. Despite their successes, they still fell short of directly addressing the disturbance attenuation property of the closed-loop system.

The objectives of robust adaptive control are to improve transient response, to accommodate unmodeled dynamics, and to reject exogenous disturbance inputs, which are the same as the objectives to motivate the study of the H^∞- optimal control problem. H^∞-optimal control was proposed as a solution to the robust control problem, where these objectives are achieved by studying only the disturbance attenuation property for the closed-loop system. The game-theoretic approach to H^∞-optimal control developed for the linear quadratic problems, offers the most promising tool to generalize the results to nonlinear systems [3]. Worst-case analysis based adaptive control design was proposed in late 1990s to address the disturbance attenuation property directly, and it is motivated by the success of the game-theoretic approach to H^∞-optimal control problems [2]. In this approach, the robust adaptive control problem is formulated as a nonlinear H^∞ control problem under imperfect state measurements. By *cost-to-come function* analysis, it is converted into an H^∞ control problem with full information measurements. This full information measurements problem is then solved using nonlinear design tools for a suboptimal solution. This design scheme has been applied to worst-case parameter identification problems [11], which has led to new classes of parametrized identifiers for linear and nonlinear systems. It has also been applied to adaptive control problems [1, 12, 14, 15, 18, 19], and the convergence properties is studied in [20]. In [14], adaptive control for a strict-feedback nonlinear systems was considered with noiseless output measurements, and more general class of nonlinear systems was studied in [1]. In [12], single-input and single output (SISO) linear systems were considered with noisy output measurements. SISO linear systems with partly measured disturbance was studied in [18], which leads to a disturbance feed-forward structure in the adaptive controller. [19] generalizes the results of [12] to the adaptive control design for SISO linear systems with zero relative degree under noisy output measurements. In [17], adaptive control for a sequentially interconnected SISO linear system was considered, and a special class of unobservable systems was also studied using the proposed approach. More recently, [16] generalized the result of [17] to adaptive control design for a linear system under simultaneous driver, plant and actuation uncertainties.

In this Chapter, we study the adaptive control design for sequentially interconnected SISO linear systems, $\mathbf{S}_1$ and $\mathbf{S}_2$(see Figure 1), under noisy output measurements and partly measured disturbance using the similar approaches as [12] and [17]. We assume that the linear systems satisfy the same assumption as [17], and the adaptive control design follows the same design method discussed above. The robust adaptive controller achieves asymptotic tracking of the reference trajectories when disturbance inputs are of finite energy. The closed-loop system is totally stable with respect to the disturbance inputs and the initial conditions. Furthermore, the closed-loop system admits a guaranteed disturbance attenuation level with respect to the exogenous disturbance inputs, where ultimate lower bound for the achievable attenuation performance level is equal to the noise intensity in the measurement channel of $\mathbf{S}_1$. The results are as same as those in [17]. In addition, the controller achieves arbitrary positive distance attenuation level with respect to the measured disturbances by proper scaling. Moreover, if the measured disturbances satisfy the assumption 2 for $\check{w}_{1,b}$ and $\check{w}_{2,b}$, the proposed controller achieves disturbance attenuation level *zero* with respect to the measured disturbances, which further leads to a stronger asymptotic tracking property, namely, the tracking error converges to zero when the unmeasured disturbances are $\mathcal{L}_2 \cap \mathcal{L}_\infty$, and the measured disturbances are $\mathcal{L}_\infty$ only.

The balance of this Chapter is organized as follows. In Section 2, we list the notations used in the Chapter. In Section 3, we present the formulation of the adaptive control problem and discuss the general solution methodology. In Section 4, we first obtain parameter identifier and state estimator using the *cost-to-come function* analysis in Subsection 4.1, then we derive the adaptive control law in Subsection 4.2. We present the main results on the robustness of the system in Section 5, and the example in Section 6. The Chapter ends with some concluding remarks in Section 7.

2. Notations

We denote $\mathbf{R}$ to be the real line; $\mathbf{R}_e$ to be the extended real line; $\mathbf{N}$ to be the set of natural numbers; $\mathbb{C}$ to be the set of complex numbers. For a function f, we say that it belongs to $\mathcal{C}$ if it is continuous; we say that it belongs to $\mathcal{C}_k$ if it is k-times continuously (partial) differentiable. For any matrix A, A' denotes its transpose. For any $b \in \mathbf{R}$, $\mathrm{sgn}(b) = \begin{cases} -1 & b < 0 \\ 0 & b = 0 \\ 1 & b > 0 \end{cases}$. For any vector $z \in \mathbf{R}^n$, where $n \in \mathbf{N}$, $|z|$ denotes $(z'z)^{1/2}$. For any vector $z \in \mathbf{R}^n$, and any $n \times n$-dimensional symmetric matrix M, where $n \in \mathbf{N}$, $|z|_M^2 = z'Mz$. For any matrix M, the vector $\overrightarrow{M}$ is formed by stacking up its column vectors. For any symmetric matrix M, $\overleftarrow{M}$ denotes the vector formed by stacking up the column vector of the lower triangular part of M. For $n \times n$-dimensional symmetric matrices M_1 and M_2, where $n \in \mathbf{N}$, we write $M_1 > M_2$ if $M_1 - M_2$ is positive definite; we write $M_1 \geq M_2$ if $M_1 - M_2$ is positive semi-definite. For $n \in \mathbf{N}$, the set of $n \times n$-dimensional positive definite matrices is denoted by $\mathcal{S}_{+n}$. For $n \in \mathbf{N} \cup \{0\}$, I_n denotes the $n \times n$-dimensional identity matrix. For any matrix M, $\|M\|_p$ denotes its p-induced norm, $1 \leq p \leq \infty$. $\mathcal{L}_2$ denotes the set of square integrable functions and $\mathcal{L}_\infty$ denotes the set of bounded functions. For any n, $m \in \mathbf{N} \cup \{0\}$, $\mathbf{0}_{n\times m}$ denotes the $n \times m$-dimensional matrix whose elements are zeros. For any $n \in \mathbf{N}$ and $k \in \{1, \cdots, n\}$, $e_{n,k}$ denotes $\left[0_{1\times(k-1)} \; 1 \; 0_{1\times(n-k)} \right]'$.

3. Problem Formulation

We consider the robust adaptive control problem for the system which is described by the block diagram in Figure 1.

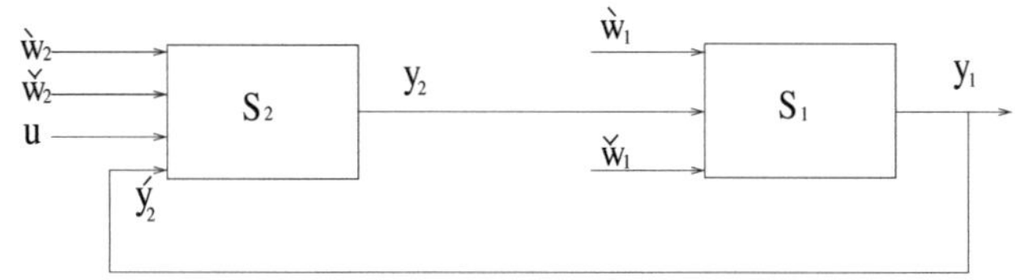

Figure 1. Diagram of two sequentially interconnected SISO linear systems.

We assume that the system dynamics for $\mathbf{S_1}$ and $\mathbf{S_2}$ are given by,

$$\dot{\grave{x}}_1 = \grave{A}_1\grave{x}_1 + \grave{B}_1 y_2 + \grave{D}_1\grave{w}_1 + \grave{\check{D}}_1\check{w}_1; \quad (1a)$$

$$y_1 = \grave{C}_1\grave{x}_1 + \grave{E}_1\grave{w}_1 \quad (1b)$$

$$\dot{\grave{x}}_2 = \grave{A}_2\grave{x}_2 + \grave{B}_2 u + \grave{A}_{2,y}\acute{y}_2 + \grave{D}_2\grave{w}_2 + \grave{\check{D}}_2\check{w}_2; \quad (1c)$$

$$y_2 = \grave{C}_2\grave{x}_2 + \grave{E}_2\grave{w}_2 \quad (1d)$$

where $\grave{x}_i$ is the n_i-dimensional state vectors with initial condition $\grave{x}_i(0) = \grave{x}_{i,0}$, $n_i \in \mathbf{N}$; u is the scalar control input; y_i is the scalar measurement output; $\grave{w}_i$ is $\grave{q}_i$-dimensional unmeasured disturbance input vector, $\grave{q}_i \in \mathbf{N}$; $\check{w}_i$ is $\check{q}_i$-dimensional measured disturbance input vector, $\check{q}_i \in \mathbf{N}$; the elements of $\check{w}_i$ are $\left[\, \check{w}_{i,1} \ \cdots \ \check{w}_{i,\check{q}_i} \,\right]'$; $\acute{y}_2 = y_1$; the matrices $\grave{A}_i$, $\grave{A}_{i,y}$, $\grave{B}_i$, $\grave{C}_i$, $\grave{D}_i$, $\grave{\check{D}}_i$, and $\grave{E}_i$ are of the appropriate dimensions, generally unknown or partially unknown, $i = 1, 2$. For subsystem $\mathbf{S_1}$, the transfer function from y_2 to y_1 is $H_1(s) = \grave{C}_1(sI_{n_1} - \grave{A}_1)^{-1}\grave{B}_1$, for subsystem $\mathbf{S_2}$, the transfer function from u to y_2 is $H_2(s) = \grave{C}_2(sI_{n_2} - \grave{A}_2)^{-1}\grave{B}_2$. All signals in the system are assumed to be continuous.

The subsystems $\mathbf{S_1}$ and $\mathbf{S_2}$ satisfy the following assumptions,

Assumption 1. *For $i = 1, 2$, the pair $(\grave{A}_i, \grave{C}_i)$ is observable; the transfer function $H_i(s)$ is known to have relative degree $r_i \in \mathbf{N}$, and is strictly minimum phase. The uncontrollable part of $\mathbf{S_1}$ (with respect to y_2) is stable in the sense of Lyapunov; any uncontrollable mode corresponding to an eigenvalue of the matrix $\grave{A}_1$ on the $j\omega$-axis is uncontrollable from $\left[\, \grave{w}_1' \ \check{w}_1' \,\right]'$. The uncontrollable part of $\mathbf{S_2}$ (with respect to u) is stable in the sense of Lyapunov; any uncontrollable mode corresponding to an eigenvalue of the matrix $\grave{A}_2$ on the $j\omega$-axis is uncontrollable from $\left[\, \grave{w}_2' \ \acute{y}_2 \ \check{w}_2' \,\right]'$.* ◇

Based on Assumption 1, for $i = 1, 2$, there exists a state diffeomorphism: $x_i = \grave{T}_i\grave{x}_i$, and a disturbance transformation: $w_i = \grave{M}_i\grave{w}_i$, such that $\mathbf{S_i}$ can be transformed into the following state space representation,

$$\dot{x}_1 = A_1 x_1 + (y_1\bar{A}_{1,211} + y_2\bar{A}_{1,212} + \sum_{j=1}^{\check{q}_1} \check{w}_{1,j}\bar{A}_{1,213j})\theta_1 + B_1 y_2 + D_1 w_1 + \check{D}_1\check{w}_1;$$

$$y_1 = C_1 x_1 + E_1 w_1$$

$$\dot{x}_2 = A_2 x_2 + (y_2\bar{A}_{2,211} + u\bar{A}_{2,212} + \sum_{j=1}^{\check{q}_2} \check{w}_{2,j}\bar{A}_{2,213j} + \acute{y}_2\bar{A}_{2,214})\theta_2 + B_2 u + A_{2,y}\acute{y}_2 + D_2 w_2 + \check{D}_2\check{w}_2;$$

$$y_2 = C_2 x_2 + E_2 w_2$$

where θ_i is the σ_i-dimensional vector of unknown parameters for the subsystem $\mathbf{S}_i$, $\sigma_i \in \mathbf{N}$; the matrices A_i, $\bar{A}_{i,211}$, $\bar{A}_{i,212}$, $\bar{A}_{i,2131}$, $\cdots$, $\bar{A}_{i,213\check{q}_i}$, $\bar{A}_{2,214}$, $A_{2,y}$, B_i, D_i, $\check{D}_i$, C_i, and E_i are known and have the following structures, $A_i = (a_{i,jk})_{n_i \times n_i}; a_{i,j(j+1)} = 1, a_{i,jk} = 0$, for $1 \le j \le r_i - 1$ and $j + 2 \le k \le n_i$; $\bar{A}_{i,212} = \left[\, \mathbf{0}_{\sigma_i \times (r_i - 1)} \ \ \bar{A}_{i,2120}' \ \ \bar{A}_{i,212r_i}' \,\right]'$, $C_i = [1 \ \mathbf{0}_{1 \times (n_i - 1)}]$, $A_{i,2120}$ is a row vector, $B_i = \left[\, \mathbf{0}_{1 \times (r_i - 1)} \ b_{i,p0} \ \cdots \ b_{i,p(n_i - r_i)} \,\right]'$, $b_{i,pj}$ $j = 0, 1, \cdots, n_i - r_i$ are constants. We denote the elements of x_1 and x_2 by $\left[\, x_{1,1} \ \cdots \ x_{1,n_1} \,\right]'$ and $\left[\, x_{2,1} \ \cdots \ x_{2,n_2} \,\right]'$, with initial conditions $x_{1,0}$ and $x_{2,0}$, respectively.

Assumption 2. *The measured disturbance $\check{w}_1$ can be partitioned as: $\check{w}_1 = \left[\check{w}'_{1,a} \; \check{w}'_{1,b} \right]'$ where $\check{w}_{1,a}$ is $\check{q}_{1,a}$ dimensional, $\check{q}_{1,a} \in \mathbf{N} \cup \{0\}$, and the transfer function from each element of $\check{w}_{1,a}$ to y_1 has relative degree less than $r_1 + r_2$; the measured disturbance $\check{w}_2$ can be partitioned as: $\check{w}_2 = \left[\check{w}'_{2,a} \; \check{w}'_{2,b} \right]'$ where $\check{w}_{2,a}$ is $\check{q}_{2,a}$ dimensional, $\check{q}_{2,a} \in \mathbf{N} \cup \{0\}$, and the transfer function from each element of $\check{w}_{2,a}$ to y_2 has relative degree less than r_2.* ◇

Based on Assumption 2, the matrix $\check{D}_i$ can be partitioned into $\left[\check{D}_{i,a} \; \check{D}_{i,b} \right]$, where $\check{D}_{i,a}$ and $\check{D}_{i,b}$ have $n_i \times \check{q}_{i,a}$- and $n_i \times \check{q}_{i,b}$-dimensional, respectively; and $\check{D}_{i,b}, \bar{A}_{i,213\,(\check{q}_{i,a}+1)}, \cdots, \bar{A}_{i,213\check{q}_i}$ have the following structure

$$\check{D}_{i,b} = \begin{bmatrix} \mathbf{0}_{(r_i-1)\times \check{q}_{i,b}} \\ \check{D}_{i,b0} \\ \check{D}_{i,br_i} \end{bmatrix}; \quad \bar{A}_{i,213\,j} = \begin{bmatrix} \mathbf{0}_{(r_i-1)\times \sigma_i} \\ \bar{A}_{i,213\,j0} \\ \bar{A}_{i,213\,jr_i} \end{bmatrix}, j = \check{q}_{i,a}+1, \cdots, \check{q}_i$$

where $\check{D}_{i,b0}$ and $\bar{A}_{i,213\,j0}$, $j = \check{q}_{i,a}+1, \cdots, \check{q}_{i,a}+\check{q}_{i,b}$, are row vectors, $i = 1, 2$.

Since we will base our design of adaptive controllers using the model (2), we call (2) the design model, and make the following two assumptions.

Assumption 3. *For $i = 1, 2$, the matrices E_i are such that $E_i E_i' > 0$.* ◇

Define $\zeta_i := (E_i E_i')^{-\frac{1}{2}}$ and $L_i := D_i E_i'$, $i = 1, 2$.

Due to the structures of A_i, $\bar{A}_{i,212}$ and B_i, the high frequency gain of the transfer function $H_i(s)$, $b_{i,0}$, is equal to $b_{i,p0} + \bar{A}_{i,2120}\theta_i$, $i = 1, 2$.

To guarantee the stability of the identified system, we make the following assumption on the parameter vectors θ_1 and θ_2.

Assumption 4. *The sign of $b_{i,0}$ is known; there exists a known smooth nonnegative radially-unbounded strictly convex function $P_i : \mathbf{R}^{\sigma_i} \to \mathbf{R}$, such that the true value $\theta_i \in \Theta_i := \{\bar{\theta}_i \in \mathbf{R}^{\sigma_i} \mid P_i(\bar{\theta}_i) \leq 1\}$; moreover, $\forall \bar{\theta}_i \in \Theta_i$, $\operatorname{sgn}(b_{i,0})(b_{i,p0} + \bar{A}_{i,2120}\bar{\theta}_i) > 0$, $i = 1, 2$.* ◇

Assumption 4 delineates *a priori* convex compact sets where the parameter vectors θ_1 and θ_2 lie in, respectively. This will guarantee the stability of the closed-loop system and the boundedness of the estimate of θ_1 and θ_2.

We make the following assumption about the reference signal, y_d.

Assumption 5. *The reference trajectory, y_d, is $r_1 + r_2$ times continuously differentiable. Define vector $Y_d := [y_d^{(0)}, \cdots, y_d^{(r_1+r_2)}]'$, where $y_d^{(0)} = y_d$, and $y_d^{(j)}$ is the jth order time derivative of y_d, $j = 1, \cdots, r_1 + r_2$; define $Y_{d0} := [y_d^{(0)}(0), \cdots, y_d^{(r_1+r_2-1)}(0)]' \in \mathbf{R}^{r_1+r_2}$. The signal Y_d is available for feedback.* ◇

The uncertainty of subsystem $\mathbf{S_1}$ is $\grave{\omega}_1 := (x_{1,0}, \theta_1, \grave{w}_{1[0,\infty)}, \check{w}_{1[0,\infty)}, Y_{d0}, y_{d[0,\infty)}^{(r_1+r_2)}) \in \grave{\mathcal{W}}_1 := \mathbf{R}^{n_1} \times \Theta_1 \times \mathcal{C} \times \mathcal{C} \times \mathbf{R}^{r_1+r_2} \times \mathcal{C}$, which comprises the initial state $x_{1,0}$, the true value of the parameters θ_1, the unmeasured disturbance waveform $\grave{w}_{1[0,\infty)}$, the measured disturbance waveform $\check{w}_{1[0,\infty)}$, the initial conditions of the reference trajectory Y_{d0}, and the waveform of the $(r_1 + r_2)$ th order derivative of the reference trajectory, $y_{d[0,\infty)}^{(r_1+r_2)}$. The uncertainty for subsystem $\mathbf{S_2}$ is $\grave{\omega}_2 := (x_{2,0}, \theta_2, \grave{w}_{2[0,\infty)}, \breve{w}_{2[0,\infty)}) \in \grave{\mathcal{W}}_2 := \mathbf{R}^{n_2} \times \Theta_2 \times \mathcal{C} \times \mathcal{C}$, which comprises the initial state $x_{2,0}$, the true value of the parameters θ_2, the unmeasured disturbance waveform $\grave{w}_{2[0,\infty)}$, and the measured disturbance waveform $\breve{w}_{2[0,\infty)}$.

Our objective is to derive a control law, which is generated by the following mapping,

$$u(t) = \mu(y_{2[0,t]}, \acute{y}_{2[0,t]}, Y_{d[0,t]}, \breve{w}_1, \breve{w}_2) \tag{3}$$

where $\mu : \mathcal{C} \times \mathcal{C} \times \mathcal{C} \times \mathcal{C} \times \mathcal{C} \to \mathbf{R}$, such that $x_{1,1}$ can asymptotically track the reference trajectory y_d, while rejecting the uncertainty $(\grave{\omega}_1, \grave{\omega}_2) \in \grave{\mathcal{W}}_1 \times \grave{\mathcal{W}}_2$, and keeping the closed-loop signals bounded. The control law μ must also satisfy that, $\forall (\grave{\omega}_1, \grave{\omega}_2) \in \grave{\mathcal{W}}_1 \times \grave{\mathcal{W}}_2$, there exists a solution $\grave{x}_{1[0,\infty)}$ and $\grave{x}_{2[0,\infty)}$ to the system (1), which yields a continuous control signal $u_{[0,\infty)}$. We denote the class of these admissible controllers by $\mathcal{M}_\mu$.

For design purposes, instead of attenuating the effect of $\left[\grave{w}_1' \; \breve{w}_1' \; \grave{w}_2' \; \breve{w}_2 \right]'$ we design the adaptive controller to attenuate the effect of $\left[w_1' \; \breve{w}_1' \; w_2' \; \breve{w}_2 \right]'$. This is done to allow our design paradigm to be carried out. This will result in a guaranteed attenuation level with respect to $\grave{\omega}_1$ and $\grave{\omega}_2$. To simplify the notation, we take the uncertainty $\omega_1 := (x_{1,0}, \theta_1, w_{1[0,\infty)}, \breve{w}_{1[0,\infty)}, Y_{d0}, y_{d[0,\infty)}^{(r_1+r_2)}) \in \mathcal{W}_1 := \mathbf{R}^{n_1} \times \Theta_1 \times \mathcal{C} \times \mathcal{C} \times \mathbf{R}^{r_1+r_2} \times \mathcal{C}$, and $\omega_2 := (x_{2,0}, \theta_2, w_{2[0,\infty)}, \breve{w}_{2[0,\infty)}) \in \mathcal{W}_2 := \mathbf{R}^{n_2} \times \Theta_2 \times \mathcal{C} \times \mathcal{C}$.

We state the control objective precisely as follows,

Definition 1. *A controller $\mu \in \mathcal{M}_\mu$ is said to achieve* disturbance attenuation level γ *with respect to* $\left[w_1' \; \breve{w}_{1,a}' \; w_2' \; \breve{w}_{2,a}' \right]'$, *and* disturbance attenuation level zero *with respect to* $\left[\breve{w}_{1,b}' \; \breve{w}_{2,b}' \right]'$, *if there exists functions* $l_1(t, \theta_1, x_1, y_{1[0,t]}, y_{2[0,t]}, \breve{w}_{1[0,t]}, \breve{w}_{2[0,t]}, Y_{d[0,t]})$, $l_2(t, \theta_2, x_2, y_{1[0,t]}, y_{2[0,t]}, \breve{w}_{1[0,t]}, \breve{w}_{2[0,t]}, Y_{d[0,t]})$, *and a known nonnegative constant* $l_0(\check{x}_{1,0}, \check{x}_{2,0}, \check{\theta}_{1,0}, \check{\theta}_{2,0})$, *such that*

$$\sup_{\grave{w}_1 \in \grave{\mathcal{W}}_1, \grave{w}_2 \in \grave{\mathcal{W}}_2} J_{\gamma t_f} \leq 0; \quad \forall t_f \geq 0 \tag{4}$$

and $l_1 \geq 0$ *and* $l_2 \geq 0$ *along the closed-loop trajectory, where*

$$J_{\gamma t_f} := J_{1,\gamma t_f} + J_{2,\gamma t_f} \tag{5a}$$

$$J_{1\gamma t_f} := \int_0^{t_f} \left((C_1 x_1 - y_d)^2 + l_1 - \gamma^2 |w_1|^2 - \gamma^2 |\breve{w}_{1,a}|^2 \right) \mathrm{d}\tau - \gamma^2 \left| \left[\theta_1' - \check{\theta}_{1,0}' \; x_{1,0}' - \check{x}_{1,0}' \right]' \right|_{\bar{Q}_{1,0}}^2 \tag{5b}$$

$$J_{2,\gamma t_f} := \int_0^{t_f} \left(l_2 - \gamma^2 |w_2|^2 - \gamma^2 |\breve{w}_{2,a}|^2 \right) \mathrm{d}\tau - l_0 - \gamma^2 \left| \left[\theta_2' - \check{\theta}_{2,0}' \; x_{2,0}' - \check{x}_{2,0}' \right]' \right|_{\bar{Q}_{2,0}}^2 \tag{5c}$$

$\check{\theta}_{i,0} \in \Theta_i$ *is the initial guess of* θ_i; $\check{x}_{i,0} \in \mathbf{R}^{n_i}$ *is the initial guess of* $x_{i,0}$; $\bar{Q}_{i,0} > 0$ *is a* $(n_i + \sigma_i) \times (n_i + \sigma_i)$*-dimensional weighting matrix, quantifying the level of confidence in the estimate* $\left[\check{\theta}_{i,0}' \; \check{x}_{i,0}' \right]'$; $\bar{Q}_{i,0}^{-1}$ *admits the structure* $\begin{bmatrix} Q_{i,0}^{-1} & Q_{i,0}^{-1} \Phi_{i,0}' \\ \Phi_{i,0} Q_{i,0}^{-1} & \Pi_{i,0} + \Phi_{i,0} Q_{i,0}^{-1} \Phi_{i,0}' \end{bmatrix}$, $Q_{i,0}$ *and* $\Pi_{i,0}$ *are* $\sigma_i \times \sigma_i$*- and* $n_i \times n_i$*-dimensional positive definite matrices, respectively,* $i = 1, 2$.

Clearly, when the inequality (4) is achieved, the squared $\mathcal{L}_2$ norm of the output tracking error $C_1x_1 - y_d$ is bounded by γ^2 times the squared $\mathcal{L}_2$ norm of the transformed disturbance input $\left[w_1' \; \breve{w}_{1,a}' \; w_2' \; \breve{w}_{2,a}' \right]'$, plus some constant. When the $\mathcal{L}_2$ norm of $\grave{w}_1$, $\grave{w}_2$, $\breve{w}_1$, and $\breve{w}_2$ are finite, the squared $\mathcal{L}_2$ norm of $C_1x_1 - y_d$ is also finite, which implies $\lim_{t\to\infty}(C_1x_1(t) - y_d(t)) = 0$, under additional assumptions.

Let ξ_i denote the expanded state vector $\xi_i = [\theta_i', x_i']'$, $i = 1, 2$, and note that $\dot{\theta}_i = 0$, we have the following expanded dynamics for system (2),

$$
\begin{aligned}
\dot{\xi}_1 &= \begin{bmatrix} \mathbf{0} & \mathbf{0} \\ y_1\bar{A}_{1,211} + y_2\bar{A}_{1,212} + \sum_{j=1}^{\breve{q}_1} \breve{w}_{1,j}\bar{A}_{1,213j} & A_1 \end{bmatrix} \xi_1 + \begin{bmatrix} \mathbf{0} \\ B_1 \end{bmatrix} y_2 + \begin{bmatrix} \mathbf{0} \\ D_1 \end{bmatrix} w_1 + \begin{bmatrix} \mathbf{0} \\ \breve{D}_1 \end{bmatrix} \breve{w}_1 \\
&=: \bar{A}_1(y_1, y_2, \breve{w}_1)\xi_1 + \bar{B}_1 y_2 + \bar{D}_1 w_1 + \bar{\breve{D}}_1 \breve{w}_1 \\
y_1 &= \begin{bmatrix} \mathbf{0} \; C_1 \end{bmatrix} \xi_1 + E_1 w_1 \\
&=: \bar{C}_1\xi_1 + E_1 w_1 \\
\dot{\xi}_2 &= \begin{bmatrix} \mathbf{0} & \mathbf{0} \\ y_2\bar{A}_{2,211} + u\bar{A}_{2,212} + \sum_{j=1}^{\breve{q}_2} \breve{w}_{2,j}\bar{A}_{2,213j} + \acute{y}_2\bar{A}_{2,214} & A_2 \end{bmatrix} \xi_2 + \begin{bmatrix} \mathbf{0} \\ B_2 \end{bmatrix} u + \begin{bmatrix} \mathbf{0} \\ A_{2,y} \end{bmatrix} \acute{y}_2 \\
&\quad + \begin{bmatrix} \mathbf{0} \\ D_2 \end{bmatrix} w_2 + \begin{bmatrix} \mathbf{0} \\ \breve{D}_2 \end{bmatrix} \breve{w}_2 \\
&=: \bar{A}_2(y_1, y_2, , \breve{w}_2, u)\xi_2 + \bar{B}_2 u + \bar{A}_{2,y}\acute{y}_2 + \bar{D}_2 w_2 + \bar{\breve{D}}_2 \breve{w}_2 \\
y_2 &= \begin{bmatrix} \mathbf{0} \; C_2 \end{bmatrix} \xi_2 + E_2 w_2 \\
&=: \bar{C}_2\xi_2 + E_2 w_2
\end{aligned}
$$

The worst-case optimization of the cost function (4) can be carried out in two steps as depicted in the following equations.

$$
\begin{aligned}
\sup_{\grave{\omega}_1 \in \grave{\mathcal{W}}_1,\, \grave{\omega}_2 \in \grave{\mathcal{W}}_2} J_{\gamma t_f} &= \sup_{\omega_m \in \mathcal{W}_m} \; \sup_{\grave{\omega}_1 \in \grave{\mathcal{W}}_1,\, \grave{\omega}_2 \in \grave{\mathcal{W}}_2 | \omega_m \in \mathcal{W}_m} J_{\gamma t_f} \\
&\leq \sup_{\omega_m \in \mathcal{W}_m} \; \sup_{\omega_1 \in \mathcal{W}_1, \omega_2 \in \mathcal{W}_2 | \omega_m \in \mathcal{W}_m} J_{\gamma t_f} \\
&= \sup_{\omega_m \in \mathcal{W}_m} \left(\sum_{i=1}^{2} \; \sup_{\omega_i \in \mathcal{W}_i | \omega_m \in \mathcal{W}_m} J_{i,\gamma t_f} \right)
\end{aligned}
\tag{6}
$$

where ω_m is the measured signals of the system, and defined as

$$
\omega_m := (y_{1[0,\infty)}, y_{2[0,\infty)}, \breve{w}_{1[0,\infty)}, \breve{w}_{2[0,\infty)}, Y_{d0}, y_{d[0,\infty)}^{(r_1+r_2)}) \in \mathcal{W}_m := \mathcal{C} \times \mathcal{C} \times \mathcal{C} \times \mathcal{C} \times \mathbf{R}^{r_1+r_2} \times \mathcal{C}.
$$

The inner supremum operators will be carried out first. We maximize over ω_i given that the measurement ω_m is available for estimator design, $i = 1, 2$. In this step, the control input, u, is a function only depended on ω_m, then u is an open-loop time function and available for the optimization. Using *cost-to-come* function analysis, we derive the dynamics of the estimators for subsystem $\mathbf{S}_1$ and $\mathbf{S}_2$ independently.

The outer supremum operator will be carried out second. In this step, we use a backstepping procedure to design the controller μ.

This completes the formulation of the robust adaptive control problem.

4. Adaptive control design

In this section, we present the adaptive control design, which involves estimation design and control design. First, we discuss estimation design.

4.1. Estimation design

In this subsection, we present the estimation design for the adaptive control problem formulated. First, we will derive the identifier of subsystem $\mathbf{S}_1$. In this step, the measurement waveform y_1, y_2 and measured disturbance $\breve{w}_1$ are assumed to be known. Then we can obtain the identifier of subsystem $\mathbf{S_1}$ from a *game-theoretic* solution methodology – *cost-to-come* function analysis.

We first set function l_1 in the definition to be $|\xi_1 - \hat{\xi}_1|^2_{\bar{Q}_1} + 2(\xi_1 - l_{1,1})'l_{1,2} + \check{l}_1$, where $\hat{\xi}_1 = [\hat{\theta}_1', \hat{x}_1']'$ is the worst-case estimate for the expanded state ξ_1, $\bar{Q}_1(y_{1[0,\tau]}, Y_{d[0,\tau]}, \breve{w}_{1[0,\tau]})$ is a matrix-valued weighting function, $l_{1,1}(y_{1[0,\tau]}, Y_{d[0,\tau]}, \breve{w}_{1[0,\tau]})$, $l_{1,2}(y_{1[0,\tau]}, Y_{d[0,\tau]}, \breve{w}_{1[0,\tau]})$ and $\check{l}_1(y_{1[0,\tau]}, Y_{d[0,\tau]}, \breve{w}_{1[0,\tau]})$ are three design functions to be introduced later, the cost function of subsystem $\mathbf{S_1}$ is then of the a linear quadratic structure.

The robust adaptive problem for $\mathbf{S_1}$ becomes an H^∞ control of affine quadratic problem, and admits a finite dimensional solution. By *cost-to-come* function analysis, we obtain the dynamics of worst-case covariance matrix $\bar{\Sigma}_1$, and state estimator $\check{\xi}_1$, which are given by

$$\begin{aligned}\dot{\bar{\Sigma}}_1 &= (\bar{A}_1 - \zeta_1^2\bar{L}_1\bar{C}_1)\bar{\Sigma}_1 + \bar{\Sigma}_1(\bar{A}_1 - \zeta_1^2\bar{L}_1\bar{C}_1)' - \bar{\Sigma}_1(\gamma^2\zeta_1^2\bar{C}_1'\bar{C}_1 - \bar{C}_1'\bar{C}_1 - \bar{Q}_1)\bar{\Sigma}_1 \\ &\quad +\gamma^{-2}\bar{D}_1\bar{D}_1' - \gamma^{-2}\zeta_1^2\bar{L}_1\bar{L}_1'; \quad \bar{\Sigma}_1(0) = \gamma^{-2}\bar{Q}_{1,0}^{-1} \end{aligned} \tag{7a}$$

$$\begin{aligned}\dot{\check{\xi}}_1 &= (\bar{A}_1 + \bar{\Sigma}_1(\bar{C}_1'\bar{C}_1 + \bar{Q}_1))\check{\xi}_1 + \bar{B}_1 y_2 + \zeta_1^2(\gamma^2\bar{\Sigma}_1\bar{C}_1' + \bar{L}_1)(y_1 - \bar{C}_1\check{\xi}_1) \\ &\quad -\bar{\Sigma}_1(\bar{C}_1' y_d + \bar{Q}_1\hat{\xi}_1 - l_{1,2}) + \breve{\bar{D}}_1\breve{w}_1; \quad \check{\xi}_1(0) = \begin{bmatrix}\check{\theta}_{1,0}' & \check{x}_{1,0}'\end{bmatrix}' \end{aligned} \tag{7b}$$

where $\bar{L}_1$ is defined as $\bar{L}_1 = \begin{bmatrix}\mathbf{0}_{1\times\sigma_1} & L_1'\end{bmatrix}'$.

We partition $\bar{\Sigma}_1$ as the same structure as

$$\bar{\Sigma}_1 = \begin{bmatrix}\Sigma_1 & \bar{\Sigma}_{1,12} \\ \bar{\Sigma}_{1,21} & \bar{\Sigma}_{1,22}\end{bmatrix} = \begin{bmatrix}\Sigma_1 & \Sigma_1\Phi_1' \\ \Phi_1\Sigma_1 & \frac{1}{\gamma^2}\Pi_1 + \Phi_1\Sigma_1\Phi_1'\end{bmatrix} \tag{8}$$

where $\Phi_1(t) := \bar{\Sigma}_{1,21}(t)(\Sigma_1(t))^{-1}$ and $\Pi_1(t) := \gamma^2(\bar{\Sigma}_{1,22}(t) - \bar{\Sigma}_{1,21}(t)(\Sigma_1(t))^{-1}\bar{\Sigma}_{1,12}(t))$, $\forall t \in [0, t_f]$. Then the weighting matrix $\bar{\Sigma}_1$ is positive definite if and only if Σ_1 and Π_1 are positive definite. To guarantee the boundedness of Σ_1, we choose weighing matrix $\bar{Q}_1$ as follows,

$$\bar{Q}_1 = \bar{\Sigma}_1^{-1}\begin{bmatrix}\mathbf{0}_{\sigma_1\times\sigma_1} & \mathbf{0}_{\sigma_1\times n_1} \\ \mathbf{0}_{n_1\times\sigma_1} & \Delta_1(t)\end{bmatrix}\bar{\Sigma}_1^{-1} + \begin{bmatrix}\epsilon_1\Phi_1'C_1'(\gamma^2\zeta_1^2 - 1)C_1\Phi_1 & \mathbf{0}_{\sigma_1\times n_1} \\ \mathbf{0}_{n_1\times\sigma_1} & \mathbf{0}_{n_1\times n_1}\end{bmatrix} \tag{9}$$

where $\Delta_1(t) = \gamma^{-2}\beta_{1,\Delta}\Pi_1(t) + \Delta_{1,1}$, with $\beta_{1,\Delta} \geq 0$ being a constant and $\Delta_{1,1}$ being an $n_1 \times n_1$-dimensional positive-definite matrix, and ϵ_1 is a scalar function defined by

$$\epsilon_1(t) := \mathrm{Tr}(\Sigma_1(t))^{-1}/K_{1,c} \quad \forall t \in [0, t_f] \tag{10a}$$

or

$$\epsilon_1(t) := 1 \quad \forall t \in [0, t_f] \tag{10b}$$

Then the dynamics of Σ_1, Φ_1, Π_1 are given as follows with initial conditions $\gamma^{-2}Q_{1,0}^{-1}$, $\Pi_{1,0}$, and $\Phi_{1,0}$ respectively,

$$\dot{\Sigma}_1 = (\epsilon_1 - 1)\Sigma_1\Phi_1' C_1' (\gamma^2\zeta_1^2 - 1)C_1\Phi_1\Sigma_1 \tag{11a}$$

$$\begin{aligned}\dot{\Pi}_1 = &(A_1 - \zeta_1^2 L_1 C_1)\Pi_1 + \Pi_1 (A_1 - \zeta_1^2 L_1 C_1)' - \zeta_1^2 L_1 L_1' - \Pi_1 C_1' (\zeta_1^2 - \gamma^{-2}) C_1 \Pi_1 \\ &+ D_1 D_1' + \gamma^2 \Delta_1\end{aligned} \tag{11b}$$

$$\dot{\Phi}_1 = A_{1,f}\Phi_1 + y_1 \bar{A}_{1,211} + y_2 \bar{A}_{1,212} + \sum_{j=1}^{\check{q}_1} \bar{A}_{1,213j}\check{w}_{1,j} \tag{11c}$$

where $A_{1,f} := A_1 - \zeta_1^2 L_1 C_1 - \Pi_1 C_1' C_1 (\zeta_1^2 - \gamma^{-2})$ is Hurwitz. By picking $\gamma \geq \zeta_1^{-1}$, we have the covariance matrix Σ_1 upper and lower bounded as summarized in the following Lemma [12].

Lemma 1. *Consider the dynamic equation (11a) for the covariance matrix Σ_1. Let $K_{1,c} \geq \gamma^2 \mathrm{Tr}(Q_{1,0})$, $Q_{1,0} > 0$, $\gamma \geq \zeta_1^{-1}$, and ϵ_1 be given by either (10b) or (10b). Then, the matrix Σ_1 is upper and lower bounded as follows, whenever Φ_1 is continuous on $[0, t_f]$,*

$$\begin{aligned}&K_{1,c}^{-1} I_{1,\sigma_1} \leq \Sigma_1(t) \leq \Sigma_1(0) = \gamma^{-2} Q_{1,0}^{-1};\\ &\gamma^2 \mathrm{Tr}(Q_{1,0}) \leq \mathrm{Tr}(\Sigma_1(t))^{-1} \leq K_{1,c}; \qquad \forall t \in [0, t_f]\end{aligned}$$

To avoid the calculation of Σ_1^{-1} online, we define $s_{1,\Sigma} = \mathrm{Tr}(\Sigma_1^{-1})$. Based on the structure of $\bar{Q}_1$, we have the following assumption to guarantee the boundedness of Σ_1 and $s_{1,\Sigma}$,

Assumption 6. *If the matrix $A_1 - \zeta_1^2 L_1 C_1$ is Hurwitz, then the desired disturbance attenuation level $\gamma \geq \zeta_1^{-1}$. In case $\gamma = \zeta_1^{-1}$, choose $\beta_{1,\Delta} \geq 0$ such that $A_1 - \zeta_1^2 L_1 C_1 + \beta_{1,\Delta}/2 I_{n_1}$ is Hurwitz. If the matrix $A_1 - \zeta_1^2 L_1 C_1$ is not Hurwitz, then the desired disturbance attenuation level $\gamma > \zeta_1^{-1}$.* ◇

This assumption implies that the achievable disturbance attenuation level γ is no smaller than ζ_1^{-1}. Under this assumption, we initialize Π_1 as the unique positive definite solution of its Riccati Differential Equation (11b), which is summarized as the following assumption.

Assumption 7. *The initial weighting matrix $\Pi_{1,0}$ is chosen as the unique positive definite solutions to the following algebraic Riccati equations:*

$$(A_1 - \zeta_1^2 L_1 C_1)\Pi_1 + \Pi_1 (A_1 - \zeta_1^2 L_1 C_1)' - \Pi_1 C_1' \zeta_1^2 C_1 \Pi_1 + D_1 D_1' - \zeta_1^2 L_1 L_1' + \gamma^2 \Delta_1 = \mathbf{0}_{n_1 \times n_1} \tag{12}$$

To guarantee the estimates parameter to be bounded and the estimate of high frequency gain to be bounded away from zero, projection function scheme is applied to modify the dynamics of $\check{\xi}_1$.

Define

$$\rho_1 := \inf\{P_1(\bar{\theta}_1) \mid \bar{\theta}_1 \in \mathbf{R}^{\sigma_1}, b_{1,p0} + \bar{A}_{1,2120}\bar{\theta}_1 = 0\} \tag{13}$$

By Assumption 4 and Lemma 2 in [19] we have $1 < \rho_1 \leq \infty$. Fix any $\rho_{1,o} \in (1, \rho_1)$, and define the open set $\Theta_{1,o} := \{\bar{\theta}_1 \in \mathbf{R}^{\sigma_1} \mid P_1(\bar{\theta}) < \rho_{1,o}\}$. Our control design will guarantee that the estimate $\check{\theta}_1$ lies in $\Theta_{1,o}$, which immediately implies $|b_{1,p0} + \bar{A}_{1,2120}\check{\theta}_1| > c_{1,0} > 0$, for some $c_{1,0} > 0$. Moreover, the convexity of P_1 implies the following inequality

$$\frac{\partial P_1}{\partial \theta_1}(\check{\theta}_1)\,(\theta_1 - \check{\theta}_1) < 0 \quad \forall \check{\theta}_1 \in \mathbf{R}^{\sigma_1} \backslash \Theta_1$$

We set $l_{1,1} = \check{\xi}_1$, and $l_{1,2} = \left[-(P_{1,r}(\check{\theta}_1))' \; \mathbf{0}_{1\times n_1} \right]'$, where

$$P_{1,r}(\check{\theta}_1) := \begin{cases} \dfrac{e^{\frac{1}{1-P_1(\check{\theta}_1)}} \left(\frac{\partial P_1}{\partial \theta_1}(\check{\theta}_1)\right)'}{\left(\rho_{1,o} - P_1(\check{\theta}_1)\right)^3} & \forall \theta_1 \in \Theta_{1,o} \backslash \Theta_1 \\ \mathbf{0}_{\sigma_1 \times 1} & \forall \theta_1 \in \Theta_1 \end{cases}$$

$$:= p_{1,r}(\check{\theta}_1) \left(\frac{\partial P_1}{\partial \theta_1}(\check{\theta}_1) \right)' \tag{14}$$

then, we obtain

$$\begin{aligned} \dot{\check{\xi}}_1 &= -\bar{\Sigma}_1 \left[(P_{1,r}(\check{\theta}_1))' \; \mathbf{0}_{1\times n_1} \right]' + \bar{A}_1 \check{\xi}_1 + \bar{\Sigma}_1 \bar{C}_1' (y_d - \bar{C}_1 \check{\xi}_1) - \bar{\Sigma}_1 \bar{Q}_1(\Phi_1, s_{1,\Sigma}) \xi_{1,c} + \bar{B}_1 y_2 \\ &\quad + \zeta_1^2 (\gamma^2 \bar{\Sigma}_1 \bar{C}_1' + \bar{L}_1)(y_1 - \bar{C}_1 \check{\xi}_1) + \bar{\check{D}}_1 \check{w}_1; \quad \check{\xi}_1(0) = \left[\check{\theta}_{1,0}' \; \check{x}_{1,0}' \right]' \end{aligned} \tag{15}$$

where $\xi_{1,c} = \hat{\xi}_1 - \check{\xi}_1$.

We summarize the equations for subsystem $\mathbf{S}_1$ as follows,

$$\begin{aligned} \mathbf{0} &= (A_1 - \zeta_1^2 L_1 C_1)\Pi_1 + \Pi_1 (A_1 - \zeta_1^2 L_1 C_1)' - \Pi_1 C_1' (\zeta_1^2 - \gamma^{-2}) C_1 \Pi_1 + D_1 D_1' - \zeta_1^2 L_1 L_1' + \gamma^2 \Delta_1 \\ \dot{\Sigma}_1 &= -(1-\epsilon_1)\Sigma_1 \Phi_1' C_1' (\gamma^2 \zeta_1^2 - 1) C_1 \Phi_1 \Sigma_1 \\ \dot{s}_{1,\Sigma} &= (\gamma^2 \zeta_1^2 - 1)(1-\epsilon_1) C_1 \Phi_1 \Phi_1' C_1' \\ \epsilon_1 &= K_{1,c}^{-1} s_{1,\Sigma} \quad \text{or} \quad 1 \\ A_{1,f} &= A_1 - \zeta_1^2 L_1 C_1 - \Pi_1 C_1' C_1 (\zeta_1^2 - \gamma^{-2}) \\ \dot{\Phi}_1 &= A_{1,f}\Phi_1 + y_1 \bar{A}_{1,211} + y_2 \bar{A}_{1,212} + \sum_{j=1}^{\check{q}_1} \bar{A}_{1,213j} \check{w}_{1,j} \\ \dot{\check{\theta}}_1 &= -\Sigma_1 P_{1,r}(\check{\theta}_1) - \Sigma_1 \Phi_1' C_1' (y_d - C_1 \check{x}_1) - \left[\Sigma_1 \; \Sigma_1 \Phi_1' \right] \bar{Q}_1 \xi_{1,c} + \gamma^2 \zeta_1^2 \Sigma_1 \Phi_1' C_1' (y_1 - C_1 \check{x}_1) \\ \dot{\check{x}}_1 &= -\Phi_1 \Sigma_1 P_{1,r}(\check{\theta}_1) + A_1 \check{x}_1 - (\gamma^{-2}\Pi_1 + \Phi_1 \Sigma_1 \Phi_1') C_1' (y_d - C_1 \check{x}_1) + B_1 y_2 + \check{D}_1 \check{w}_1 \\ &\quad - \left[\Phi_1 \Sigma_1 \; \gamma^{-2}\Pi_1 + \Phi_1 \Sigma_1 \Phi_1' \right] \bar{Q}_1 \xi_{1,c} + (y_1 \bar{A}_{1,211} + y_2 \bar{A}_{1,212} + \sum_{j=1}^{\check{q}_1} \check{w}_{1,j} \bar{A}_{1,213j}) \check{\theta}_1 \\ &\quad + \zeta_1^2 (\Pi_1 C_1' + \gamma^2 \Phi_1 \Sigma_1 \Phi_1' C_1' + L_1)(y_1 - C_1 \check{x}_1) \end{aligned}$$

This completes the estimation design of $\mathbf{S_1}$.

Next, we will derive the estimator for subsystem $\mathbf{S_2}$. In this step, the measurements waveform ω_m is assumed to be known. Since the control input, u, is a causal function of ω_m, then it is known. Again, we will apply the *cost-to-come function* methodology to derive the estimator. We briefly summarize the estimation design for $\mathbf{S_2}$ as follows.

Set function l_2 in definition to be $|\xi_2 - \hat{\xi}_2|^2_{\bar{Q}_2} + 2(\xi_2 - \check{\xi}_2)' l_{2,2} + \check{l}_2$, where $\hat{\xi}_2 = [\hat{\theta}_2', \hat{x}_2']'$ is the worst-case estimate for the expanded state ξ_2, $\check{\xi}_2$ is the estimate of ξ_2, $\bar{Q}_2$ is a matrix-valued weighting function, $l_{2,2}$ and $\check{l}_2$ are two design functions to be introduced later, the cost function of subsystem $\mathbf{S_2}$ is then of a linear quadratic structure. By *cost-to-come* function analysis, we obtain the dynamics of worst-case covariance matrix $\bar{\Sigma}_2$, and state estimator $\check{\xi}_2$. We partition $\bar{\Sigma}_2$ as $\bar{\Sigma}_2 = \begin{bmatrix} \Sigma_2 & \bar{\Sigma}_{2,12} \\ \bar{\Sigma}_{2,21} & \bar{\Sigma}_{2,22} \end{bmatrix}$ and introduce $\Phi_2 := \bar{\Sigma}_{2,21}\Sigma_2^{-1}$ and $\Pi_2 := \gamma^2(\bar{\Sigma}_{2,22} - \bar{\Sigma}_{2,21}\Sigma_2^{-1}\bar{\Sigma}_{2,12})$, then the weighting matrix $\bar{\Sigma}_2$ is positive definite if and only if Σ_2 and Π_2 are positive definite. To guarantee the boundedness of Σ_2, we choose weighing matrix $\bar{Q}_2$ as follows,

$$\bar{Q}_2 = \begin{bmatrix} -\Phi_2' \\ I_{n_2} \end{bmatrix} \gamma^4 \Pi_2^{-1} \Delta_2 \Pi_2^{-1} \begin{bmatrix} -\Phi_2' \\ I_{n_2} \end{bmatrix}' + \begin{bmatrix} \epsilon_2 \Phi_2' C_2' \gamma^2 \zeta_2^2 C_2 \Phi_2 & \mathbf{0}_{\sigma_2 \times n_2} \\ \mathbf{0}_{n_2 \times \sigma_2} & \mathbf{0}_{n_2 \times n_2} \end{bmatrix} \tag{16}$$

where $\Delta_2(t) = \gamma^{-2}\beta_{2,\Delta}\Pi_2(t) + \Delta_{2,1}$, with $\beta_{2,\Delta} \geq 0$ being a constant and $\Delta_{2,1}$ being an $n_2 \times n_2$- dimensional positive-definite matrix, and ϵ_2 is a scalar function defined by $\epsilon_2 = K_{2,c}^{-1}\mathrm{Tr}(\Sigma_2^{-1})$ or $\epsilon_2 = 1$. $K_{2,c} \geq \gamma^2 \mathrm{Tr}(Q_{2,0})$ is a design constant, $Q_{2,0}$ is an $\sigma_2 \times \sigma_2$-dimensional positive-definite matrix. Then the dynamics of Σ_2, Φ_2, Π_2 are given as follows,

$$\dot{\Sigma}_2 = (\epsilon_2 - 1)\Sigma_2 \Phi_2' C_2' \gamma^2 \zeta_2^2 C_2 \Phi_2 \Sigma_2; \quad \Sigma_2(0) = \frac{\gamma^{-2}}{Q_{2,0}} \tag{17a}$$

$$\begin{aligned} \dot{\Pi}_2 = {} & (A_2 - \zeta_2^2 L_2 C_2 + \beta_{2,\Delta}/2 I_{n_2})\Pi_2 + \Pi_2 (A_2 - \zeta_2^2 L_2 C_2 + \beta_{2,\Delta}/2 I_{n_2})' - \Pi_2 C_2' \zeta_2^2 C_2 \Pi_2 + D_2 D_2' \\ & - \zeta_2^2 L_2 L_2' + \gamma^2 \Delta_{2,1}; \quad \Pi_2(0) = \Pi_{2,0} \end{aligned} \tag{17b}$$

$$\dot{\Phi}_2 = A_{2,f}\Phi_2 + y_2 \bar{A}_{2,211} + u\bar{A}_{2,212} + \sum_{j=1}^{\check{q}_2} \bar{A}_{2,213j}\check{w}_{2,j} + \acute{y}_2 \bar{A}_{2,214}; \qquad \Phi_2(0) = \Phi_{2,0} \tag{17c}$$

where $A_{2,f} := A_2 - \zeta_2^2 L_2 C_2 - \Pi_2 C_2' C_2 \zeta_2^2$ is Hurwitz. By Lemma [12], we have the covariance matrix Σ_2 upper and lower bounded as follows, $K_{2,c}^{-1} I_{\sigma_2} \leq \Sigma_2(t) \leq \Sigma_2(0) = \gamma^{-2} Q_{2,0}^{-1}$, $\gamma^2 \mathrm{Tr}(Q_{2,0}) \leq \mathrm{Tr}(\Sigma_2(t))^{-1} \leq K_{2,c}$, whenever it exists on $[0, t_f]$ and Φ_2 is continuous on $[0, t_f]$. To avoid the calculation of Σ_2^{-1} online, we define $s_{2,\Sigma} = \mathrm{Tr}(\Sigma_2^{-1})$.

To guarantee the estimates parameter to be bounded and the estimate of high frequency gain to be bounded away from zero without persistently exciting signals, we introduce the following soft projection design on the parameter estimate.

Define $\rho_2 := \inf\{P_2(\bar{\theta}_2) \mid \bar{\theta}_2 \in \mathbf{R}^{\sigma_2}, b_{2,p0} + \bar{A}_{2,2120}\bar{\theta}_2 = 0\}$, we have $1 < \rho_2 \leq \infty$. Fix any $\rho_{2,o} \in (1, \rho_2)$, we define the open set $\Theta_{2,o} := \{\bar{\theta}_2 \mid P_2(\bar{\theta}) < \rho_{2,o}\}$. Our control design will guarantee that the estimate $\check{\theta}_2$ lies in $\Theta_{2,o}$, which immediately implies $|b_{2,p0} + \bar{A}_{2,2120}\check{\theta}_2| > c_{2,0} > 0$, for some $c_{2,0} > 0$. Moreover, the convexity of P_2 implies the following inequality: $\frac{\partial P_2}{\partial \theta_2}(\check{\theta}_2)(\theta_2 - \check{\theta}_2) < 0 \quad \forall \check{\theta}_2 \in \mathbf{R}^{\sigma_2} \backslash \Theta_2$. To incorporate the modifier to the estimates dynamics, we introduce $l_{2,2} = [-(P_{2,r}(\check{\theta}_2)' \ \mathbf{0}_{1\times n_2}]'$, where

$$P_{2,r}(\check{\theta}_2) := \begin{cases} \frac{\exp\left(\frac{1}{1-P_2(\check{\theta}_2)}\right)}{\left(\rho_{2,o}-P_2(\check{\theta}_2)\right)^3} \left(\frac{\partial P_2}{\partial \theta_2}(\check{\theta}_2)\right)' & \forall \theta_2 \in \Theta_{2,o}\backslash\Theta_2 \\ 0_{\sigma_2 \times 1} & \forall \theta_2 \in \Theta_2 \end{cases}$$

$$:= p_{2,r}(\check{\theta}_2)\left(\frac{\partial P_2}{\partial \theta_2}(\check{\theta}_2)\right)'$$

and the dynamics of $\check{\xi}_2$ is then given as follows,

$$\dot{\check{\xi}}_2 = -\bar{\Sigma}_2\left[(P_{2,r}(\check{\theta}_2))' \ \mathbf{0}_{1\times n_2}\right]' + \bar{A}_2\check{\xi}_2 + \bar{B}_2 u + \zeta_2^2\left(\gamma^2\bar{\Sigma}_2\bar{C}_2' + \bar{A}_{2,y}\acute{y}_2 + \bar{L}_2\right)(y_2 - \bar{C}_2\check{\xi}_2)$$
$$+\check{\bar{D}}_2\check{w}_2 - \bar{\Sigma}_2\bar{Q}_2(\hat{\xi}_2 - \check{\xi}_2)$$

where $\check{\xi}_2 = [\check{\theta}_2' \ \check{x}_2]'$ with initial condition $[\check{\theta}_{2,0}' \ \check{x}_{2,0}']'$, and $\bar{L}_2$ is defined as $\bar{L}_2 = [\mathbf{0}_{1\times\sigma_2} \ L_2']'$. This completes the estimation design of $\mathbf{S_2}$.

Associated with the above identifier and estimator of subsystem $\mathbf{S}_i$, $i = 1, 2$, we introduce the value function $W_i : \mathbf{R}^{n_i+\sigma_i} \times \mathbf{R}^{n_i+\sigma_i} \times \mathcal{S}_{+(n_i+\sigma_i)} \to \mathbf{R}$ and the time derivative are as follows

$$W_i(\xi_i, \check{\xi}_i, \bar{\Sigma}_i) = |\theta_i - \check{\theta}_i|^2_{\bar{\Sigma}_i^{-1}} + \gamma^2|x_i - \check{x}_i - \Phi_i(\theta_i - \check{\theta}_i)|^2_{\Pi_i^{-1}} \tag{18}$$

$$\dot{W}_1 = -|x_{1,1} - y_d|^2 - \gamma^4|x_1 - \hat{x}_1 - \Phi_1(\theta_1 - \hat{\theta}_1)|^2_{\Pi_1^{-1}\Delta_1\Pi_1^{-1}} + |C_1\check{x}_1 - y_d|^2$$
$$-\epsilon_1(\gamma^2\zeta_1^2 - 1)|\theta_1 - \hat{\theta}_1|^2_{\Phi_1'C_1'C_1\Phi_1} - \gamma^2\zeta_1^2|y_1 - C_1\check{x}_1|^2 + \gamma^2|w_1|^2 - \gamma^2|w_1 - w_{1,*}|^2$$
$$+2(\theta_1 - \check{\theta}_1)'P_{1,r}(\check{\theta}_i) + |\xi_{1,c}|^2_{\bar{Q}_1} \tag{19}$$

$$\dot{W}_2 = -\gamma^4|x_2 - \hat{x}_2 - \Phi_2(\theta_2 - \hat{\theta}_2)|^2_{\Pi_2^{-1}\Delta_2\Pi_2^{-1}} - \epsilon_2\gamma^2\zeta_2^2|\theta_2 - \hat{\theta}_2|^2_{\Phi_2'C_2'C_2\Phi_2} + |\xi_{2,c}|^2_{\bar{Q}_2}$$
$$-\gamma^2\zeta_2^2|y_2 - C_2\check{x}_2|^2 + \gamma^2|w_2|^2 - \gamma^2|w_2 - w_{2,*}|^2 + 2(\theta_2 - \check{\theta}_2)'P_{2,r}(\check{\theta}_2) \tag{20}$$

where $w_{i,*}$ is the worst-case disturbance, given by $w_{i,*} : \mathbf{R} \times \mathbf{R}^{n_i+\sigma_i} \times \mathbf{R}^{n_i+\sigma_i} \times \mathcal{S}_{+(n_i+\sigma_i)} \longrightarrow \mathbf{R}$

$$w_{i,*}(\xi_i, \check{\xi}_i, \bar{\Sigma}_i, w_i) = \zeta_i^2 E_i'(y_i - \bar{C}_i\xi_i) + \gamma^{-2}(I_{q_i} - \zeta_i^2 E_i'E_i)\bar{D}_i'\bar{\Sigma}_i^{-1}(\xi_i - \check{\xi}_i); \quad i = 1, 2$$

We note that (18) holds when $\Sigma_i > 0$ and $\theta_i \in \Theta_{i,0}$, and the last term in $\dot{W}_i$ is nonpositive, zero on the set Θ_i and approaches $-\infty$ as $\check{\theta}_i$ approaches the boundary of the set $\Theta_{i,o}$, which guarantees the boundedness of $\check{\theta}_i$, $i = 1, 2$.

Then (5) can be equivalently written as, $i = 1, 2$:

$$J_{1,\gamma t_f} = \int_0^{t_f}\left(|C_1\check{x}_1 - y_d|^2 + |\xi_{1,c}|^2_{\bar{Q}_1} + \check{l}_1 - \gamma^2\zeta_1^2|y_1 - C_1\check{x}_1|^2 - \gamma^2|w_1 - w_{1,*}|^2 - \gamma^2|\check{w}_{1,a}|^2\right)d\tau$$
$$-l_{1,0} - |\xi_1(t_f) - \check{\xi}_1(t_f)|^2_{(\bar{\Sigma}_1(t_f))^{-1}}$$

$$J_{2,\gamma t_f} = \int_0^{t_f}\left(|\xi_{2,c}|^2_{\bar{Q}_2} + \check{l}_2 - \gamma^2\zeta_2^2|y_2 - C_2\check{x}_2|^2 - \gamma^2|w_2 - w_{2,*}|^2 - \gamma^2|\check{w}_{2,a}|^2\right)d\tau$$
$$-l_{2,0} - |\xi_2(t_f) - \check{\xi}_2(t_f)|^2_{(\bar{\Sigma}_2(t_f))^{-1}}$$

This completes the identification design step.

4.2. Control design

In this section, we describe the controller design for the uncertain system under consideration. Note that, we ignored some terms in the cost function (5) in the identification step, since they are constant when y_1, y_2, $\check{w}_1$, $\check{w}_2$ and $\acute{y}_2$ are given. In the control design step, we will include such terms. Then, based on the cost function (5), the controller design is to guarantee that the following supremum is less than or equal to zero for all measurement waveforms,

$$
\begin{aligned}
&\sup_{\check{w}_1 \in \check{\mathcal{W}}_1, \check{w}_2 \in \check{\mathcal{W}}_2} J_{\gamma t_f} \\
&\leq \sup_{\omega_m \in \mathcal{W}_m} \left(\sup_{\omega_1 \in \mathcal{W}_1 | \omega_m \in \mathcal{W}_m} J_{1,\gamma t_f} + \sup_{\omega_2 \in \mathcal{W}_2 | \omega_m \in \mathcal{W}_m} J_{2,\gamma t_f} \right) \\
&\leq \sup_{\omega_m \in \mathcal{W}_m} \left\{ \int_0^{t_f} \left(|C_1 \check{x}_1 - y_d|^2 + \sum_{i=1}^{2} \left(|\xi_{i,c}|^2_{\bar{Q}_i} + \check{l}_i - \gamma^2 \zeta_i^2 |y_i - C_i \check{x}_i|^2 - \gamma^2 |\check{w}_{i,a}|^2 \right) \right) \mathrm{d}\tau \right\} \qquad (21)
\end{aligned}
$$

where function $\check{l}_1(\tau, y_{1[0,\tau]}, Y_{d[0,\tau]}, \check{w}_1)$ is part of the weighting function $l_1(\tau, \theta_1, x_1, y_{1[0,\tau]}, Y_{d[0,\tau]}, \check{w}_1)$, and $\check{l}_2(\tau, y_{2[0,\tau]}, Y_{d[0,\tau]}, \check{w}_2)$ is part of the weighting function $l_2(\tau, \theta_2, x_2, y_{2[0,\tau]}, Y_{d[0,\tau]}, \check{w}_2)$ to be designed, which are constants in the identifier design step and are therefore neglected.

By equation (21), we observe that the cost function is expressed in term of the states of the estimator we derived, whose dynamics are driven by the measurement y_1, y_2, $\check{w}_1$, $\check{w}_2$, $\acute{y}_2$, the reference trajectory y_d, the input u, and the worst-case estimate for the expanded state vector $\hat{\xi}_1$ and $\hat{\xi}_2$, which are signals we either measure or can construct. This is then a nonlinear H^∞-optimal control problem under full information measurements. Since $\acute{y}_2 = y_1$ in the adaptive system under consideration, we can equivalently deal with the following transformed variables instead of considering y_1, y_2, $\check{w}_1$, $\check{w}_2$, and $\acute{y}_2$ as the maximizing variable,

$$
v = \begin{bmatrix} \zeta_1 (y_1 - C_1 \check{x}_1) \\ \check{w}_{1,a} \\ \check{w}_{1,b} \\ \hline \zeta_2 (y_2 - C_2 \check{x}_2) \\ \check{w}_{2,a} \\ \check{w}_{2,b} \end{bmatrix} = \begin{bmatrix} v_1 \\ v_2 \end{bmatrix}
$$

where $v_i = \left[\zeta_i (y_i - C_i \check{x}_i) \; \check{w}'_{i,a} \; \check{w}'_{i,b} \right]'$, $i = 1, 2$.

By the special structure of the system, we define $v_{i,a} = \left[\zeta_i (y_i - C_i \check{x}_i) \; \check{w}'_{i,a} \right]'$, $i = 1, 2$, $v_a = \left[v'_{1,a} \; v'_{2,a} \right]'$, and we will attenuate disturbance v_a, and cancel the disturbance $\check{w}_{1,b}$ and $\check{w}_{2,b}$. In view of $y_2 = \zeta_2^{-1} e'_{\check{q}_a+2, \check{q}_{1,a}+2} v_a + \check{x}_{2,1}$, we will treat $\check{x}_{2,1}$ as the virtual control input of subsystem $\mathbf{S_1}$, where $\check{q}_a = \check{q}_{1,a} + \check{q}_{2,a}$.

For $i = 1, 2$, we introduce the matrix $M_{i,f} := \left[A_{i,f}^{n_i - 1} p_{i,n_i} \; \cdots \; A_{i,f} p_{i,n_i} \; p_{i,n_i} \right]$, where p_{i,n_i} is a n_i-dimensional vector such that the pair $(A_{i,f}, p_{i,n_i})$ is controllable. We note that $\acute{y}_2 = y_1$, then the following $3n_1+4n_2+\check{q}_1+\check{q}_2$-dimensional prefiltering system for $y_1, y_2, u, \check{w}_1, \check{w}_2$, and $\acute{y}_2$ generates the Φ_1 and Φ_2 online:

$$
\begin{aligned}
\dot{\eta}_1 &= A_{1,f}\eta_1 + p_{1,n_1}y_1; \\
\dot{\eta}_{\check{w}_1,j} &= A_{1,f}\eta_{\check{w}_1,j} + p_{1,n_1}\check{w}_{1,j}; \quad \eta_{\check{w}_i,j}(0) = \eta_{\check{w}_i,j0}, i = 1,\cdots,\check{q}_i \\
\dot{\lambda}_1 &= A_{1,f}\lambda_1 + p_{1,n_1}y_2; \quad \lambda_1(0) = \lambda_{1,0} \\
\Phi_1 &= \left[A_{1,f}^{n_1-1}\eta_1 \ \cdots \ A_{1,f}\eta_1 \ \eta_1 \right] M_{1,f}^{-1}\bar{A}_{1,211} + \left[A_{1,f}^{n_1-1}\lambda_1 \ \cdots \ A_{1,f}\lambda_1 \ \lambda_1 \right] M_{1,f}^{-1}\bar{A}_{1,212} \\
&\quad + \sum_{j=1}^{\check{q}_1} \left[A_{1,f}^{n_1-1}\eta_{\check{w}_1,j} \ \cdots \ A_{1,f}\eta_{\check{w}_1,j} \ \eta_{\check{w}_1,j} \right] M_{1,f}^{-1}\bar{A}_{i,213j} \\
\dot{\eta}_2 &= A_{2,f}\eta_2 + p_{2,n_2}y_2; \\
\dot{\eta}_{\check{w}_2,j} &= A_{2,f}\eta_{\check{w}_2,j} + p_{2,n_2}\check{w}_{2,j}; \quad \eta_{\check{w}_2,j}(0) = \eta_{\check{w}_2,j0}, j = 1,\cdots,\check{q}_i \\
\dot{\lambda}_2 &= A_{2,f}\lambda_2 + p_{2,n_2}u; \quad \lambda_1(0) = \lambda_{1,0} \\
\dot{\eta}_{2,y} &= A_{2,f}\eta_{2,y} + p_{2,n_2}\acute{y}_2; \eta_{2,y}(0) = \eta_{2,y0} \\
\Phi_2 &= \left[A_{2,f}^{n_2-1}\eta_1 \ \cdots \ A_{2,f}\eta_2 \ \eta_2 \right] M_{2,f}^{-1}\bar{A}_{2,211} + \left[A_{2,f}^{n_2-1}\lambda_2 \ \cdots \ A_{2,f}\lambda_2 \ \lambda_2 \right] M_{2,f}^{-1}\bar{A}_{2,212} \\
&\quad + \sum_{j=1}^{\check{q}_2} \left[A_{2,f}^{n_2-1}\eta_{\check{w}_2,j} \ \cdots \ A_{2,f}\eta_{\check{w}_2,j} \ \eta_{\check{w}_2,j} \right] M_{2,f}^{-1}\bar{A}_{2,213j} \\
&\quad + \left[A_{2,f}^{n_2-1}\eta_{2,y} \ \cdots \ A_{2,f}\eta_{2,y} \ \eta_{2,y} \right] M_{2,f}^{-1}\bar{A}_{2,214}
\end{aligned}
$$

The variables to be designed at this stage include $\check{x}_{2,1}$, u, $\xi_{1,c}$, and $\xi_{2,c}$. Note that the structures of A_1 and A_2 in the dynamics is in strict-feedback form, we will use the backstepping methodology, see [9], to design the control input u, which will guarantee the global boundedness of the closed-loop system states and the asymptotic convergence of the tracking error. Since there are the nonnegative definite weighting on $\xi_{1,c}$ and $\xi_{2,c}$ in the cost function (21), we can not use integrator backstepping to design feedback law for $\xi_{1,c}$ and $\xi_{2,c}$. Hence, we set $\xi_{1,c} = \xi_{2,c} = 0$ in the backstepping procedure. After the completion of the backstepping procedure, we will then optimize the choice of $\xi_{1,c}$ and $\xi_{2,c}$ based on the value function obtained. Note that Σ_1, Π_1, $s_{1,\Sigma}$, $\check{\theta}_1$, Σ_2, Π_2, $s_{2,\Sigma}$, and $\check{\theta}_2$ are always bounded by the design in Section 4.1. Since Φ_1 is driven by control y_2, and Φ_2 is explicitly driven by u, they can not be stabilized in conjunction with $\check{x}_1$ and $\check{x}_2$ in the backstepping design. We will assume they are bounded and prove later they are indeed so under the derived control law.

We carry out the backstepping design for subsystem $\mathbf{S_1}$ first, and treat $\check{x}_{2,1}$ as the virtual control input of subsystem $\mathbf{S_1}$ in view of $y_2 = \zeta_2^{-1}e'_{\check{q}_a+2,\check{q}_{1,a}+2}v_a + \check{x}_{2,1}$. To stabilize η_1, we introduce variable $\eta_{1,d}$, which satisfies $\dot{\eta}_{1,d} = A_{1,f}\eta_{1,d} + p_{1,n_1}y_d$ with initial condition $\eta_{1,d}(0) = \eta_{1,d0}$, and is the reference trajectory for η_1 to track. Choosing value function $V_{1,0} := |\eta_1 - \eta_{1,d}|^2_{Z_1}$, where Z_1 is the solution to an algebraic Riccati equation. Treating $\check{x}_{1,1}$ as the virtual control input, we complete the step 0 with the virtual control law $\alpha_{1,0} = y_d$, which will guarantee the $\dot{V}_{1,0} \le 0$ under $\check{x}_{1,1} = \alpha_{1,0}$. At step 1, we introduce $z_{1,1} := \check{x}_{1,1} - y_d$, and choose value function $V_{1,1} = V_{1,0} + \frac{1}{2}z_{1,1}^2$. Treating $\check{x}_{1,2}$ as the virtual control input, we end the step 1 with the virtual control law $\alpha_{1,1}$, which guarantees $\dot{V}_{1,1} \leq 0$ under $\check{x}_{1,2} = \alpha_{1,1}$. Define the variable $z_{1,2} = \check{x}_{1,2} - \alpha_{1,1}$ for step 2. Repeating the backstepping procedure until step r_1, the virtual control input $\check{x}_{2,1}$ will appear in the dynamic of $\dot{z}_{1,r_1}$. Using the similar procedure as previous steps, we can derive the robust adaptive controller α_{1,r_1} such that $\dot{V}_{1,r_1} \leq 0$ under $\check{x}_{2,1} := \alpha_{1,r_1}$. This completes the control design for subsystem $\mathbf{S_1}$.

To stabilize η_2, we introduce variable $\eta_{2,d}$ as below,

$$\dot{\eta}_{2,d} = A_{2,f}\eta_{2,d} + p_{2,n_2}\alpha_{1,r_1} + p_{2,n_2}e'_{\check{q}_a+2,\check{q}_{1,a}+2}\nu_{1,r_1}; \eta_{2,d}(0) = \eta_{2,d0}$$

and is the reference trajectory for η_2 to track, where ν_{1,r_1} is a function obtained after step r_1. Choosing value function $V_{2,0} := |\eta_2 - \eta_{2,d}|^2_{Z_2} + V_{1,r_1}$, where Z_2 is the solution to an algebraic Riccati equation. We complete the step $r_1 + 1$ with the virtual control law $\alpha_{2,0} = \alpha_{1,r_1}$, which will guarantee the $\dot{V}_{2,0} \leq 0$ under $\check{x}_{2,1} = \alpha_{2,0}$. Repeating the backstepping procedure until step $r_1 + r_2 + 1$, the virtual control input u will appear in the dynamic of $\dot{z}_{2,r_2}$. Introduce $V_{2,r_2} = \sum_{j=1}^{2}(|\tilde{\eta}_j|^2_{Z_j} + \sum_{k=1}^{r_j}\frac{1}{2}z^2_{j,k})$, we then can derive the robust adaptive controller μ such that $\dot{V}_{2,r_2} \leq 0$ under $u := \mu$. Later, we will show that the control law μ will guarantee the boundedness of the closed-loop system states and the asymptotic convergence of tracking error.

For the closed-loop adaptive nonlinear system, we have the following value function, $U = W_1 + W_2 + V_{2,r_2}$, and its time derivative is given by

$$\begin{aligned}
\dot{U} = -|x_{1,1} - y_d|^2 &-\sum_{j=1}^{2}\Big(\gamma^4|x_j-\hat{x}_j-\Phi_j(\theta_j-\hat{\theta}_j)|^2_{\Pi_j^{-1}\Delta_j\Pi_j^{-1}} + \epsilon_j(\gamma^2\zeta_j^2 - 1)|\theta_j - \hat{\theta}_j|^2_{\Phi_j'C_j'C_j\Phi_j} \\
&-2(\theta_j - \check{\theta}_j)'P_{j,r}(\check{\theta}_j) + |\tilde{\eta}_j|^2_{Y_j} + \sum_{k=1}^{r_j}\beta_{j,k}z^2_{j,k} - \gamma^2|w_j|^2 + \gamma^2|w_j - w_{j,opt}|^2 - \gamma^2|\check{w}_{j,a}|^2 \\
&+\gamma^2|\check{w}_{j,a} - \check{w}_{j,opt}|^2\Big) - \epsilon_2|\theta_2 - \hat{\theta}_2|_{\Phi_2'C_2'C_2\Phi_2} - \frac{1}{4}\left|\varsigma_{1,(r_1+r_2)}\right|^2_{\bar{Q}_1} - \frac{1}{4}|\varsigma_{2,r_2}|^2_{\bar{Q}_2} \\
&+\left|\xi_{1,c} + \frac{1}{2}\varsigma_{1,(r_1+r_2)}\right|^2_{\bar{Q}_1} + \left|\xi_{2,c} + \frac{1}{2}\varsigma_{2,r_2}\right|^2_{\bar{Q}_2}
\end{aligned}$$

where ς_{1,r_1+r_2} and ς_{2,r_2} are functions obtained after step $r_1 + r_2 + 1$, $w_{1,opt}$ and $w_{2,opt}$ are the worst case disturbance with respect to the value function U, which are given by

$$w_{1,opt} = \zeta_1 E_1' e_{2,1}' \nu_{1,r_1} + \gamma^{-2}(I_{q_1} - \zeta_1^2 E_1' E_1)\bar{D}_1'\bar{\Sigma}_1^{-1}(\xi_1 - \check{\xi}_1) + \zeta_1^2 E_1' C_1(\check{x}_1 - x_1)$$

$$w_{2,opt} = \zeta_2 E_2' e_{2,2}' \nu_{2,r_2} + \gamma^{-2}(I_{q_2} - \zeta_2^2 E_2' E_2)\bar{D}_2'\bar{\Sigma}_2^{-1}(\xi_2 - \check{\xi}_2) + \zeta_2^2 E_2' C_2(\check{x}_2 - x_2)$$

$$\check{w}_{1,opt} = \left[\mathbf{0}'_{1\times(2+\check{q}_{1,a}+\check{q}_{2,a})}\ e_{(2+\check{q}_{1,a}+\check{q}_{2,a}),1} \cdots e_{(2+\check{q}_{1,a}+\check{q}_{2,a}),\check{q}_{1,a}}\ \mathbf{0}'_{(1+\check{q}_{2,a})\times(2+\check{q}_{1,a}+\check{q}_{2,a})}\right]' \nu_{1,r_1}$$

$$\check{w}_{2,opt} = \left[\mathbf{0}'_{(2+\check{q}_{1,a})\times(2+\check{q}_{1,a}+\check{q}_{2,a})}\ e_{(2+\check{q}_{1,a}+\check{q}_{2,a}),1} \cdots e_{(2+\check{q}_{1,a}+\check{q}_{2,a}),\check{q}_{2,a}}\right]' \nu_{2,r_2}$$

where ν_{1,r_1} and ν_{2,r_2} are functions obtained after backstepping design.

Then the optimal choice for the variable $\tilde{\xi}_{i,c}$ and $\hat{\xi}_i, i = 1,2$, are:

$$\tilde{\xi}_{1,c*} = -\frac{1}{2}\varsigma_{1,r_1+r_2} \Longleftrightarrow \hat{\xi}_{1,*} = \check{\xi}_1 - \frac{1}{2}\varsigma_{1,r_1+r_2};$$
$$\tilde{\xi}_{2,c*} = -\frac{1}{2}\varsigma_{2,r_2} \Longleftrightarrow \hat{\xi}_{2,*} = \check{\xi}_2 - \frac{1}{2}\varsigma_{2,r_2}$$

which yields that the closed-loop system is dissipative with storage function U and supply rate with optimal choice for $\hat{\xi}_i, i = 1,2$:

$$-|x_{1,1} - y_d|^2 + \gamma^2|w_1|^2 + \gamma^2|w_2|^2 + \gamma^2|\breve{w}_{1,a}|^2 + \gamma^2|\breve{w}_{2,a}|^2$$

This completes the adaptive controller design step. We will discuss the robustness and tracking properties of the proposed adaptive control laws.

5. Main result

In this Section, we present the main result by stating two theorems.

For the adaptive control law, with the optimal choice of $\xi_{i,c*}$, the closed-loop system dynamics are

$$\dot{X} = F(X, y_d^{(r_1+r_2)}) + G(X)\left[w_1' \; w_2'\right] + G_{\breve{w}}(X)\left[\breve{w}_1' \; \breve{w}_2'\right]; X(0) = X_0 \tag{22}$$

where F, G and G_M are smooth mapping of $\mathcal{D} \times \mathbf{R}$, $\mathcal{D}$ and $\mathcal{D}$, respectively; and the initial condition $X_0 \in \mathcal{D}_0 := \{X_0 \in \mathcal{D} \mid \theta_i \in \Theta_i, \check{\theta}_{i,0} \in \Theta_i, \Sigma_i(0) = \gamma^{-2}Q_{i,0}^{-1} > 0, \mathrm{Tr}\left((\Sigma_i(0))^{-1}\right) \leq K_{i,c}, s_{i,\Sigma}(0) = \gamma^2\mathrm{Tr}(Q_{i,0}); \quad i = 1,2\}$. And the value function U satisfies an Hamilton-Jacobi-Isaacs equation, $\forall X \in \mathcal{D}, \forall y_d^{(r_1+r_2)} \in \mathbf{R}$.

$$\frac{\partial U}{\partial X}(X)F(X, y_d^{(r_1+r_2)}) + \frac{1}{4\gamma^2}\frac{\partial U}{\partial X}(X)\left[G(X) \; G_{\breve{w}}(X)\right]\left[G(X)' \; G_{\breve{w}}(X)'\right]'\left(\frac{\partial U}{\partial X}(X)\right)'$$
$$+Q(X, y_d^{(r_1+r_2)}) = 0;$$

where $Q : \mathcal{D} \times \mathbf{R} \to \mathbf{R}$ is smooth and given by

$$Q(X, y_d^{(r_1+r_2)}) = |x_{1,1} - y_d|^2 + \sum_{j=1}^{2}\Big(\gamma^4|x_j - \hat{x}_j - \Phi_j(\theta_j - \hat{\theta})_j|^2_{\Pi_j^{-1}\Delta_j\Pi_j^{-1}}$$
$$+\epsilon_j(\gamma^2\zeta_j^2 - 1)|\theta_j - \hat{\theta}_j|^2_{\Phi_j'C_j'C_j\Phi_j} - 2(\theta_j - \check{\theta}_j)'P_{j,r}(\check{\theta}_j) + |\tilde{\eta}_j|^2_{Y_j} + \sum_{k=1}^{r_j}\beta_{j,k}z_{j,k}^2\Big)$$
$$+\frac{1}{4}\left|\varsigma_{1,(r_1+r_2)}\right|^2_{\bar{Q}_1} + \frac{1}{4}|\varsigma_{2,r_2}|^2_{\bar{Q}_2} + \epsilon_2|\theta_2 - \hat{\theta}_2|_{\Phi_2'C_2'C_2\Phi_2}$$

The closed-loop adaptive system possesses a strong stability property, which will be stated precisely in the following theorem.

Theorem 1. *Consider the robust adaptive control problem formulated and assumptions in Section 3. The robust adaptive controller μ with the optimal choice of $\xi_{i,c}$, achieves the following strong robustness properties for the closed-loop system.*

1. *Given $c_w \geq 0$, and $c_d \geq 0$, there exists a constant $c_c \geq 0$ and compact sets $\Theta_{1,c} \subset \Theta_{1,o}$, and $\Theta_{2,c} \subset \Theta_{2,o}$ such that for any uncertainty $(x_{1,0}, \theta_1, \grave{w}_{1,[0,\infty)}, \check{w}_{1,[0,\infty)}, Y_{d0}, y^{(r_1+r_2)}_{d[0,\infty)}) \in \grave{\mathcal{W}}_1$ and $(x_{2,0}, \theta_2, \grave{w}_{2,[0,\infty)}, \check{w}_{2,[0,\infty)}) \in \grave{\mathcal{W}}_2$ with $|x_{1,0}| \leq c_w; |x_{2,0}| \leq c_w; |\grave{w}_1(t)| \leq c_w; |\grave{w}_2(t)| \leq c_w; |\check{w}_1(t)| \leq c_w; |\check{w}_2(t)| \leq c_w; |Y_d(t)| \leq c_d;\quad \forall t \in [0,\infty)$ all closed-loop state variables x_1, $\check{x}_1$, $\check{\theta}_1$, Σ_1, $s_{1,\Sigma}$, η_1, $\eta_{1,d}$, $\Phi_{1,u}$, x_2, $\check{x}_2$, $\check{\theta}_2$, Σ_2, $s_{2,\Sigma}$, η_2, $\eta_{2,d}$, $\Phi_{2,u}$ are bounded as follows, $\forall t \in [0,\infty)$,*

$$|x_i(t)| \leq c_c; |\check{x}_i(t)| \leq c_c; \check{\theta}_i(t) \in \Theta_{i,c}; |\eta_i(t)| \leq c_c; |\eta_{i,\grave{w},1}(t)| \leq c_c; \cdots |\eta_{i,\grave{w},\grave{q}_i}| \leq c_c;$$

$$|\eta_{i,d}(t)| \leq c_c; |\Phi_{i,u}(t)| \leq c_c; K_{i,c}^{-1} I \leq \Sigma_i(t) \leq \gamma^{-2} Q_{i,0}^{-1};\quad \gamma^2 \mathrm{Tr}(Q_{i,0}) \leq s_{i,\Sigma}(t) \leq K_{i,c};\quad i=1,2$$

The inputs are also bounded $|u(t)| \leq c_u$, and $\hat{\xi}_1 \leq c_u$, $\hat{\xi}_2 \leq c_u$, $\forall t \in [0,\infty)$, for some constant $c_u \geq 0$. Furthermore, there exists constant $c_\lambda \geq 0$ such that $|\lambda_{i,0}(t)| \leq c_\lambda$, $|\lambda_i(t)| \leq c_\lambda$, $i = 1,2$, and $|\eta_{2,y}(t)| \leq c_\lambda$, $\forall t \geq 0$.

2. *For any uncertainty $(x_{1,0}, \theta_1, \grave{w}_{1,[0,\infty)}, \check{w}_{1,[0,\infty)}, Y_{d0}, y^{(r_1+r_2)}_{d[0,\infty)}) \in \grave{\mathcal{W}}_1$, and $(x_{2,0}, \theta_2, \grave{w}_{2,[0,\infty)}, \check{w}_{2,[0,\infty)}) \in \grave{\mathcal{W}}_2$ the controller $\mu \in \mathcal{M}$ achieves disturbance attenuation level γ with respect to w_1 and w_2, arbitrary disturbance attenuation level $\check{\gamma}$ with respect to $\check{w}_{1,a}$ and $\check{w}_{2,a}$, and disturbance attenuation level zero with respect to $\check{w}_{1,b}$ and $\check{w}_{2,b}$, .*

3. *For any uncertainty $(x_{1,0}, \theta_1, \grave{w}_{1[0,\infty)}, \check{w}_{1[0,\infty)}, Y_{d0}, y^{(r_1+r_2)}_{d[0,\infty)}) \in \grave{\mathcal{W}}_1$, and $(x_{2,0}, \theta_2, \grave{w}_{2[0,\infty)}, \check{w}_{2[0,\infty)}) \in \grave{\mathcal{W}}_2$ with $\grave{w}_{1[0,\infty)} \in \mathcal{L}_2 \cap \mathcal{L}_\infty$, $\grave{w}_{2[0,\infty)} \in \mathcal{L}_2 \cap \mathcal{L}_\infty$, $\check{w}_{1,a[0,\infty)} \in \mathcal{L}_2 \cap \mathcal{L}_\infty$, $\check{w}_{2,a[0,\infty)} \in \mathcal{L}_2 \cap \mathcal{L}_\infty$ $\check{w}_{1,b[0,\infty)} \in \mathcal{L}_\infty$, $\check{w}_{2,b[0,\infty)} \in \mathcal{L}_\infty$, and $Y_{d[0,\infty)} \in \mathcal{L}_\infty$, the noiseless output of the system, $x_{1,1}$, asymptotically tracks the reference trajectory, y_d, i.e.,*

$$\lim_{t\to\infty} (x_{1,1}(t) - y_d(t)) = 0$$

4. *The ultimate lower bound on the achievable performance level is only relevant to the Subsystem $\mathbf{S_1}$, i.e., $\gamma \geq \zeta_1^{-1}$ or $\gamma > \zeta_1^{-1}$.*

Proof For the first statement, fix $c_w \geq 0$, and $c_d \geq 0$ consider any uncertainty $(x_{1,0}, x_{2,0}, \theta_1, \theta_2, \grave{w}_{1,[0,\infty)}, \grave{w}_{2,[0,\infty)}, \check{w}_{1,[0,\infty)}, \check{w}_{2,[0,\infty)}, y^{(r_1+r_2)}_{d[0,\infty)})$ that satisfies:

$$|x_{1,0}| \leq c_w; |x_{2,0}| \leq c_w; |\grave{w}_1(t)| \leq c_w; |\grave{w}_2(t)| \leq c_w; |\check{w}_1(t)| \leq c_w; |\check{w}_2(t)| \leq c_w; |Y_d(t)| \leq c_d; \forall t \in [0,\infty)$$

We define $[0, T_f)$ to be the maximal length interval on which the closed system (22) has a solution that lies in $\mathcal{D}$. Note that we have Σ_1, Σ_2, $s_{1,\Sigma}$ and $s_{2,\Sigma}$ are uniformly upper bounded and uniformly bounded away from 0 as desired by Section 4.

Introduce the vector of variables

$$X_e := \left[\, \check{\theta}_1' \;\check{\theta}_2' \;(\check{x}_1 - \Phi_1\tilde{\theta}_1)' \;(\check{x}_2 - \Phi_2\tilde{\theta}_2)' \;\tilde{\eta}_1' \;\tilde{\eta}_2' \;z_{1,1} \cdots z_{1,r_1} \;z_{2,1} \cdots z_{2,r_2} \,\right]'$$

and two nonnegative and continuous functions defined on $\mathbf{R}^{2n_1+2n_2+\sigma_1+\sigma_2+r_1+r_2}$

$$U_M(X_e) := \sum_{i=1}^{2} K_{i,c}|\tilde{\theta}_i|^2 + \sum_{i=1}^{2} \gamma^2 |\tilde{x}_i - \Phi_i\tilde{\theta}_i|^2_{\Pi_i^{-1}} + \sum_{i=1}^{2} |\tilde{\eta}_i|^2_{Z_i} + \sum_{j=1}^{r_1} \gamma_{1,j} z_{1,j}^2 + \sum_{j=1}^{r_2} \gamma_{2,j} z_{2,j}^2$$

$$U_m(X_e) := \sum_{i=1}^{2} \gamma^2 |\tilde{\theta}_i|^2_{Q_{i,0}} + \sum_{i=1}^{2} \gamma^2 |\tilde{x}_i - \Phi_i\tilde{\theta}_i|^2_{\Pi_i^{-1}} + \sum_{i=1}^{2} |\tilde{\eta}_i|^2_{Z_i} + \sum_{j=1}^{r_1} \gamma_{1,j} z_{1,j}^2 + \sum_{j=1}^{r_2} \gamma_{2,j} z_{2,j}^2$$

then, we have

$$U_m(X_e) \le U(t,X_e) \le U_M(X_e), \quad \forall (t,X_e) \in [0,T_f) \times \mathbf{R}^{2(n_1+n_2)+\sigma_1+\sigma_2+r_1+r_2}$$

Since $U_m(X_e)$ is continuous, nonnegative definite and radially unbounded, then $\forall \alpha \in \mathbf{R}$, the set $S_{1\alpha} := \{X_e \in \mathbf{R}^{2(n_1+n_2)+\sigma_1+\sigma_2+r_1+r_2} \mid U_m(X_e) \le \alpha\}$ is compact or empty. Since $|\check{w}_1(t)| \le c_w$, and $|\check{w}_2(t)| \le c_w$, $\forall t \in [0,\infty)$, there exists a constant $c > 0$ such that we have the following inequality for the derivative of U:

$$\dot{U} \le -\sum_{i=1}^{2}\left(\frac{\gamma^4}{2}|x_i - \check{x}_i - \Phi_i(\theta_i - \check{\theta}_i)|^2_{\Pi_i^{-1}\Delta_i\Pi_i^{-1}} - 2(\theta_i - \check{\theta}_i)' P_{i,r}(\check{\theta}_i) + |\tilde{\eta}_i|^2_{Y_i} + \sum_{j=1}^{r_i} c_{i,\beta_j} z_{i,j}^2\right) + c$$

Since $-\sum_{i=1}^{2}(\frac{\gamma^4}{2}|x_i + \check{x}_i - \Phi_i(\theta_i - \check{\theta}_i)|^2_{\Pi_i^{-1}\Delta_i\Pi_i^{-1}} + |\tilde{\eta}_i|^2_{Y_i} - 2(\theta_i - \check{\theta}_i)' P_{i,r}(\check{\theta}_i) + \sum_{j=1}^{r_i} c_{\beta_{i,j}} z_{i,j}^2)$ will tend to $-\infty$ when X_e approaches the boundary of $\Theta_{1,o} \times \Theta_{2,o} \times \mathbf{R}^{2(n_1+n_2)+r_1+r_2}$, then there exists a compact set $\Omega_1(c_w) \subset \Theta_{1,o} \times \Theta_{2,o} \times \mathbf{R}^{2(n_1+n_2)+r_1+r_2}$, such that $\dot{U} < 0$ for $\forall X_e \in \Theta_{1,o} \times \Theta_{2,o} \times \mathbf{R}^{2(n_1+n_2)+r_1+r_2} \backslash \Omega_1$.

Then we have $U(t,X_e(t)) \le c_1$, and $X_e(t)$ is in the compact set $S_{1c_1} \subseteq \mathbf{R}^{2(n_1+n_2)+\sigma_1+\sigma_2+r_1+r_2}$, $\forall t \in [0,T_f)$. It follows that the signal X_e is uniformly bounded, namely, $\tilde{\theta}_1$, $\tilde{\theta}_2$, $\tilde{x}_1 - \Phi_1\tilde{\theta}_1$, $\tilde{x}_2 - \Phi_2\tilde{\theta}_2$, $\tilde{\eta}_1$, $\tilde{\eta}_2$, $z_{1,1}, \cdots, z_{1,r_1}$ and $z_{2,1}, \cdots, z_{2,r_2}$ are uniformly bounded.

Based on the dynamics of $\eta_{1,d}$, we have $\eta_{1,d}$ is uniformly bounded. Since $\tilde{\eta}_1 = \eta_1 - \eta_{1,d}$ is uniformly bounded, then η_1 is also uniformly bounded. Furthermore, there is a particular linear combination of the components of η_1, denoted by $\eta_{1,L}$,

$$\dot{\eta}_1 = A_{1,f}\eta_1 + p_{1,n_1}y_1$$

$$\eta_{1,L} = T_{1,L}\eta_1$$

which is strictly minimum phase and has relative degree 1 with respect to y_1. Then the signal $\eta_{1,L}$ has relative degree $r_1 + 1$ with respect to the input y_2, and is uniformly bounded. The composite system of η_1 and $\check{x}_1$ with input $\check{w}_1$ and y_2 and output $\eta_{1,L}$ may serve as a reference system in the application of bounding Lemma [12].

Note $\Phi_1 = \Phi_{1,y} + \Phi_{1,u}$ and $\Phi_{1,y}$ is uniformly bounded. To prove Φ_1 is bounded, we need to prove $\Phi_{1,u}$ is uniformly bounded. Define the following equations to separate $\Phi_{1,u}$ into two part:

$$\Phi_{1,u} = \Phi_{1,u_s} + \lambda_{1,b}\bar{A}_{1,2120} \tag{23a}$$

$$\dot{\lambda}_{1,b} = A_{1,f}\lambda_{1,b} + e_{n_1,r_1}y_2; \quad \lambda_{1,b}(0) = \mathbf{0}_{n_1\times 1} \tag{23b}$$

$$\dot{\Phi}_{1,u_s} = A_{1,f}\Phi_{1,u_s} + y_2\begin{bmatrix}\mathbf{0}_{r_1\times\sigma_1} \\ \bar{A}_{1,212r_1}\end{bmatrix}; \quad \Phi_{1,u_s}(0) = \Phi_{1,u0} \tag{23c}$$

We observe that the relative degree for each element of Φ_{1,u_s1} is at least $r_1 + 1$ with respect to the input y_2, and is the output of a stable linear system. Take $\eta_{1,L}$ and y_2 as output and input of the reference system, we conclude $\Phi_{1,u_s 1}$ is uniformly bounded by bounding Lemma. Because the first row element of $\tilde{x}_1 - \Phi_1\tilde{\theta}_1$ is:

$$\tilde{x}_{1,1} - \Phi_{1,u_s 1}\tilde{\theta}_1 - \lambda_{1,b1}\bar{A}_{1,2120}\tilde{\theta}_1 - \eta_1' T_{1,1}\tilde{\theta}_1$$

we can conclude that $\tilde{x}_{1,1} - \lambda_{1,b1}\bar{A}_{1,2120}\tilde{\theta}_1$ is uniformly bounded in view of the boundedness of $\tilde{x}_1 - \Phi_1\tilde{\theta}_1$, $\tilde{\theta}_1$, $\Phi_{1,u_s 1}$, and η_1. Since $z_{1,1} = \check{x}_{1,1} - y_d$, and $z_{1,1}, y_d$ are both uniformly bounded, we have that $\check{x}_{1,1}$ is also uniformly bounded.

Notice that $A_{1,f} = A_1 - \zeta_1^2 L_1 C_1 - \Pi_1 C_1' C_1 (\zeta_1^2 - \gamma^{-2})$, we generated the signal $x_{1,1} - b_{1,0}\lambda_{1,b1}$ by:

$$\begin{aligned}
\dot{x}_1 - b_{1,0}\dot{\lambda}_{1,b} &= A_{1,f}(x_1 - b_{1,0}\lambda_{1,b}) + \begin{bmatrix} \mathbf{0}_{r_1\times 1} \\ \bar{A}_{1,212r_1}\theta_1 \end{bmatrix} y_2 + \bar{A}_{1,211}\theta_1 y_1 + D_1\grave{M}_1\check{w}_1 + (\zeta_1^2 L_1 \\
&\quad + \Pi_1 C_1' (\zeta_1^2 - \frac{1}{\gamma^2}))(y_1 - E_1\grave{M}_1\check{w}_1) + \begin{bmatrix} \mathbf{0}_{1\times r_1} & b_{1,p1} & \cdots & b_{1,p\,n_1-r_1} \end{bmatrix}' y_2 \\
&\quad + \sum_{j=1}^{\check{q}_1} \bar{A}_{1,213,j}\check{w}_{1,j}\theta_1 + \check{D}_1\check{w}_1 \\
x_{1,1} - b_{1,0}\lambda_{1,b1} &= C_1(x_1 - b_{1,0}\lambda_{1,b})
\end{aligned}$$

Now we will separate the above dynamics into y_1 dependent and y_2 dependent parts by the linearity of the system, $x_{1,1} - b_{1,0}\lambda_{1,b1} := x_{1,u1} + x_{1,y1}$, which are respectively given by,

$$\begin{aligned}
\dot{x}_{1,u} &= A_{1,f}x_{1,u} + \begin{bmatrix} \mathbf{0}_{r_1\times 1} \\ \bar{A}_{1,212r_1}\theta \end{bmatrix} y_2 + \begin{bmatrix} \mathbf{0}_{1\times r_1} & b_{1,p1} & \cdots & b_{1,p\,n_1-r_1} \end{bmatrix}' y_2 \\
x_{1,u1} &= C_1 x_{1,u} \\
\dot{x}_{1,y} &= A_{1,f}x_{1,y} + (\zeta_1^2 L_1 + \Pi_1 C_1' (\zeta_1^2 - \frac{1}{\gamma^2}))(y_1 - E_1\grave{M}_1\check{w}_1) + \bar{A}_{1,211}\theta_1 y_1 + D_1\grave{M}_1\check{w}_1 \\
&\quad + \sum_{j=1}^{\check{q}_1} \bar{A}_{1,213,j}\check{w}_{1,j}\theta_1 + \check{D}_1\check{w}_1 \\
x_{1,y1} &= C_1 x_{1,y}
\end{aligned}$$

We observe that the signal $x_{1,u1}$ has relative degree at least $r_1 + 1$ with respect to y_2, take $\eta_{1,L}$ and y_2 as output and input of the reference system, we conclude $x_{1,u1}$ is uniformly bounded by bounding Lemma . Since $x_{1,y1}$ has relative degree at least 1 with respect to y_1, take $\eta_{1,L}$ and y_1 as output and input of the reference system, we conclude $x_{1,y1}$ is uniformly bounded by bounding Lemma. Then, $x_{1,1} - b_{1,0}\lambda_{1,b1}$ is uniformly bounded. It follows that $\check{x}_{1,1} - \lambda_{1,b1}(b_{1,p0} + A_{1,2120}\check{\theta}_1)$ is also uniformly bounded. Since $\check{x}_{1,1}$ is uniformly bounded and $\check{\theta}_1$ is uniformly bounded away from 0, we have $\lambda_{1,b1}$ is uniformly bounded. That further imply $\Phi_{1,1}$, i.e., $C_1\Phi_1$, is uniformly bounded. Furthermore, since $x_{1,1} - b_{1,0}\lambda_{1,b1}$ and $\check{w}_1$ are bounded, we have that the signals of $x_{1,1}$ and y_1 are uniformly bounded.

Next, we need to prove the existence of a compact set $\Theta_{1,c} \subset \Theta_{1,o}$ such that $\check{\theta}_1(t) \in \Theta_{1,c}$, $\forall t \in [0, T_f)$. First introduce the function

$$\Upsilon_1 := U + (\rho_{1,o} - P_1(\check{\theta}_1))^{-1} P_1(\check{\theta}_1)$$

We notice that, when $\check{\theta}_1$ approaches the boundary of $\Theta_{1,o}$, $P_1(\check{\theta}_1)$ approaches $\rho_{1,o}$. Then Υ_1 approaches ∞ as X_e approaches the boundary of $\Theta_{1,o} \times \Theta_{2,o} \times \mathbf{R}^{2(n_1+n_2)+r_1+r_2}$. There exist some constant $c > 0$ such that the following inequalities hold.

$$\begin{aligned}\dot{\Upsilon}_1 &= \dot{U} + (\rho_{1,o} - P_1(\check{\theta}_1))^{-2}\rho_{1,o}\frac{\partial P_1}{\partial \theta_1}(\check{\theta}_1)\dot{\check{\theta}}_1 \\ &\leq -\sum_{i=1}^{2}(\frac{\gamma^4}{2}|x_{i,1} - \check{x}_{i,1} - \Phi_i(\theta_i - \check{\theta}_i)|^2_{\Pi_i^{-1}\Delta_i\Pi_i^{-1}} - 2(\theta_i - \check{\theta}_i)'P_{i,r}(\check{\theta}_i) + |\tilde{\eta}_i|^2_{\Upsilon_i} + \sum_{j=1}^{r_i} c_{i,\beta_j} z_{i,j}^2) \\ &\quad - \left|\left(\frac{\partial P_1}{\partial \theta_1}(\check{\theta}_1)\right)'\right|^2 (\rho_{1,o} - P_1(\check{\theta}_1))^{-4}\left(K_{1,c}^{-1}\rho_{1,o}p_{1,r}(\check{\theta}_1)(\rho_{1,o} - P_1(\check{\theta}_1))^2 - c\right) + c\end{aligned}$$

Since $\dot{\Upsilon}_1$ will tend to $-\infty$ when X_e approaches the boundary of $\Theta_{1,o} \times \Theta_{2,o} \times \mathbf{R}^{2(n_1+n_2)+r_1+r_2}$, then there exists a compact set $\Omega_{1,2}(c_w) \subset \Theta_{1,o} \times \Theta_{2,o} \times \mathbf{R}^{2(n_1+n_2)+r_1+r_2}$, such that $\forall X_e \in \Theta_{1,o} \times \Theta_{2,o} \times \mathbf{R}^{2(n_1+n_2)+r_1+r_2} \backslash \Omega_{1,2}$, $\dot{\Upsilon}_1(X_e) < 0$.

Then there exists a compact set $\Theta_{1,c} \subset \Theta_{1,o}$, such that $\check{\theta}_1(t) \in \Theta_{1,c}$, $\forall t \in [0, T_f)$. Moreover, $\Upsilon_1(t, X_e(t)) \leq c_2$, and $X_e(t)$ is in the compact set $S_{1,2c_2} \subseteq \Theta_{1,o} \times \Theta_{2,o} \times \mathbf{R}^{2(n_1+n_2)+r_1+r_2}$, $\forall t \in [0, T_f)$.

To derive the uniformly boundedness of the closed-loop system states, we separate the relative degree, r_1, into two cases: $r_1 = 1$, and $r_1 \geq 2$. First, we consider the case 1: $r_1 = 1$.

Taking $x_{1,1}$ and y_2 as the output and input of the reference system, we note that $x_{1,1}$ is strictly minimum phase and has relative degree r_1 with respect to input y_2. Since the state x_1 can be viewed as stably filtered output signals of y_2 and y_1, it is uniformly bounded. Since λ_1 is also some stably filtered signals of y_1 and y_2, it is uniformly bounded. It further implies Φ is uniformly bounded. Then we can conclude $\check{x}_1$ is uniformly bounded from the boundedness of $\tilde{x}_1 - \Phi_1\tilde{\theta}_1$. This further implies that the inputs $\check{x}_{2,1}$ and $\hat{\xi}_1$ are uniformly bounded.

Case 2: $r_1 \geq 2$. Considering the canonical form (78) in [12] for the true system (1), we denote the elements of $\bar{x}$ by $\left[\, \bar{x}_{11} \;\cdots\; \bar{x}_{1r_1} \,\right]'$. We will use the mathematical induction to derive the boundedness of $\Phi_{1,u_s i}$, $\tilde{x}_{1,i} - \lambda_{1,bi}\bar{A}_{1,2120}\tilde{\theta}_1$, $\check{x}_{1,i}$, $x_{1,i} - b_{1,0}\lambda_{1,bi}$, $\lambda_{1,bi}$, $\Phi_{1,u\,i}$, $x_{1,i}$, $\bar{x}_{1,1i}$, $\forall i = \{1, \cdots, r_1\}$. For the boundedness of $\bar{x}_{1i}$, we will show that $\bar{x}_{1i}$ is a linear combination of $x_{1,1}$, $\cdots$, $x_{1,i}$, $\bar{x}_3$, and $\bar{x}_4$, i.e.,

$$\bar{x}_{1i} = \tilde{a}_{1,1i}x_{1,1} + \cdots + \tilde{a}_{1,i-1\,i}x_{1,i-1} + x_{1,i} + \tilde{T}_{1,i3}\bar{x}_3 + \tilde{T}_{1,i4}\bar{x}_4; \quad 1 \leq i \leq r_1 \tag{24}$$

where $\tilde{a}_{1,1i}$, $\cdots$, $\tilde{a}_{1,i-1\,i}$ are constants, $\tilde{T}_{1,i3}$, $\tilde{T}_{1,i4}$ are constant matrices, and $\bar{x}_3$ and $\bar{x}_4$ are defined at (78) in [12].

1°: We have deduced that η_1, $\eta_{1,d}$, $\eta_{1,L}$, $\Phi_{1,u_s 1}$, $\tilde{x}_{1,1} - \lambda_{1,b1}\bar{A}_{1,2120}\tilde{\theta}_1$, $\check{x}_{1,1}$, $x_{1,1} - b_{1,0}\lambda_{1,b1}$, $\lambda_{1,b1}$, $\Phi_{1,u\,1}$, and $x_{1,1}$ are uniformly bounded in $[0, T_f)$. $\bar{x}_{11}$ is bounded in view of $x_{1,1} - \grave{C}_3\bar{x}_3 - \grave{C}_4\bar{x}_4$.

2°: We assume that $\Phi_{1,u_s i}$, $\tilde{x}_{1,i} - \lambda_{1,bi}\bar{A}_{1,2120}\tilde{\theta}_1$, $\check{x}_{1,i}$, $x_{1,i} - b_{1,0}\lambda_{1,bi}$, $\lambda_{1,bi}$, $\Phi_{1,u\,i}$, $x_{1,i}$, and $\bar{x}_{1i}$ are bounded, and

$$\bar{x}_{1i} = \tilde{a}_{1,1i}x_{1,1} + \cdots + \tilde{a}_{1,i-1\,i}x_{1,i-1} + x_{1,i} + \tilde{T}_{1,i3}\bar{x}_3 + \tilde{T}_{1,i4}\bar{x}_4; \quad \forall i \in \{1, \cdots k\} \tag{25}$$

where $1 \leq k < r_1$.

3°: First, we need to show that $\Phi_{1,u_s k+1}$, $\tilde{x}_{1,k+1} - \lambda_{1,bk+1}\bar{A}_{1,2120}\tilde{\theta}_1$, $\check{x}_{1,k+1}$, $x_{1,k+1} - b_{1,0}\lambda_{1,bk+1}$, $\lambda_{1,bk+1}$, $\Phi_{1,uk+1}$, $x_{1,k+1}$, and $\bar{\check{x}}_{1k+1}$ are bounded.

From equation (23c), we note that every element of $\Phi_{1,u_s k+1}$ has relative degree of at least $r_1 - k + 1$ with respect to y_2, and is the output of a stable linear system. Since the boundedness of $\bar{\check{x}}_{11}, \cdots, \bar{\check{x}}_{1k}$, we conclude $\Phi_{1,u_s k+1}$ is uniformly bounded by Lemma 11 in [12], where the reference system has input y_2 and output y_1.

Note that $k+1$st row element of $\tilde{x}_1 - \Phi_1\tilde{\theta}_1$ is

$$\tilde{x}_{1,k+1} - \Phi_{1,u_s k+1}\tilde{\theta}_1 - \lambda_{1,bk+1}\bar{A}_{1,2120}\tilde{\theta}_1 - \eta_1' T_{1,k+1}\tilde{\theta}_1$$

We can conclude that $\tilde{x}_{1,k+1} - \lambda_{1,bk+1}\bar{A}_{1,2120}\tilde{\theta}_1$ is uniformly bounded in view of the boundedness of $\tilde{x}_1 - \Phi_1\tilde{\theta}_1$, $\tilde{\theta}_1$, $\Phi_{1,u_s k+1}$, and η_1. Since the boundedness of y_d, $s_{1,\Sigma}$, η_1, $\eta_{1,d}$, Σ_1, $\check{x}_{1,1}$, $y_d^{(1)}$, $\Phi_{1,u1}, \cdots$ $\check{x}_{1,k}$, $y_d^{(k)}$, $\Phi_{1,uk}$, and $\check{\theta}_1(t) \in \Theta_{1,c}$, $\forall t \in [0, T_f)$, $\alpha_{1,k}$ is bounded. Since $z_{1,k+1} = \check{x}_{1,k+1} - \alpha_{1,k}$, and $z_{1,k+1}$ is uniformly bounded, we have that $\check{x}_{1,k+1}$ is also uniformly bounded.

The signal $x_{1,k+1} - b_{1,0}\lambda_{1,bk+1}$ is generated by:

$$\begin{aligned}
\dot{x}_1 - b_{1,0}\dot{\lambda}_{1,b} &= A_{1,f}(x_1 - b_{1,0}\lambda_{1,b}) + \begin{bmatrix} \mathbf{0}_{r_1\times 1} \\ \bar{A}_{1,212r_1}\theta_1 \end{bmatrix} y_2 + \bar{A}_{1,211}\theta_1 y_1 + D_1\grave{M}_1\check{w}_1 + (\zeta_1^2 L_1 \\
&\quad + \Pi_1 C_1'(\zeta_1^2 - \frac{1}{\gamma^2}))(y_1 - E_1\grave{M}_1\check{w}_1) + \begin{bmatrix} \mathbf{0}_{1\times r_1} & b_{1,p1} & \cdots & b_{1,pn_1-r_1} \end{bmatrix}' y_2 \\
&\quad + \sum_{j=1}^{\check{q}_1} \bar{A}_{1,213,j}\check{w}_{1,j}\theta_1 + \check{D}_1\check{w}_1 \\
x_{1,k+1} - b_{1,0}\lambda_{1,bk+1} &= e_{n_1,k+1}'(x_1 - b_{1,0}\lambda_{1,b})
\end{aligned}$$

Now we will separate the above dynamics into y_1 dependent and y_2 dependent parts by the linearity of the system, $x_{1,k+1} - b_{1,0}\lambda_{1,bk+1} := x_{1,uk+1} + x_{1,yk+1}$, which are respectively given by,

$$\begin{aligned}
\dot{x}_{1,u} &= A_{1,f}x_{1,u} + \begin{bmatrix} \mathbf{0}_{r_1\times 1} \\ \bar{A}_{1,212r_1}\theta_1 \end{bmatrix} y_2 + \begin{bmatrix} \mathbf{0}_{1\times r_1} & b_{1,p1} & \cdots & b_{1,pn_1-r_1} \end{bmatrix}' y_2 \\
x_{1,uk+1} &= e_{n_1,k+1}' x_{1,u} \\
\dot{x}_{1,y} &= A_{1,f}x_{1,y} + (\zeta_1^2 L_1 + \Pi_1 C_1'(\zeta_1^2 - \frac{1}{\gamma^2}))(y_1 - E_1\grave{M}_1\check{w}_1) + \bar{A}_{1,211}\theta_1 y_1 + D_1\grave{M}_1\check{w}_1 \\
&\quad + \sum_{j=1}^{\check{q}_1} \bar{A}_{1,213,j}\check{w}_{1,j}\theta_1 + \check{D}_1\check{w}_1 \\
x_{1,yk+1} &= e_{n_1,k+1}' x_{1,y}
\end{aligned}$$

We observe that the signal $x_{1,uk+1}$ has relative degree at least $r_1 - k + 1$ with respect to y_2. Since $\bar{\check{x}}_{1,11}, \cdots, \bar{\check{x}}_{1,1k}$ are uniformly bounded, we conclude $x_{1,uk+1}$ is uniformly bounded by Lemma 11 in [12], where the reference system with input y_2 and output y_1. We conclude $x_{1,yk+1}$ is uniformly bounded since y_1 is bounded. Then, $x_{1,k+1} - b_{1,0}\lambda_{1,bk+1}$ is uniformly bounded. It follows that $\check{x}_{1,k+1} - \lambda_{1,bk+1}(b_{1,p0} + \bar{A}_{1,2120}\check{\theta}_1)$ is also uniformly bounded. Since

$\check{x}_{1,k+1}$ is uniformly bounded and $b_{1,p0} + \bar{A}_{1,2120}\check{\theta}_1$ is uniformly bounded away from 0, we have $\lambda_{1,bk+1}$ is uniformly bounded. That further imply $\Phi_{1,uk+1}$ is uniformly bounded. Furthermore, since $x_{1,k+1} - b_{1,0}\lambda_{1,bk+1}$ and $\lambda_{1,bk+1}$ are bounded, we have that the signals of $x_{1,k+1}$ is uniformly bounded.

Next, we need to show $\bar{\tilde{x}}_{1,1\,k+1}$ is satisfied equation (24). Comparing the design model (2) and the canonical form (78) in [12], we have $\bar{\tilde{C}}\bar{\tilde{x}} = C_1x_1$. It further implies

$$\bar{\tilde{C}}_1\bar{\tilde{A}}_1^k\bar{\tilde{x}}_1 = C_1(A_1 + \bar{A}_{1,211}\theta_1C_1)^k x_1$$

Hence, we have

$$\bar{\tilde{x}}_{1\,k+1} = \tilde{a}_{1,1\,k+1}x_1 + \cdots + \tilde{a}_{1,k\,k+1}x_{1,k} + x_{1,k+1} + \tilde{T}_{1,k+13}\bar{\tilde{x}}_3 + \tilde{T}_{1,k+14}\bar{\tilde{x}}_4 \tag{26}$$

where $\tilde{a}_{1,1\,k+1}, \cdots, \tilde{a}_{1,k\,k+1}$ are constants, and $\tilde{T}_{1,k+13}$, $\tilde{T}_{1,k+14}$ are constant matrices.

Then, we have the boundedness of $\bar{\tilde{x}}_{1\,k+1}$. Thus, we can conclude the boundedness of $\Phi_{1,u_s\,i}$, $\tilde{x}_{1,i} - \lambda_{1,b\,i}\bar{A}_{1,2120}\check{\theta}_1$, $\check{x}_{1,i}$, $x_{1,i} - b_{1,0}\lambda_{1,b\,i}$, $\lambda_{1,b\,i}$, $\Phi_{1,u\,i}$, $x_{1,i}$, and $\bar{\tilde{x}}_{1\,i}$, $\forall i \in \{1, \cdots r_1\}$.

Since the state x_1 can be viewed as stably filtered output signals of y_2 and y_1, it is uniformly bounded. Also, η_1, λ_1 are some stably filtered signals of y_2 and y_1, they are uniformly bounded. It further implies Φ_1 is uniformly bounded. Then we can conclude $\check{x}_1$ is uniformly bounded from the boundedness of $\tilde{x}_1 - \Phi_1\tilde{\theta}_1$. This further implies that the control input $\check{x}_{2,1}$ is uniformly bounded. Therefore, it follows $T_f = \infty$ and the complete system states are uniformly bounded on $[0, \infty)$.

The boundedness of closed-loop state variables of $\mathbf{S_2}$ can be proven with the similar line of reasoning above. Thus, we have established statement 1 in all cases.

We define $l_0 = l_{1,0} + l_{2,0} = V_{2,r_2}(X_1(0), X_2(0))$, and

$$l_1 + l_2 := \sum_{i=1}^{2}\left(\gamma^4|x_i - \hat{x}_i - \Phi_i(\theta_i - \hat{\theta}_i)|^2_{\Pi_i^{-1}\Delta_i\Pi_i^{-1}} + |\tilde{\eta}_i|^2_{\Upsilon_i} - 2(\theta_i - \check{\theta}_i)'P_{i,r}(\check{\theta}_i) + \sum_{j=1}^{r_i}\beta_{i,j}z_{i,j}^2\right)$$
$$+\frac{1}{4}|\varsigma_{1,(r_1+r_2)}|^2_{\bar{Q}_1} + \frac{1}{4}|\varsigma_{2,r_2}|^2_{\bar{Q}_2} + \epsilon_1(\gamma^2\zeta_1^2 - 1)|\theta_1 - \hat{\theta}_1|^2_{\Phi_1'C_1'C_1\Phi_1} + \epsilon_2|\theta_2 - \hat{\theta}_2|^2_{\Phi_2'C_2'C_2\Phi_2}$$

$$\sup_{\grave{w}_1\in\grave{\mathcal{W}}_1, \grave{w}_2\in\grave{\mathcal{W}}_2}\left\{\int_0^{t_f}\left((x_{1,1} - y_d)^2 + l_1 + l_2 - \sum_{i=1}^{2}\gamma^2|w_i|^2 - \sum_{i=1}^{2}\gamma^2|w_{i,a}|^2\right)d\tau\right.$$
$$\left. - \sum_{i=1}^{2}\gamma^2\left|\left[\theta_i' - \check{\theta}_{i,0}'\ x_{i,0}' - \check{x}_{i,0}'\right]\right|^2_{\bar{Q}_{i,0}} - l_0\right\}$$
$$\leq \sup_{\grave{w}_1\in\grave{\mathcal{W}}_1, \grave{w}_2\in\grave{\mathcal{W}}_2}\left\{\int_0^{t_f}\left((x_{1,1} - y_d)^2 + l_1 + l_2 - \sum_{i=1}^{2}\gamma^2|w_i|^2 - \sum_{i=1}^{2}\check{\gamma}^2|\frac{\gamma}{\check{\gamma}}\grave{w}_{i,a}|^2\right)d\tau - l_0\right.$$
$$\left. -\gamma^2\sum_{i=1}^{2}\left|\left[\theta_i' - \check{\theta}_{i,0}'\ x_{i,0}' - \check{x}_{i,0}'\right]\right|^2_{\bar{Q}_{i,0}} + \int_0^{t_f}\dot{U}d\tau - U(t) + U(0)\right\}$$
$$\leq -U(t) \leq 0$$

then, we establish the second statement.

For the third statement, we consider the following inequality,

$$\int_0^\infty \dot{U}\mathrm{d}\tau \leq \int_0^\infty (-|x_1 - y_d|^2 + \gamma^2|\check{w}_{1,a}|^2 + \gamma^2|\grave{M}_1\grave{w}_1|^2 + \gamma^2|\check{w}_{2,a}|^2 + \gamma^2|\grave{M}_2\grave{w}_2|^2)\mathrm{d}\tau$$

it follows that

$$\int_0^\infty |x_1 - y_d|^2\mathrm{d}\tau \leq \int_0^\infty \left(\gamma^2|\check{w}_{1,a}|^2 + \gamma^2|\grave{M}_1\grave{w}_1|^2 + \gamma^2|\check{w}_{2,a}|^2 + \gamma^2|\grave{M}_2\grave{w}_2|^2\right)\mathrm{d}\tau + U(0) < +\infty$$

By the first statement, we notice that

$$\sup_{0 \leq t < \infty} |\dot{x}_1 - \dot{y}_d| < \infty.$$

Then, we have

$$\lim_{t \to \infty} |x_1(t) - y_d(t)| = 0$$

For the last statement, it's easy to establish by Section 4.

This complete the proof of the theorem. ◇

6. Example

In this section, we present one example to illustrate the main results of this Chapter. The designs were carried out using MATLAB symbolic computation tools, and the closed-loop systems were simulated using SIMULINK.

Consider the following linear systems with zeros initial conditions:

$$\dot{\grave{x}}_1 = \grave{x}_1 + \grave{x}_2 + \grave{x}_3 + 0.1\grave{w}_1; \tag{27a}$$

$$\dot{\grave{x}}_2 = (1 + \theta_1)\grave{x}_3 + (1 + \theta_2)\check{w}_1 \tag{27b}$$

$$\dot{\grave{x}}_3 = -\grave{x}_1 - \grave{x}_3 + \grave{x}_4 + u + \check{w}_2 \tag{27c}$$

$$\dot{\grave{x}}_4 = \grave{x}_1 + (2 + \theta_3)u + 0.1\grave{w}_2 + \check{w}_2 \tag{27d}$$

$$y = \grave{x}_1 + 0.1\grave{w}_1 \tag{27e}$$

where θ_1, θ_2 and θ_3 are three unknown parameters with true value 0s. The coefficient terms, 0.1 and 1, reflect the *a priori* knowledge that the disturbances $\grave{w}_1$ and $\grave{w}_2$ are weak in power relative to that of the disturbance $\check{w}_1$ and $\check{w}_2$. We note that (27) is an unobservable system. We can decompose (27) into the following two SISO linear systems, $\mathbf{S_1}$ and $\mathbf{S_2}$, sequentially interconnected with additional output measurement,

$$\dot{x}_{11} = x_{11} + x_{12} + y_2 + w_{11}; \tag{28a}$$

$$\dot{x}_{12} = (1 + \theta_1)y_2 + (1 + \theta_2)\check{w}_1 + w_{12}; \tag{28b}$$

$$y_1 = x_{11} + w_{13}; \tag{28c}$$

$$\dot{x}_{21} = -x_{21} + x_{22} + u - y_1 + \check{w}_2 + w_{21}; \tag{28d}$$

$$\dot{x}_{22} = (2 + \theta_3)u + y_1 + \check{w}_2 + w_{22}; \tag{28e}$$

$$y_2 = x_{21} + w_{23} \tag{28f}$$

where

$$x_1 = \begin{bmatrix} x_{11} \\ x_{12} \end{bmatrix} = \begin{bmatrix} \check{x}_1 \\ \check{x}_2 \end{bmatrix}; x_2 = \begin{bmatrix} x_{21} \\ x_{22} \end{bmatrix} = \begin{bmatrix} \check{x}_3 \\ \check{x}_4 \end{bmatrix};$$

$$w_1 = \begin{bmatrix} w_{11} \\ w_{12} \\ w_{13} \end{bmatrix} = \begin{bmatrix} 0.1\check{w}_1 - \check{w}_3 \\ -(1-\theta_1)\check{w}_3 \\ 0.1\check{w}_1 \end{bmatrix}; w_2 = \begin{bmatrix} w_{21} \\ w_{22} \\ w_{23} \end{bmatrix} = \begin{bmatrix} 0.1\check{w}_1 \\ -0.1\check{w}_1 + 0.1\check{w}_2 \\ \check{w}_3 \end{bmatrix}$$

Here $\check{w}_3$ is the measurement disturbance of the state $\check{x}_3$. It is easy to check that $\mathbf{S_1}$ and $\mathbf{S_2}$ in (28) satisfied the assumptions 1–5.

For the adaptive control design, we set the desired disturbance attenuation level $\gamma = 10$. We select the true value of the parameters in subsystem $\mathbf{S_1}$ and subsystem $\mathbf{S_2}$ are zeros, and belong to the interval $[-1, 1]$. The projection function $P_1(\theta_1)$ and $P_2(\theta_2)$ are chosen as $P_1(\theta_1) = 0.5(\theta_1^2 + \theta_2^2)$, $P_2(\theta_2) = \theta_3^2$. The reference trajectory, y_d, is generated by the following linear system $\dot{x}_{d,1} = -x_{d,2}$, $\dot{x}_{d,2} = x_{d,1} - x_{d,2} + d$, $y_d = x_{d,1}$ with zeros initial condition, where d is the command input signal. The objective is to achieve asymptotic tracking of $\check{x}_1$ to the reference trajectory y_d.

For design and simulation parameters of $\mathbf{S}_i$, $(i = 1, 2)$, we select

$$\check{x}_{1,0} = \begin{bmatrix} 0.2 \; 0 \end{bmatrix}'; \; \check{x}_{2,0} = \begin{bmatrix} 0.1 \; 0 \end{bmatrix}'; \check{\theta}_{1,0} = \begin{bmatrix} 0.5 \; -0.5 \end{bmatrix}'; \; \check{\theta}_{2,0} = -1/2; Q_{i,0} = 0.001 I_2;$$

$$K_{i,c} = 0.2; \; \Delta_i = I_2; \; p_{i,n_i} = e_{2,2}; \; \Phi_{i,0} = \mathbf{0}_{2\times 1}; \; \rho_{i,o} = 2; \; \beta_{i,\Delta} = 0; \; \epsilon_i = K_{i,c}^{-1} s_{i,\Sigma}; \; \lambda_{i,0} = \mathbf{0}_{2\times 1};$$

$$\beta_{i,1} = 0.5; \; \eta_{i,0} = \mathbf{0}_{2\times 1}; \; Z_1 = \begin{bmatrix} 0.0893 & -0.0081 \\ -0.0081 & 0.0097 \end{bmatrix}; \; Z_2 = \begin{bmatrix} 0.1094 & -0.0099 \\ -0.0099 & 0.0099 \end{bmatrix}$$

We present one set of simulation results in this example to illustrate the regulatory behavior of the adaptive controller. We set $d(t) = 0.4\sin(0.1t) + \sin(0.6t)$, $\check{w}_1(t) = 0$, $\check{w}_2(t) = 0$, $\check{w}_3(t) = 0$, $\check{w}_1(t) = \sin(12t + \frac{\pi}{9}) + 0.8\sin(3t)$, and $\check{w}_2(t) = 3\sin(3t + \frac{\pi}{3})$. The results are shown in Figure 2(a)–(f). To illustrate that the proposed controller can improve the system performance by incorporating the measurements and/or the estimation of the significant external disturbances into the control design, the simulation results based on [17] are presented in Figure 2(c)(d), where the measured disturbances $\check{w}_1$ and $\check{w}_2$ are treated as arbitrary disturbances and θ_3 is treated as constant in control design. We observe that the output tracking error asymptotically converges to zero and the parameter estimates asymptotically converge to its true value 0 in (a) and (b) even if there is a non-zero measured disturbance in the system. But the parameter estimates doesn't converge to the true value,

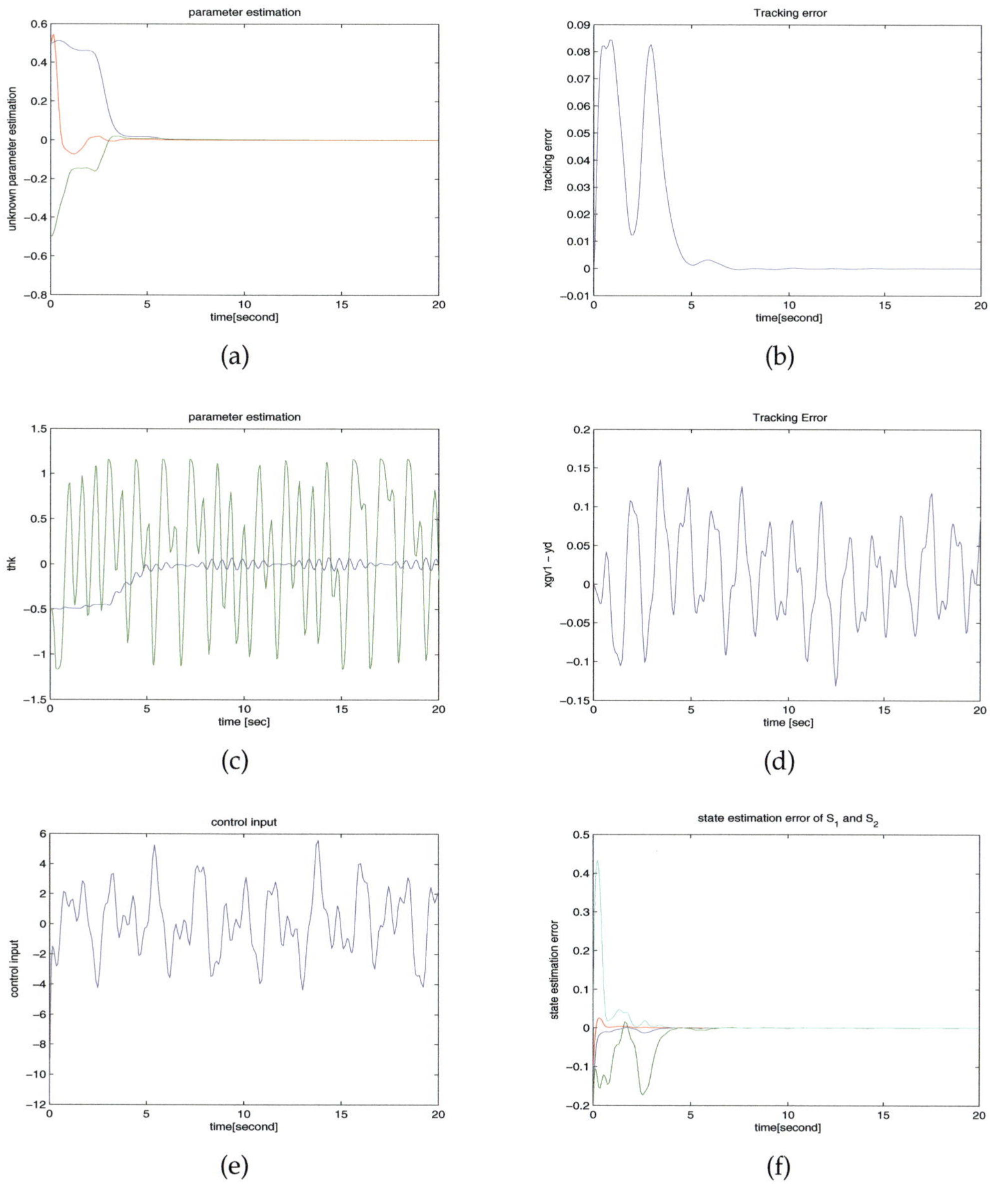

Figure 2. System response for Example under command input $d(t) = 0.4\sin(0.1t) + \sin(0.6t)$, $\breve{w}_1 = 0$, $\breve{w}_2 = 0$, $\breve{w}_3 = 0$, $\breve{w}_1(t) = \sin(12t + \frac{\pi}{9}) + 0.8\sin(3t)$, and $\breve{w}_2(t) = 3\sin(3t + \frac{\pi}{3})$. (a) Parameter estimate; (b) Tracking error; (c) Parameter estimate(based on [17]); (d) Tracking error(based on [17]); (e) control input; (f) State estimation error;

and the tracking error doesn't converge to zero in (c) and (d). State estimation error, $x_1 - \check{x}_1$ and $x_2 - \check{x}_2$, converge to zero in (f), and the transient performance behaves well as in (e).

7. Conclusions

In this Chapter, we present the game-theoretical approach based adaptive control design for a special class of MIMO linear systems, which is composed of two sequentially interconnected SISO linear systems, $\mathbf{S}_1$ and $\mathbf{S}_2$. We assume the subsystem under studied subject to noisy output measurements, unknown initial state conditions, linear unknown parametric uncertainties, measured and unmeasured additive exogenous disturbance input uncertainties. Our design objective is to address the asymptotical tracking, the transient response and robustness of the closed-loop system, which are the same as the objectives to motivate the study of the H^{∞}- optimal control problem. In view of the similar solution between H^{∞} optimal control design and *zero sum differential game*, we convert the original adaptive control design problem into a zero-sum game with soft constraints on the disturbance input uncertainties and the unknown initial state uncertainties, which incorporates the measures of transient response, disturbance attenuation, and asymptotic tracking into a single game-theoretic cost function and formulates the design problem as a nonlinear H^{∞} control problem under imperfect state measurements. A game-theoretical approach, *cost-to-come function* analysis, is then applied to obtain the finite dimensional estimators of $\mathbf{S_1}$ and $\mathbf{S_2}$ independently, which is also converted the control design as an H^{∞} control problem with full information measurements. The integrator backstepping methodology is finally applied on this full information measurements problem to obtain a suboptimal solution. The controller achieves the same result as [17], namely the total stability of the closed-loop system, the desired disturbance attenuation level, and asymptotic tracking of the reference trajectory when the disturbance is of finite energy and uniformly bounded. In addition, the proposed controller may achieve arbitrary positive disturbance attenuation level with respect to the measured disturbances by proper scaling. The contribution of the measurements of part of the disturbance inputs is that we can design an adaptive controller with disturbance feedforward structure with respect to $\check{w}_{1,b}$ and $\check{w}_{2,b}$ to eliminate their effect on the squared $\mathcal{L}_2$ norm of the tracking error. Moreover, the asymptotic tracking is achieved even if the measured disturbances are only uniformly bounded without requiring them to be of finite energy.

Author details

Sheng Zeng
Critical Care R&D Engineering, Carefusion Corporation, CA, 92887, USA

Emmanuel Fernandez
School of Electronics & Computing Systems, University of Cincinnati, OH, 45221-0030, USA

8. References

[1] Arslan, G. & Başar, T. [2001]. Disturbance attenuating controller design for strict-feedback systems with structurally unknown dynamics, *Automatica* 37(8): 1175–1188.

[2] Başar, T. & Bernhard, P. [1995]. *H^∞-Optimal Control and Related Minimax Design Problems: A Dynamic Game Approach*, 2nd edn, Birkhäuser, Boston, MA.

[3] Didinsky, G. [1994]. *Design of minimax controllers for nonlinear systems using cost-to-come methods*, PhD thesis, University of Illinois, Urbana, IL.

[4] Goodwin, G. C. & Mayne, D. Q. [1987]. A parameter estimation perspective of continuous time adaptive control, *Automatica* 23: 57–70.

[5] Goodwin, G. C. & Sin, K. S. [1984]. *Adaptive Filtering, Prediction and Control*, Prentice-Hall, Englewood Cliffs.

[6] Ioannou, P. A. & Sun, J. [1996]. *Robust Adaptive Control*, Prentice Hall, Upper Saddle River, NJ.

[7] Isidori, A. [1995]. *Nonlinear Control Systems*, 3rd edn, Springer-Verlag, London.

[8] Kanellakopoulos, I., Kokotović, P. V. & Morse, A. S. [1991]. Systematic design of adaptive controllers for feedback linearizable systems, *IEEE Transactions on Automatic Control* 36: 1241–1253.

[9] Krstić, M., Kanellakopoulos, I. & Kokotović, P. V. [1995]. *Nonlinear and Adaptive Control Design*, Wiley, New York, NY.

[10] Kumar, P. R. [1985]. A survey of some results in stochastic adaptive control, *SIAM Journal on Control and Optimization* 23(3): 329–380.

[11] Pan, Z. & Başar, T. [1996]. Parameter identification for uncertain linear systems with partial state measurements under an H^∞ criterion, *IEEE Transactions on Automatic Control* 41: 1295–1311.

[12] Pan, Z. & Başar, T. [2000]. Adaptive controller design and disturbance attenuation for SISO linear systems with noisy output measurements, *CSL report*, University of Illinois at Urbana-Champaign, Urbana, IL.

[13] Rohrs, C. E., Valavani, L., Athans, M. & Stein, G. [1985]. Robustness of continuous-time adaptive control algorithms in the presence of unmodeled dynamics, *IEEE Transactions on Automatic Control* 30: 881–889.

[14] Tezcan, I. E. & Başar, T. [1999]. Disturbance attenuating adaptive controllers for parametric strict feedback nonlinear systems with output measurements, *Journal of Dynamic Systems, Measurement and Control, Transactions of the ASME* 121(1): 48–57.

[15] Zeng, S. [2010]. Adaptive controller design and disturbance attenuation for a general class of sequentially interconnected siso linear systems with noisy output measurements, *Proceedings of the 49th IEEE Conference on Decision and Control(CDC)*, Atlanta, GA, pp. 2608–2613.

[16] Zeng, S. [2011]. Worst-case analysis based adaptive control design for siso linear systems with plant and actuation uncertainties, *Proceedings of the 50th IEEE Conference on Decision and Control and European Control Conference (CDC-ECC)*, Orlando, FL, pp. 6349–6354.

[17] Zeng, S. & Fernandez, E. [2010]. Adaptive controller design and disturbance attenuation for sequentially interconnected siso linear systems under noisy output measurements, *IEEE Transactions on Automatic Control* 55: 2123–2129.

[18] Zeng, S. & Pan, Z. [2009]. Adaptive controller design and disturbance attenuation for SISO linear systems with noisy output measurements and partly measured disturbances, *International Journal of Control* 82(2): 310–334.

[19] Zeng, S., Pan, Z. & Fernandez, E. [2010]. Adaptive controller design and disturbance attenuation for SISO linear systems with zero relative degree under noisy output measurements, *International Journal of Adaptive Control and Signal Processing* 24: 287–310.

[20] Zhao, Q., Pan, Z. & Fernandez, E. [2009]. Convergence analysis for reduced-order adaptive controller design of uncertain siso linear systems with noisy output measurements, *International Journal of Control* 82(11): 1971–1990.

Chapter 6

Models for Highway Cost Allocation

Alberto Garcia-Diaz and Dong-Ju Lee

Additional information is available at the end of the chapter

http://dx.doi.org/10.5772/53927

1. Introduction

Historically, *equity* has been one of the most important principles applied to formulate tax policy. It has been considered when raising revenues and allocating funds for maintenance, capital improvements, operating programs, and services to the public. The problem of determining how the total cost of a shared facility or service should be divided fairly and rationally is common both in public and private enterprises. The theory of cooperative games is widely used for allocating these costs. Examples of this include but are not limited to public utilities providing telephone services, electricity, water, and transport; public works projects designed to serve different constituencies; access fees or user charges for airports, highways, bridges or waterways; internal accounting rules to allocate overhead costs in private companies [1-4].

The purpose of a *Highway Cost Allocation* (HCA) study is to determine the fair share that each class of road user (vehicle class) should pay for the construction, maintenance, operation, improvement, and related costs of highways, roads, bridges, and streets in a highway network, such as those managed by state Departments of Transportation in the U.S.A. Particular emphasis should be placed on criteria and methods for allocating costs among vehicle classes using a common highway facility (road or bridge, for example) in a just, equitable, fair, and reasonable manner. Cost allocation is ultimately concerned with fairness. Through a comparison of revenues (user fees paid) and cost responsibilities, this study will estimate current equity and recommend alternatives to bring about a closer match between payments and cost responsibilities for each vehicle class.

A significant objective of HCA studies is to analyze highway-related costs attributable to different highway users as a basis for evaluating the equity and efficiency of user charges. Ideally, the costs incurred by the various user groups should be in proportion to the damage they contribute to the highway system. The cost of supporting a highway infrastructure may be deemed fair if there is an equitable distribution of costs and revenues among the various groups of highway users. With this assumption, equity is achieved when each group's percentage of total assigned costs is equal to the percentage of the revenues contributed by that group. This chapter focuses exclusively on highway cost allocation, specifically the allocation of pavement and bridge costs.

Highway users are concerned about the fairness of road-use charges and demand that these be allocated equitably among the various vehicle classes occasioning the total cost. Although the word *equity* conveys the general intent of any cost allocation procedure, there are many possible ways to formulate a cost allocation objective to measure equity. In general, there exists no perfect cost allocation method. This is why there is a rich menu of cost allocation methods each intended to reflect the problem-specific logical, historical, political, economic, as well as mathematical analysis.

Costs associated with highway construction, maintenance, and operation can be divided into several categories. Because the impact of different vehicle classes on the costs is different, each of the cost categories should be allocated among the various user groups or vehicle classes in a different manner. These cost categories are:

a. Costs associated with new pavement construction.
b. Costs associated with pavement maintenance, rehabilitation, and reconstruction.
c. Costs associated with new bridge construction.
d. Costs associated with bridge maintenance, rehabilitation, and reconstruction.
e. Costs associated with system enhancement.
f. Other highway-related costs.

In addition to this Introduction, this chapter is organized according to six additional sections. Section 2 briefly describes several traditional and non-traditional procedures for highway cost allocation and outlines some important properties of game-theory-based procedures. Section 3 presents a conceptual framework for conducting a highway cost allocation study for a transportation agency, such as a U.S. state Department of Transportation. Section 4 discusses the application of the *nucleolus method* in highway cost allocation combining it with the concept of statistical cost effect to determine a unique solution from multiple optimal solutions. Section 5 describes a new procedure for allocating highway costs having one component due to pavement thickness and another one due to traffic capacity (measured in terms of lanes). Section 6 develops a procedure for bridge cost allocation that integrates both game theory concepts and the traditional incremental approach. Two numerical examples are designed to illustrate the proposed procedures.

2. Highway cost allocation procedures and properties

2.1. Traditional HCA Methods

During the last three decades, several methods have been developed for the purpose of allocating the total cost of a transportation facility among all the vehicle classes using it. Most procedures that can be used to achieve this goal can be grouped as either *incremental* or *proportional* allocation procedures, or a combination of these two. The proportional and incremental methods have been used by the Federal Highway Administration [5][6] and by several state Departments of Transportation. In the **Incremental Method** a highway facility is initially designed to accommodate only the vehicles with lowest axle weight, and then it is sequentially redesigned as the additional vehicle classes are included in increasing order of axle weights. As the process of adding vehicle classes continues, after each inclusion the marginal or incremental cost is charged to the most recently included class. This method satisfies two of the three fundamental properties: completeness and marginality, sometimes marginality but this not guaranteed. Furthermore, this method is not *consistent* because the cost allocated to each vehicle class depends on the *order* in which vehicle classes are included in the analysis. As the name suggests, the **Proportional Method** distributes costs proportionally among vehicle classes according to a specified measure. The cost allocator could be vehicle-miles of travel (VMTs), 18,000 lb. *equivalent single-axle loads* (ESALs), or some other measure. While this procedure may not satisfy marginality and rationality, it does satisfy the completeness principle.

2.2. Non-traditional HCA methods

Several non-traditional allocation methods have been developed based on concepts from the theory of cooperative games by Neumann and Morgenstern [7].The application of *non-atomic game theory* to cost allocation was proposed by Castaño-Pardo and Garcia-Diaz [4]. This approach is different from the analysis of the game in which entire vehicle classes are considered as players; instead, each vehicle passage is considered as a player. Such a game obviously has a large number of players, and the decisions of a single player are irrelevant to the total outcome of the game. The value of this non-atomic game is utilized to find the solution to the problem of pavement cost allocation.

The **Generalized Method** is based on concepts from the theory of cooperative games [7], and was proposed for conducting highway cost allocation by Villarreal and Garcia-Diaz [8]. The method satisfies completeness, marginality, and rationality because these principles are forcibly satisfied due to constraints in its mathematical formulation. In essence the method guarantees that every vehicle class will be allocated a lower cost in the *grand coalition* (consisting of all vehicle classes), as compared to any other *smaller coalition* (one with fewer vehicle classes than the grand coalition). This method is known in the game theory literature as the *Nucleolus Method*. Its conditions are considered of primary importance in a large number of applications (as in public utility pricing, for example). The solution procedure is actually an application of *Linear Programming* (*LP*). Sometimes the linear programming solution may not be unique and then there is the need to introduce a tie-breaker rule.

The **Shapley Value** [9] is the average marginal cost for a vehicle class considering all possible permutations of the vehicles in the grand coalition. For example, if there are three vehicle classes, represented by 1, 2, 3, the following permutations are possible: 123, 132, 213, 231, 321, and 312. If we calculate the marginal cost for each vehicle and the compute the average for the six permutations, this average marginal cost is known as the Shapley value. The Shapley value, primarily due to its simplicity and mathematical properties, is one of the most widely studied and used joint cost allocation solution concepts. It represents the average marginal cost contribution each vehicle class i would make to the grand coalition if it were to form one vehicle class at a time. Thus the average or expected cost assessment is

$$x_i = \sum_{\substack{i \in S \\ S \subseteq N}} \frac{(|S|-1)!(|N|-|S|)!}{|N|!} C^i(S) \tag{1}$$

where $|S|$ and $|N|$ represent the cardinality of sets S and N, $C^i(S)$ represents the marginal cost contribution of i relative to S, which can readily be computed using $C^i(S) = C(S)-C(S-i)$ if $i \in \mathbf{S}$, and where the sum is computed over all subsets S containing vehicle class i. For example, for the cost game given by $C(1)=7$, $C(2)=8$, $C(3)=8$, $C(1,2)=10$, $C(1,3)=10$, $C(2,3)=15$ and $C(1,2,3)=17$, the Shapley value allocation is calculated as shown below:

$$x_1 = \frac{0!2!}{3!}(7-0) + \frac{1!1!}{3!}(10-8) + \frac{1!1!}{3!}(10-8) + \frac{2!0!}{3!}(17-15) = \frac{11}{3} = 3.67$$

$$x_2 = \frac{0!2!}{3!}(8-0) + \frac{1!1!}{3!}(10-7) + \frac{1!1!}{3!}(15-8) + \frac{2!0!}{3!}(17-10) = \frac{20}{3} = 6.67$$

$$x_3 = \frac{0!2!}{3!}(8-0) + \frac{1!1!}{3!}(10-7) + \frac{1!1!}{3!}(15-8) + \frac{2!0!}{3!}(17-10) = \frac{11}{3} = 6.67$$

The **Aumann-Shapley Value** [10,11] is a procedure that considers two types of costs. The first cost is for ESALs (pavement thickness) and the second cost is for highway-lanes (traffic capacity). The total cost allocated to a vehicle class is the sum of these two costs. This procedure allows the consideration of the number of lanes as being variable and depending on the composition of the traffic using a highway. In particular, it addresses two seemingly conflicting objectives: lighter vehicles require less pavement thickness and more lanes while heavier vehicles require fewer lanes but thicker pavements. This method calculates a cost per ESAL and a cost per lane. Then it allocates the number of available lanes among the vehicle classes using the Shapley value (which is the average incremental number of lanes over all possible orderings of the vehicle classes). Since the ESALs are given as data, then the cost allocated to a vehicle class can be calculated as the sum of the *ESALs cost* plus the *lanes cost*.

2.3. Desirable HCA properties

In order to explain some desirable properties of Highway Cost Allocation (HCA) procedures we will consider a highway facility such as a pavement or a bridge. First, **completeness** is the property that highway costs (construction, rehabilitation, maintenance) are fully paid for by all participating vehicle classes. Second, **rationality** is the property that each vehicle class is guaranteed a lower cost by participating in the *grand coalition* (group consisting of all vehicle classes). The fundamental observation is that if a highway facility is designed for the grand coalition, the cost share of each vehicle class would be smaller than the share paid by the vehicle class in a smaller coalition for which an alternative facility can be designed and for which the cost is available. **Marginality** means that each vehicle class should pay at least the incremental cost incurred by including it in the grand coalition. D**emand monotonicity** is a property that implies that the cost-share of a player does not decrease when the player increases its level of demand. **Additivity** means that the allocated costs can be divided into two corresponding components if a cost function can be divided into two distinct and independent cost components. The **dummy** property means that a cost allocation should be equal to zero for a player that does not contribute to any coalition. Some of these properties will be further addressed in Sections 4 and 5.

3. Overview of a highway cost allocation study

Figure 1 outlines a typical framework of a highway cost allocation study for a transportation agency, such as a State Department of Transportation. Instead of directly allocating a total cost at the state level, a more equitable approach is to divide the total cost on the basis of three *classification attributes* known as climatic region, highway system, and highway location. For each of these three attributes several choices must be identified. As an example, a state may be divided into one to four climatic regions depending on the climatic factors affecting pavement performance, the highways may be classified into at least two highway systems to include state and federal highways as a minimum, and the locations may be classified into at least two major classes to accommodate urban and rural highways.

For any *cost classification*, i.e. one choice of each region, highway system and location, the corresponding total cost to be allocated among vehicle classes is first calculated or estimated by dividing the state total among all classifications according to well-known cost allocators, such as vehicle miles of travel (VMTs) or vehicle loadings measured in terms of *18,000 lb*

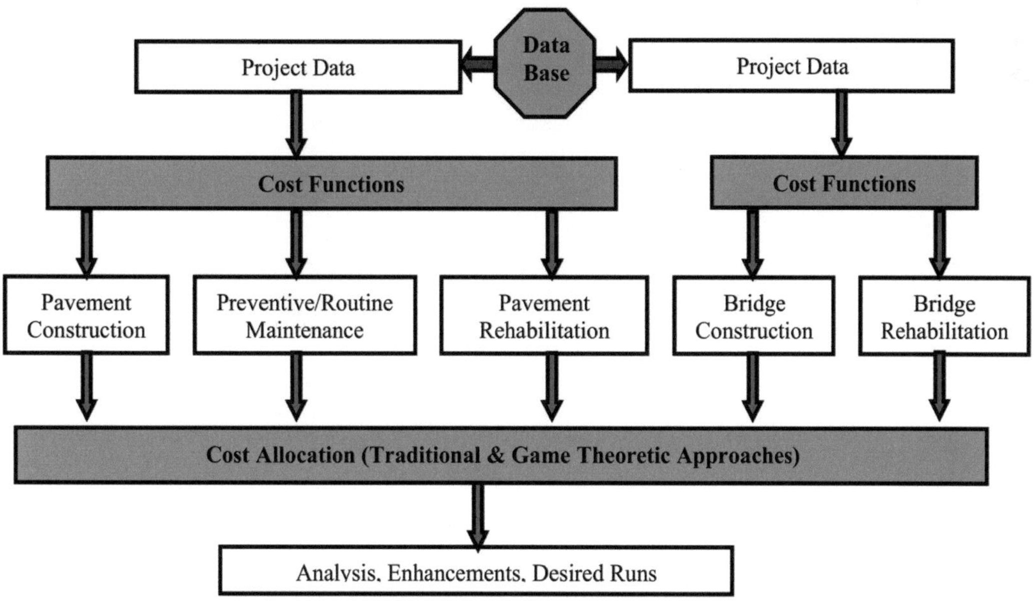

Figure 1. Framework for HCA Study.

Equivalent Single-Axle Load applications (ESALs). To divide the cost for any *cost classification* among vehicle classes, we need to find a cost function, known in game theory as the *characteristic function*, that provides a cost in $/mile for any specified number of ESALs. The characteristic function can be determined by statistical regression analysis using data on expenditures and traffic volumes extracted from several representative highway projects. The characteristic function allows the use of game-theoretic procedures that require costs estimates for coalitions or groups of vehicle classes. In particular, the Shapley value, Generalized Method, and A-S value methods require the use of a characteristic function.

3.1. Vehicle classes

Vehicle classes are viewed as players in a cooperative game. The object of a highway cost location procedure is to fairly divide the construction, rehabilitation or maintenance cost of a transportation facility, such as a highway or a bridge, among these users or players. The following vehicle classes are typically included in highway cost allocation studies:

1. Motorcycles
2. Passenger cars
3. Other Two-Axle, Four-Tire Single Unit Vehicles
4. Buses
5. Two-Axle, Six-Tire, Single-Unit Trucks
6. Three-Axle Single-Unit Trucks
7. Four or More Axle Single-Unit Trucks
8. Four or Fewer Axle Single-Trailer Trucks
9. Five-Axle Single-Trailer Trucks
10. Six or More Axle Single-Trailer Trucks
11. Five or fewer Axle Multi-Trailer Trucks
12. Six-Axle Multi-Trailer Trucks
13. Seven or More Axle Multi-Trailer Trucks

3.2. Database description

The database includes the information of traffic levels and costs of relevant pavement maintenance or rehabilitation projects for different data classifications. Typically the database has data for all classifications formed with the following attributes:

1. *Climactic Regions.* Since the performance of a pavement is affected by climatic conditions, it is customary to divide a large geographic area into smaller homogeneous climatic regions.
2. *Highway Systems.* In a number of studies two to three highway systems are included when defining the scope of the study. In a number of U.S. states at least Interstate Highways, US highways, and State highways/roads are included.
3. *Highway Locations.* There are two primary types of locations considered in a number of studies: urban and rural areas.

For each of the resulting classifications or combinations of climactic region, highway system, and location, at least three (or four) projects are extracted from the database and used to estimate cost relationships (characteristic functions) that can be used to estimate costs for different levels of ESALs. Typically traffic data available will include the following:

a. Annual Average Daily Traffic (AADT) and Equivalent Single-Axle Loads (ESALs).
b. The distribution of vehicles on the road (proportion of passenger cars, single-axle trucks, etc.)

In order to generate data for a more detailed level of classification, the following information can be used:

a. Vehicle Miles Traveled (VMT)
b. Required number of lanes for various combinations of vehicle classes

Since each treated or constructed pavement has a specific service life and all the vehicles traveled in its service life should pay the maintenance or construction cost, the *Equivalent Annual Cost* (*EAC*) of the project in its service life is calculated and used as the cost of that specific project. *EAC* is the cost per year of owning and operating an asset over its entire lifespan. This cost is calculated for the following highway work activities:

1. *Pavement maintenance*: typically both routine and preventive maintenance activities are included in this cost component. Routine maintenance activities are needed to repair cracks of different types, fill pot holes and correct other signs of pavement distress. Preventive maintenance is done mostly applying thin seal coats, micro surfacing, fog sealing, chip sealing, etc.
2. *Pavement rehabilitation*: pavement rehabilitation activities include conventional hot mixed asphalt overlay with or without milling. Generally, thicker overlays will be used for high traffic level roads and thus the cost will be also higher.
3. *Pavement construction*: new pavement construction includes the subgrade, base layer and surface layer.

4. Generalized method

Let N be the set (grand coalition) of all vehicle classes using a highway. Let $C(N)$ be the cost per mile of this highway (construction, rehabilitation or maintenance). Furthermore, let R_i the cost paid by vehicle class $i \in N$. The *completeness* property can be formulated as

$$\sum_{i \in N} R_i = C(N) \tag{2}$$

Now, let us consider a subset (coalition) of vehicle classes, $S \in N$, and let $C(S)$ be the cost per mile of a highway designed specifically to accommodate only the vehicle classes in S. The *rationality* property can be formulated as

$$\sum_{i \in S} R_i \leq C(S) \quad \textit{for all} \;\; S \subset N \tag{3}$$

Furthermore, the *marginality* property implies that

$$\sum_{i \in S} R_i \geq C(N) - C(N - S) \quad \text{for all } S \subset N \tag{4}$$

It can be proved that if the completeness property (2) is held then the rationality and marginality properties (3) and (4) are equivalent. From (3) it is concluded that the savings enjoyed by a coalition S when joining the grand coalition are given by

$$C(S) - \sum_{i \in S} R_i \tag{5}$$

To maximize these savings, we maximize t, where

$$C(S) - \sum_{i \in S} R_i \geq t$$

which can be rewritten as

$$\sum_{i \in S} R_i \leq C(S) - t \tag{6}$$

As an illustration, for $\mathbf{N} = \{1,2,3\}$, the *LP* model for the Generalized Method is formulated in (7)-(15).

$$\text{Maximize } t \tag{7}$$

Subject to

$$R_1 \leq C_1 - t \tag{8}$$

$$R_2 \leq C_2 - t \tag{9}$$

$$R_3 \leq C_3 - t \tag{10}$$

$$R_1 + R_2 \leq C_{12} - t \tag{11}$$

$$R_1 + R_3 \leq C_{13} - t \tag{12}$$

$$R_2 + R_3 \leq C_{23} - t \tag{13}$$

$$R_1 + R_2 + R_3 = C_{123} \tag{14}$$

$$R_1, R_2, R_3, t \geq 0 \tag{15}$$

Constraints (8)-(10) correspond to highways (pavements) designed to accommodate single-vehicle-class coalitions. Constraints (11)-(13) correspond to two-vehicle-class coalitions. Constraint (14) corresponds to the grand coalition. Each coalition has a level of traffic loadings, measured in ESALs, for a specified design period (typically 20 years). The highway cost per mile should be strictly increasing as the number of ESALs increases. Under this assumption, Constraints (8)-(14) define a feasible region called the *core of the game* when $t=0$. If W_1 and W_2 are measured in ESALS then the core exists if C_1, C_2, C_3, C_{12}, C_{13}, C_{23} and C_{123} satisfy the following condition

$$C(W_1+W_2) \leq C(W_1) + C(W_2) \tag{16}$$

It can be proved that a typical non-decreasing cost function satisfying (16) is the one represented in Figure 2. In this figure, W is the total number of standard loads (ESALs) for the grand coalition and $C(W)$ is the cost to be allocated. In a number of highway cost allocation studies functions like the one shown in this figure are found using regression analysis from cost data for a set of highway projects available in the database of the study.

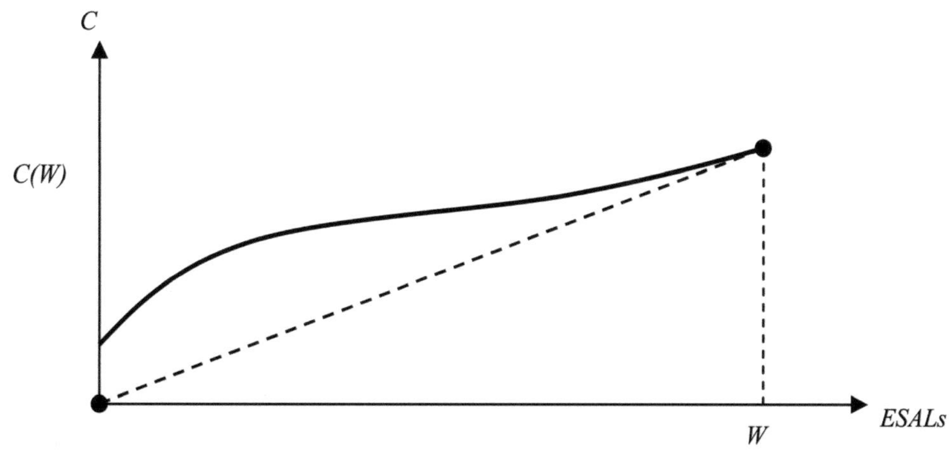

Figure 2. Cost function.

Figure 3(a) shows the feasible region for the above formulation when $t = 0$. Figure 3(b) shows the effect of increasing the value of the variable t. It is noted in this figure that as the value of t increases, the feasible region gets smaller, becoming either a point or a line when t reaches its maximum value. A solution represented by one point indicates a unique solution. The line represents infinitely many optimal solutions, a case already indicated in Section 2.

When the model formulated in (7)-(15) has infinitely many optimal solutions an additional condition must be considered to select a unique solution. The solution procedure can, therefore, be divided into two phases, with the second one needed only to break the tie among multiple solutions in the first phase.

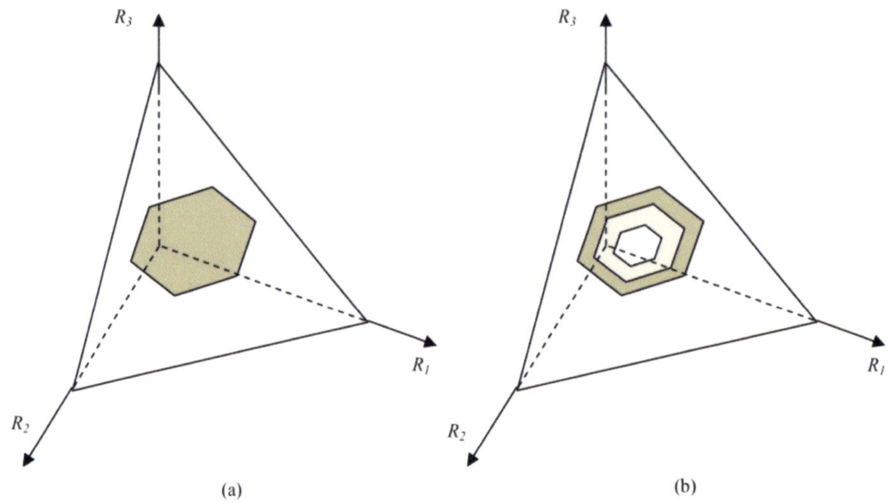

Figure 3. Feasible region.

4.1. Phase 1 of generalized method

$$Maximize\ t \tag{17}$$

Subject to

$$\sum_{i \in N} R_i = C(N) \tag{18}$$

$$\sum_{i \in S} R_i \leq C(S)\text{-}t \text{ for all } S \subset N \tag{19}$$

$$R_i,\ t \geq 0 \text{ for all } i \in N \tag{20}$$

4.2. Phase 2 of generalized method

Villarreal-Cavazos and Garcia-Diaz [13] proposed to break the tie among multiple solutions using the concept of statistical *cost effect* of vehicle classes. This is defined as the difference in average cost between all coalitions *including* a given vehicle class and all coalitions *not including* the class. If E_i is the cost effect of vehicle class *i*, the *relative effect* is defined as

$$e_i = \frac{E_i}{\sum_{i \in N} E_i} \text{ for all } i \in N$$

Also, the *relative cost allocated* to vehicle class *i* is defined as

$$r_i = \frac{R_i}{\sum_{i \in N} R_i} \text{ for all } i \in N$$

A unique solution is obtained from the solution to the *non-linear* model formulated in (21)-(24).

Minimize

$$\sum_{i \in N} | r_i - e_i | \tag{21}$$

Subject to

$$\sum_{i \in N} R_i = C(N) \tag{22}$$

$$\sum_{i \in S} R_i \le C(S) - t^* \quad \text{for all } S \subset N \tag{23}$$

$$R_i,\ t \ge 0 \text{ for all } i \in N \tag{24}$$

where t^* is the optimal value obtained for t in Phase 1. The model formulated in (21)-(24) can be *linearized* as shown in (25)-(30).

Minimize

$$\sum_{i \in N} (L_i + H_i) \tag{25}$$

Subject to

$$r_i + L_i - H_i = e_i \quad \text{for all } i \in N \tag{26}$$

$$\sum_{i \in S} R_i \le C(S) - t^* \quad \text{for all } S \subset N \tag{27}$$

$$\sum_{i \in N} R_i = C(N) \tag{28}$$

$$(\sum_{i \in N} R_i) \times r_i = R_i \quad \text{for all } i \in N \tag{29}$$

$$R_i,\ t \ge 0 \text{ for all } i \in N \tag{30}$$

It is noted that by *LP* optimality conditions,

$$L_i = \begin{cases} e_i - r_i \ ; \ r_i < e_i, \\ 0 \ ; \ r_i \ge e_i, \end{cases} \tag{31}$$

$$H_i = \begin{cases} 0 \ ; \ r_i \le e_i, \\ r_i - e_i \ ; \ r_i > e_i. \end{cases} \tag{32}$$

4.3. Statistical cost effects

A grand coalition consisting of the set of vehicle classes {1,2,3} is considered again to illustrate the calculation of the relative cost effects of the classes. First we regard each vehicle class as a *two-level factor*. The levels can be represented by the signs – and +, where – means that the vehicle class is not in a coalition and + indicates that it is in the coalition. Moreover, the number of level combinations for three two-level factors is equal to $2^3 = 8$. These eight combinations are listed in Table 1. Now, it is noted that combinations 2-8 represent the 7 coalitions that can be formed with the three vehicle classes being considered. The last column in the table shows the highway cost for each coalition. Combination 1 corresponds to an *empty coalition.* Its cost can be viewed as the environmental cost, that is, the cost needed to have a facility able to withstand the impact of climatic conditions alone, not considering the impact of vehicle loadings. In HCA studies this cost can be regarded as a specified fraction of C_{123}.

Combination	X_1	X_2	X_3	Cost
1	-	-	-	C_o
2	+	-	-	C_1
3	-	+	-	C_2
4	+	+	-	C_{12}
5	-	-	+	C_3
6	+	-	+	C_{13}
7	-	+	+	C_{23}
8	+	+	+	C_{123}

Table 1. Level combinations

The effect of factor X_i, for example, as previously indicated, is the difference in average cost between the coalitions including vehicle class i and those not including it. Based on this definition, the cost effects of the three vehicle classes are obtained as follows using the results shown in Table 1:

$$E_1 = \frac{C_1 + C_{12} + C_{13} + C_{123}}{4} - \frac{C_o + C_2 + C_3 + C_{23}}{4}$$

$$E_2 = \frac{C_2 + C_{12} + C_{23} + C_{123}}{4} - \frac{C_o + C_1 + C_3 + C_{13}}{4}$$

$$E_3 = \frac{C_3 + C_{13} + C_{23} + C_{123}}{4} - \frac{C_o + C_1 + C_2 + C_{12}}{4}$$

Once $E_1, E_2, \ldots, E_n$ are calculated their values are used to define the relative cost effects

$$e_i = \frac{E_i}{\sum_{i \in N} E_i} \quad \text{for all } i \in N$$

and we can formulate the tie-breaking constraints (26) needed in the second phase of the generalized method.

5. Separation of pavement thickness and traffic capacity costs

The proposed approach [12] distributes traffic-related costs in a more fair way than any other method by considering both traffic loads and traffic capacity. Furthermore, the development of a new cost allocation methodology considering allows us to analyze the impact of traffic capacity costs. The two concepts used in the proposed methodology to allocate costs among vehicle classes, according to traffic load and capacity requirements, are known as the Shapley value and the Aumann-Shapley value. In essence the Aumann-Shapley value determines an *average cost per ESAL* and an *average cost per lane* (per mile). The Shapley value allocates the total number of lanes of a highway among the vehicle classes. With these results, it is then possible to calculate costs per mile for each vehicle class by adding the cost due to ESALs (pavement thickness) and the cost due to lanes (capacity).

Two types of players will be considered. $\mathbf{E} = \{1,2,\ldots,q_1\}$ and $\mathbf{L} = \{1,2,\ldots,q_2\}$ are sets of players of type 1 and type 2, respectively. Thus, $\mathbf{M} = \mathrm{E} \cup \mathbf{L}$ is the set of all players. Now, let $P(\mathbf{M})$ be the set of all subsets or *coalitions* formed with the elements of **M**. Furthermore, let **N** be the set of natural numbers and $\mathbf{R}^+$ be the set of positive real numbers. Let C: $P(\mathbf{M}) \rightarrow \mathbf{R}^+$ be a real-valued cost function known as the *characteristic function*. Finally, let $x(q_1,q_2;C)$ be allocated costs yielded by a cost allocation method, $x_1(q_1,q_2;C)$ be the cost allocated to player 1 and $x_2(q_1,q_2;C)$ be the allocated cost to player 2. With these conventions, four important definitions are given below.

5.1. Definitions

Definition 1

If $x_1(q_1,q_2;C) + x_2(q_1,q_2;C) = C(\mathbf{M})$, then the method x is called *complete*.

Definition 2

If $x(q_1,q_2;C_1+C_2) = x(q_1,q_2;C_1) + x(q_1,q_2;C_2)$, where C_1 and C_2 are non-decreasing cost functions, then the method x is called *additive*. If a cost function can be divided into two distinct and independent cost components, then the *allocated costs* can be divided into two corresponding components.

Definition 3

If $C(S) - C(S\backslash\{i\}) = 0$ for any $i \in S$, $i \in N$, and $S \subseteq N$, then $x_i(q_1,q_2;C) = 0$. In this case, the method x is called *dummy*. If any player does not contribute to any coalition, then the cost allocated to it is zero.

Definition 4

If $x_1(q_1,q_2;C) \geq x_1(q_1-1,q_2;C)$, then the cost allocation method x is called *demand monotonic* for any $q_1>2$. Similarly, if $x_2(q_1,q_2;C) \geq x_2(q_1,q_2-1;C)$, then the cost allocation method x is called *demand monotonic* for any q_2>2. The cost-share of a player should not decrease when the player increases its demand.

Friedman [13] shows that the A-S value is *complete, additive,* and *dummy.* In addition, Friedman and Moulin [14] show that the A-S value does not satisfy the demand monotonicity property for general non-decreasing cost functions. Lee & Garcia-Diaz [15] show that demand monotonicity will be held in the following cases.

5.2. Pavement and capacity costs allocation

Assume the *log concave* cost function formulated in (33):

$$C(e,l) = l\left(a + b\, e^{r}\right) \tag{33}$$

where e is the number of *ESALs*, $l \in \mathbf{N}$ is the number of lanes, $C(e,l)$ is the cost in dollars per lane-mile, and a, b, and r are non-negative parameters. For this function the following results can be proved:

a. Demand monotonicity for the number of lanes.
b. Demand monotonicity for the number of *ESALs* $r \geq 0.32$.

In this chapter we use a *compact form* developed for the discrete A-S value [15]. This compact form allows the use of the A-S value in realistic applications with a large number of players, where the computational work becomes excessive without using the form. This section states some fundamental results regarding the demand monotonicity of the log concave characteristic function. The proposed approach [12] is composed of the following three steps.

Step 1. Traffic-related pavement cost separation

To separate traffic-related pavement costs into the costs for traffic load and the costs for traffic capacity, the discrete A-S value is used. Suppose that there are m types of players and q_i players of a type i. Further, let

$$Q = \sum_i q_i,\ T = \sum_i t_i,\ T' = \sum_i t'_i$$

and

$$t'_i = q_i - t_i.$$

There are two formulas for the discrete A-S value. A formula by Moulin [11] is shown in (34), where $i = 1,\ldots, m$:

$$x_i(q;C) = \frac{q_1!\ldots q_m!}{Q} \sum_{t\in[0,q]} \frac{T!}{t_1!\ldots t_m!} \frac{T'!}{t'_1!\ldots t'_m!} \left(\frac{t_i}{T} - \frac{t'_i}{T'}\right) C(t). \tag{34}$$

Another formula by Redekop [16] is given in (35):

$$x_i(q;C) = \sum_{\substack{t\in[0,q]\\ t_i>0}} q_i \times \frac{\binom{q_i-1}{t_i-1}\left(\prod_{j\neq i}\binom{q_j}{t_j}\right)}{T\binom{Q}{T}}[C(t)-C(t_1,t_2,...,t_i-1,...,t_m)]. \tag{35}$$

The cost per ESAL and the cost per lane are calculated by averaging since all the players of the same type are identical. Thus, the cost per ESAL (C_e) and the cost per lane (C_l) can be calculated as follows, where i = e or l:

$$C_i = \frac{x_i(q;C)}{q_i} \tag{36}$$

There are two types of players, namely, ESALs and lanes. Furthermore, let q_1 be the total number of players for ESALs, and q_2 be the total number of players for lanes. Then, the cost per lane and the cost per ESAL can be calculated from Redekop's formula as shown in (37) and (38).

$$C_l = \frac{q_1!(q_2-1)!}{(q_1+q_2)!}\sum_{t_1=0}^{q_1}\sum_{t_2=1}^{q_2}\frac{(t_1+t_2-1)!}{t_1!(t_2-1)!}\frac{(q_1-t_1+q_2-t_2)!}{(q_1-t_1)!(q_2-t_2)!}\{C(t_1,t_2)-C(t_1,t_2-1)\} \tag{37}$$

$$C_e = \frac{(q_1-1)!q_2!}{(q_1+q_2)!}\sum_{t_2=0}^{q_1}\sum_{t_1=1}^{q_2}\frac{(t_1+t_2-1)!}{(t_1-1)!t_2!}\frac{(q_1-t_1+q_2-t_2)!}{(q_1-t_1)!(q_2-t_2)!}\{C(t_1,t_2)-C(t_1-1,t_2)\} \tag{38}$$

If the cost increment remains the same when t_1 (or t_2) is fixed and t_2 (or t_1) is increased by 1 the A-S value can be determined using the simplified compact form formulated in (39).

$$x_{t_2}(q,C) = \frac{q_2}{q_1+1}\sum_{t_1=0}^{q_1} C(t_1,1) \tag{39}$$

Step 2. Lane assignment

Since the A-S value satisfies the *completeness property*, the sum of costs for traffic capacity and traffic load for the grand coalition equals the total cost for that coalition. The sum of ESALs over all vehicle classes is equal to the number of ESALs for the grand coalition (q_1), but the sum of the lanes required for each vehicle class is greater than or equal to the lanes required for the grand coalition (q_2). Hence, to calculate cost responsibilities for each vehicle class, the number of lanes for the grand coalition should be assigned to the vehicle classes. The Shapley value will be used to determine the number of lanes assigned to vehicle class i (L_i). The ith Shapley value for n players is determined using (1), with i = 1, …, n

$$L_i = \sum_{s=1}^{n}\frac{(s-1)!(n-s)!}{n!}\sum_{\substack{S\subseteq N:i\in S\\ |S|=s}}\left(F(S)-F(S-i)\right) \tag{40}$$

Step 3. Cost allocation

Costs are allocated to each vehicle class in proportion to the number of ESALs and the number of lanes, that is

$$x_i(E_i, L_i) = E_i C_e + L_i C_l \tag{41}$$

where

$x_i(E_i, L_i)$: Cost allocated to vehicle class i
E_i : ESALs for vehicle class i
C_e: Cost per ESAL
L_i : Number of lanes assigned to vehicle class i
C_l: Cost per lane

5.3. An example

The proposed approach is now illustrated using a simple example. Suppose that there are 3 vehicles: two automobiles (A), one pickup truck (P), and one 5-axle-trailer truck (T). Furthermore, there is 1 *base lane,* 2 additional lanes, and a total of 4 ESALs. These loads are divided into 1 ESAL for two automobiles, 1 ESALs for one pickup truck, and 2 ESALs for one 5 axle-trailer truck. The numbers of additional lanes required by each vehicle coalition are in shown in Table 2.

COALITION	{A}	{P}	{T}	{A,P}	{A,T}	{P,T}	{A,P,T}
Number of additional lanes	1	1	0	2	2	1	2

Table 2. Number of additional lanes required by each vehicle coalition

The cost in $/mile as a function of the number of ESALs and the number of lanes is assumed to be $C(e,l) = l\left(2 + 3\sqrt{e}\right)$. To calculate the A-S value for cost per ESAL (C_e) and cost per lane (C_l), Table 3 will be used. All possible 6!/2!4! = 15 inclusion sequences are shown in this table, where an E stands for one unit of ESALs and an L for one unit of lanes. The gray-colored column is for the *base lane.*

A *base lane* is first included in any possible sequence, and then either E or L is included. The average marginal costs, C_e and C_l, for including E or L in each sequence can be calculated from Table 3. The A-S values (C_e and C_l) can be also calculated by using the formulas shown in Step 1. The calculated values for C_e and C_l are 2.66 and 5.68, respectively.

To calculate number of lanes assigned to each vehicle class by the Shapley value, we first determine the total number of possible sequences as 3! = 6. The average marginal number of lanes, L_i, for including A, P, or T in each sequence is calculated from Table 4. The Shapley value for L_i can be also calculated by using formulas shown in Step 2.

Sequences	Inclusion Sequences						
1	L	E	E	E	E	L	L
2	L	E	E	E	L	E	L
3	L	E	E	E	L	L	E
4	L	E	E	L	E	E	L
5	L	E	E	L	E	L	E
6	L	E	E	L	L	E	E
7	L	E	L	E	E	E	L
8	L	E	L	E	E	L	E
9	L	E	L	E	L	E	E
10	L	E	L	L	E	E	E
11	L	L	E	E	E	L	E
12	L	L	E	E	E	E	L
13	L	L	E	E	L	E	E
14	L	L	E	L	E	E	E
15	L	L	L	E	E	E	E

Table 3. All possible inclusion sequences for the A-S value

Sequences	Including sequences			Marginal number of lanes		
1	A	P	T	1	1	0
2	A	T	P	1	1	0
3	P	A	T	1	1	0
4	P	T	A	1	0	1
5	T	A	P	0	2	0
6	T	P	A	0	1	1

Table 4. All possible inclusion sequences for the Shapley value

The Shapley values for the three vehicle classes are:

$$L_A = \frac{1}{6}(1+1+1+1+2+1) = 1.67$$

$$L_P = \frac{1}{6}(1+0+1+1+0+1) = 0.67$$

$$L_T = \frac{1}{6}(0+1+0+0+0+0) = 0.16\,.$$

The cost for the base lane is 2. This cost may be allocated proportionally by ESALs or, perhaps more appropriately, by vehicle miles of travel (VMT), since this cost is a non-load-related cost. Cost responsibilities for the three vehicle classes are shown in Table 5, where the base lane cost has been allocated proportionally according to ESALs.

Vehicle Classes	Load costs (E_iC_e)	Capacity costs (L_iC_l)	Costs for base lane (proportional)	Cost responsibilities
Automobile	1×2.66	1.17×5.68	2×0.50	10.30
Pickup truck	1×2.66	0.67×5.68	2×0.25	6.97
5-ax-trailer truck	2×2.66	0.16×5.68	2×0.25	6.73

Table 5. Cost responsibility calculation for each vehicle class

6. Separation of bridge construction and traffic capacity costs

A cost function is needed to estimate the bridge construction cost for the gross vehicle weight associated with any coalition of vehicle classes. This cost function can be developed by determining the cost of the bridge required by a coalition as a percentage of the cost of a *baseline bridge.* To accommodate all possible coalitions, the range of gross vehicle weight can be divided into an adequate number of intervals or categories. Results for nine categories of gross weight ranging from 5,000 lb to 108,000 lb are shown in Table 6. This table was built using a study by Moses [17] and the 1997 Federal Highway Cost allocation Study [6]. The table shows the required bridge cost for each gross vehicle weight category as a percentage of the cost of a baseline HS20 bridge which has a weight carrying capacity of 72,000 lb. The results for each gross vehicle weight category are the coordinates of one point of the bridge cost function.

Gross Vehicle Weight (1000 lb)	5	10	20	30	40	54	**72**	90	108
Bridge Cost Percentage	80.78	82.61	86.52	90.43	95.80	94.59	**100**	105	110

Table 6. Bridge cost percentages considering a baseline HS20 bridge

6.1. Bridge cost allocation procedure

The proposed model for the relationship between cost per lane-mile and the gross vehicle weight to be applied is formulated as

$$Y = l_i\ (a_i + b_i X) \tag{42}$$

where Y is the cost in dollars per lane-mile, l_i is the number of lanes of bridge type i, X is the gross vehicle weight in kips, and a_i and b_i are known parameters (to be estimated using regression analysis). Depending on the required number of lanes, more than one cost function can be formulated to determine accurate bridge construction cost estimates. A short-span structured bridge may be proper for a bridge with one lane in each direction, while a longer-span structured bridge may be so for a bridge with more lanes

The bridge construction cost allocation procedure is outlined below [18]. The procedure is essentially the same one developed in Section 5. In the case of bridges, however, there is an additional step (referred to as Step 2 below) to apply the incremental method of highway cost allocation.

Step 1. Traffic-related pavement cost separation

This step is identical Step 1 in the methodology described in Section 5 of this chapter.

Step 2. Traffic-load cost allocation

The cost per unit of weight (C_e) was obtained in *Step 1*. The traffic-load cost can be allocated to each weight group in vehicle class by using the incremental method, as indicated below:

a. The lightest vehicle group is first considered. The unit of weight (C_e) is allocated to each vehicle class in this group and all heavier groups according to average daily traffic (ADT).
b. The next light group is considered. The marginal cost equal to C_e is allocated to each vehicle class in this group and all heavier groups according to ADT.
c. If the heaviest group is considered, then go to d. Otherwise, continue to b.
d. If a vehicle class i has several weight groups, then sum up the cost for those weight groups.

Step 3. Lane assignment

Again, this procedure is identical to the Step 2 of the methodology described in Section 5.

Step 4. Cost allocation

This procedure is also identical to the Step 3 of the methodology described in Section 5.

6.2. An example

A simple hypothetical numerical example is presented in this section to illustrate and clarify the application of the proposed method. It is assumed that there are 3 vehicles: automobile {A}, pickup truck {P}, and 5-ax-trailer truck {T}. Also, it is assumed that 1 *base* lane is required. The number of *additional* lanes is the same in Table 1. The total vehicle weight is distributed along four intervals: 0-10 kips, 11-20 kips, and 21-30 kips. The percentages of total ADT due to vehicles of each class, for the given weight intervals, are: {A} belongs to the 0-10 kip interval with 65 % of ADT; {P} belongs to the 0-10 kip interval with 20 % of ADT and to the 11-20 kip interval with 5 percent of ADT; {T} belongs to the11-20 kips interval with 5 percent of ADT and to the 21-30 kip interval with 5 percent of ADT. The cost functions for this example are formulated below:

$$C(k,l) = l(1+2k) \quad l = 1$$

$$C(k,l) = l(2+3k) \quad l \geq 2$$

The following results are obtained in each step.

Step 1. Bridge construction cost separation

To calculate the A-S value for the cost per unit of weight (10 kips in this example) and the cost per lane the sequences shown in Table 7 can be used. It is noted that the total number of sequences is 5!/(3!2!) = 10. In Table 7 letter K represents one unit of weight (10 kips) and letter L represents one unit of lanes. A gray-shaded column is used for the base lane. A base lane is first included in any possible sequence, and then either a K or an L is included. The average marginal costs (or the A-S value) C_k and C_l can be calculated by using Table 7. The calculated values are $C_k = 170/30 = 5.67$ and $C_l = 150/20 = 7.5$.

Sequence	Including Sequence						Marginal Cost					
1	L	K	K	K	L	L	1	2	2	2	15	11
2	L	K	K	L	K	L	1	2	2	11	6	11
3	L	K	K	L	L	K	1	2	2	11	8	9
4	L	K	L	K	K	L	1	2	7	6	6	11
5	L	K	L	K	L	K	1	2	7	6	8	9
6	L	K	L	L	K	K	1	2	7	5	9	9
7	L	L	K	K	K	L	1	3	6	6	6	11
8	L	L	K	K	L	K	1	3	6	6	8	9
9	L	L	K	L	K	K	1	3	6	5	9	9
10	L	L	L	K	K	K	1	3	2	9	9	9

Table 7. Sequences and marginal cost for calculation of A-S value

Step 2. Traffic-load cost allocation:

$$E_T:\ 5.67\times\frac{5+5}{65+20+5+5+5}+5.67\times\frac{5+5}{5+5+5}+5.67\times\frac{5}{5}=10$$

$$E_A:\ 5.67\times\frac{65}{65+20+5+5+5}=3.7$$

$$E_P:\ 5.67\times\frac{20+5}{65+20+5+5+5}+5.67\times\frac{5}{5+5+5}=3.3$$

Step 3. Lane assignment:

See Table 3. L_A=1.17, L_P=0.67, L_T=0.16

Step 4. Cost allocation:

The value (cost) of parameter *a* for the base lane is 2. This cost is allocated proportionally by ADTs in this example. The total cost allocations for the three vehicle classes are shown in Table 8.

Vehicle Classes	Load costs (E_i)	Capacity costs (L_iC_l)	Costs for base lanes (proportional)	Cost responsibilities
Automobile	3.7	1.17×7.5	1×0.65	13.12
Pickup truck	3.3	0.67×7.5	1×0.25	8.58
5-ax-trailer truck	10	0.16×7.5	1×0.10	11.30

Table 8. Cost allocations for vehicle classes

Author details

Alberto Garcia-Diaz
Department of Industrial & Information Engineering, The University of Tennessee, USA

Dong-Ju Lee
Department of Industrial & Systems Engineering, Kongju National University, South Korea

7. References

[1] Young H P (1985) Methods and Principles of Cost Allocation In: Young, H P (Eds.), Cost Allocation: Methods, Principles, Applications. Elsevier Science Ltd., North Holland, NY, pp. 3-29.

[2] Dror M (1990) Cost Allocation: The Traveling Salesman, Binpacking, and the Knapsack. Applied Mathematics and Computation 35: 191-207.

[3] Hartman, B C, Dror M (1996) Cost Allocation in Continuous-Review Inventory Models. Naval Research Logistics 43: 549-561.

[4] Castano-Pardo A, Garcia-Diaz A (1995) Highway Cost Allocation: An Application of the Theory of Nonatomic Games. Transportation Research Part A 29: 187-203.

[5] Federal Highway Administration (1982) Final Report on the Federal Highway Cost Allocation Study, U.S. Department of Transportation, Washington, D.C.

[6] Federal Highway Administration (U.S. Department of Transportation) (1997) 1997 Federal Highway Cost Allocation Study Final Report.

[7] Neumann J, Morgenstern O (1944) Theory of Games and Economic Behavior, Princeton University Press.

[8] Villarreal-Cavazos A, Garcia-Diaz A (1985) Development and Application of New Highway Cost Allocation Procedures. Transportation Research Record 1009: 34-41.

[9] Shapley L S (1953) A Value for n-person Games. In: Kuhn, H.W., Tucker, A.W. (Eds.), Contributions to the Theory of Games, Vol. II, Annual Mathematics Study 28. Princeton University Press, Princeton, NJ, pp. 307-317.

[10] Aumann R J, Shapley L S (1974) Values of Non-Atomic Games. Princeton University Press, Princeton, NJ.

[11] Moulin H (1995) On Additive Methods to Share Joint Costs. The Japanese Economic Review; 46: 303-332.

[12] Lee D (2002) Game-Theoretic Procedures for Determining Pavement Thickness and Traffic Lane Costs in Highway Cost Allocation. Ph. D. Dissertation. College Station, TX: Texas A&M University.

[13] Friedman E J (2004) Strong Monotonicity in Surplus Sharing. Economic theory. 23: 643-658.

[14] Friedman E, Moulin H (1999) Three Methods to Share Joint Costs or Surplus. Journal of Economic Theory. 87: 275-312.

[15] Lee D, Garcia-Diaz A, Lee C (2012) "Demand Monotonicity Analysis of the Discrete Aumann-Shapley Value Based on a Compact Form Used in Highway Cost Allocation" *Working Paper.*

[16] Redekop J (2000) Increasing marginal cost and the monotonicity of Aumann-Shapley pricing. Working Paper, Department of Economics, University of Waterloo, Waterloo, Ontario, Canada.
[17] Moses F (1989) Effects on Bridges of Alternative Truck Configurations and Weights, NCHRP Contract No. HR 2-16 (b)m, Transportation Research Board, Washington, D.C.
[18] Lee D, Garcia-Diaz A (2007) Procedure for Bridge Construction Cost Allocation Based on Game Theory. In Transportation Research Record: Journal of the Transportation Research Board, No., Transportation Research Board of the National Academies, Washington, D.C., 1996: 100-105.

Chapter 7

A Game Theoretic Analysis of Price-QoS Market Share in Presence of Adversarial Service Providers

Mohamed Baslam, Rachid El-Azouzi, Essaid Sabir,
Loubna Echabbi and El-Houssine Bouyakhf

Additional information is available at the end of the chapter

http://dx.doi.org/10.5772/54380

1. Introduction

Recently, the selfish behavior of customers and Service Providers (SPs) in telecommunications systems has been widely analyzed using game theory with all its powerful solution concepts. It was shown in several works that customer's selfish behavior leads to a network collapse, where a typical prisonerís dilemma situation arises. Despite of the bounty of works and efforts investigated in analyzing market share game, this filed is still an ideal tool to understand interaction among SPs and customers. Indeed, it is common in the literature to assume a single decision action (e.g., cost) through which an equilibrium would be computed. Yet, in order to take into account Quality of Service (QoS), it is necessary to incorporate into the model more than one decision parameter. A simple example is to include both price and some measure of QoS (e.g., delay, throughput, loss probability, etc.). Other multi-criteria models may incorporate, for example, delay and reliability, the latter representing the QoS, price or delay and jitter, etc.

The competition in terms of prices and QoS among SPs entails the formation of non-cooperative games. We consider multiple SPs (players of the game), where each one seeks to maximize its own revenue, whereby the whole system of SPs would have no incentive to deviate from the Nash equilibrium[1] point, i.e., the vector of equilibrium strategies. Yet, such equilibrium point should first mathematically exist. In this chapter, we present a general model for computing a bi-criteria Nash equilibrium for multiple SPs. We shall then analyze the interactions between SPs who won't attract more clients and maximize their respective profits. We address the important problem of Nash Equilibrium characterization with two-component action, when the two components of each provider are the service price and a measure of QoS. Our model is mainly inspired from, [6], where the authors studied a

[1] A Nash equilibrium is a strategy profile where no player has sensitive to deviate unilaterally from its current strategy.

non-cooperative game for pricing problem considering QoS as an extra decision parameter. The authors build a Markovian model to derive the behavior of customers depending on the strategic actions of the SPs. In contrast to this chapter, we base our study on the concepts of demand for the services of a given SP (defined by linear function that depends on the vectors of prices and QoSs), which is a commonly used function in research related to competitive network and equilibrium models, [9], [5], to calculate the reputation of an SP in the market.

We focus our studies on the non-cooperative games in terms of stable solutions, which are the pure strategy Nash equilibria of the game. We do not consider mixed strategy equilibria, because our environment requires a concrete strategy rather than a randomized strategy, which would be the result of a mixed strategy. Hence, when using the term Nash equilibrium we mean pure strategy exact Nash equilibria unless mentioned otherwise.

We note that the most fundamental assumption in relative works of game theory is rationality. It implies that every player is motivated by increasing his own payoff, i.e. every player is looking to maximize his own utility. John V. Neumann and Morgenstern justified the idea of maximizing the expected payoff in their work in [23]. In this context, all information concerning the game is known to all players, i.e., there is complete information. So, we consider that all players are said to be rational and intelligent. A rational person is one who acts in such a way as to maximize his or her expected payoff or utility as economists would say. An intelligent person is one who can deduce what his or her opponent will do when acting rationally. In fact, humans use a propositional calculus in reasoning, the propositional calculus concerns truth functions of propositions, which are logical truths (statements that are true in virtue of their form). For this reason, the assumption of rational behaviour of players in telecommunications systems is more justified, as the players are usually devices programmed to operate in certain ways. However, there are previous studies that have shown that humans do not always act rationally [10].

Related Works :

Applying game theory in telecommunications problems is an active research area, in which game-theoretic models have been developed and studied in the last decades, [1, 2, 6, 7, 9, 16, 18]. These models are interested in pricing issues, they proposed non-cooperative game formulations to analyze behaviours of players that selfishly decide their strategies to maximize their respective profits. Other works consider the criteria of price as an implicit parameter, which is determined as a function of the degree of saturation on the network. Typically in these approaches, the price is a shadow price. For more details on those approaches see, [14, 15, 24]. Nonetheless, the price of anarchy has been studied in a large and diverse number of games, e.g., in areas like wireless ad-hoc networks [8, 13], routing and congestion [4, 19], network creation [3], or facility location [22]. In our model, we do not take into account network topology, but rather the effective service proposed by each SP as a single entity. In other words, the price and QoS proposed by an SP will not depend on the source or destination, distance, etc. that underlies the request of each user. After we have proved existence of Nash equilibrium, we propose a joint price and QoS algorithm which allows to learn the equilibrium price and QoS strategies decided by SPs. This is a simple algorithm implementation with lower computational complexity.

Organization :

The rest of the chapter is organized as follows : in Section 2 we describe the system model and introduce a new demand and utility functions. In Section 3 we formulate the joint price and QoS problem as a non-cooperative game, and investigate existence and uniqueness of a Nash equilibrium solution. Then, we present numerical results obtained from simulations that exploits our joint price and QoS algorithm in Section 4 . Conclusions and future guidelines are drawn in Section 5.

2. Problem modeling

In this chapter, we formulate the interaction among service providers (SPs) as a non-cooperative game. Each SP chooses the Quality of Service to guarantee (it depends on the amount of requested bandwidth) and the corresponding price.

We consider a system with N service providers. Let p_i and q_i be, respectively, the tariff/pricing policy and the QoS guaranteed by SP-i. Now, each customer seeks to subscribe to the operator which allows him to meet a QoS sufficient to satisfy his/her needs, at suitable price. We consider that behaviors of customer's has been handled by a simple function so called demand functions, see equation (1). This later depends on the price and QoS strategies of all SPs. From a tagged SP's point of view, the question is to set the best pricing strategy and the best QoS (amount of bandwidth to request from the network owner). SPs are supposed to know the effect of their policy on the customer's subscription policy. Whereas from customer's point of view, the question is to find the SP that has the best price-QoS tradeoff conditions.

2.1. Demand model

For simplicity, we consider that the demand function D_i for services of the tagged SP-i is linear with respect to the set price p_i and the promised QoS q_i, see, [9]. This demand function depends also on prices $\mathbf{p}_{-i}$ and QoS $\mathbf{q}_{-i}$ set by the competitors. Namely, the demand function of SP-i depends on $\mathbf{p} = [p_1, .., p_N]$ and $\mathbf{q} = [q_1, .., q_N]$. Eventually, D_i is decreasing w.r.t. p_i and increasing w.r.t. p_j, $j \neq i$. Whereas it is increasing w.r.t q_i and decreasing w.r.t. q_j, $j \neq i$. Then, the demand functions w.r.t services of SP-i can be written as follows:

$$D_i(\mathbf{p}, \mathbf{q}) = D_i^0 - \alpha_i^i p_i + \beta_i^i q_i + \sum_{j, j \neq i} \left[\alpha_i^j p_j - \beta_i^j q_j \right], \quad \forall i \in \{1, .., N\}. \tag{1}$$

where D_i^0 is a positive constant used to insure non-negative demands over the feasible region. While α_i^j and β_i^j are positive constants representing respectively the sensitivity of service provider i to price and QoS of service provider j.

2.2. Utility model

The total revenue of SP-i is $D_i(\mathbf{p}, \mathbf{q}) p_i$. We assume that we have a single network owner, this latter charges each SP-i a cost ϑ_i per unit of requested bandwidth. In order to insure the customers loyalty, the amount of bandwidth μ_i required by SP-i should depend on $D_i(.)$ and on the QoS q_i it wishes to offer to its customers. Therefore, the net profit of SP-i is simply the difference between the total revenue and the fee paid to the network owner:

$$U_i(\mathbf{p},\mathbf{q}) = D_i(\mathbf{p},\mathbf{q})p_i - F_i(q_i, D_i), \quad \forall i \in \{1,..,N\}.$$

where $F_i(q_i, D_i)$ is the fee paid by SP-i (investment of SP-i) :

$$F_i = \vartheta_i \mu_i \left(q_i, D_i\right)$$

where μ_i is the amount of bandwidth required by SP-i, such that ϑ_i is a cost per unit of requested bandwidth We assume that the QoS corresponds to the expected delay, also we consider the Kleinrock delay which is a common delay used in Networking Games, so :

$$q_i = \frac{1}{\sqrt{Delay_i}} = \sqrt{\mu_i - D_i}$$

that mean that:

$$\mu_i = q_i^2 + D_i$$

While, the utility function of the SP-i is given by the following formula:

$$U_i(\mathbf{p},\mathbf{q}) = D_i(\mathbf{p},\mathbf{q})\left(p_i - \vartheta_i\right) - \vartheta_i q_i^2, \quad \forall i \in \{1,..,N\}. \tag{2}$$

3. A non-cooperative game formulation

For a precise formulation of a non-cooperative game, we have to specify (i) the number of players, (ii) the possible actions available to each player, and any constraints that may be imposed on them, (iii) the objective function of each player which she attempts to optimize. Here we will consider formulation of games where items (i)-(iii) above are relevant.

Let $\mathbf{G} = [\mathcal{N}, \{P_i, Q_i\}, \{U_i(.)\}]$ denote the non-cooperative price and QoS game (NPQG), where $\mathcal{N} = \{1,..,N\}$ is the index set identifying the SPs, P_i is the price strategy set of SP-i, Q_i is the QoS strategy set of SP-i, and $U_i(.)$ is the utility function. Each SP-i selects a price $p_i \in P_i$ and a QoS measure $q_i \in Q_i$. Let the price vector $\mathbf{p} = (p_1,..,p_N)^T \in P^N = P_1 \times P_2 \times ... \times P_N$, QoS vector $\mathbf{q} = (q_1,..,q_N)^T \in Q^N = Q_1 \times Q_2 \times ... \times Q_N$ (where T represents the transpose operator). The utility of SP-i when it decides the strategy price p_i to allocate the QoS q_i is given in equation (2). We assume that the strategy spaces P_i and Q_i of each SP are compact and convex sets with maximum and minimum constraints, For any given user i we consider strategy spaces the closed intervals $Pi = [\underline{p_i}, \overline{p_i}]$ and $Qi = [\underline{q_i}, \overline{q_i}]$.

In order to maximize their utilities, each SP-i decides a price p_i and QoS q_i. Formally, the NPQG problem can be expressed as:

$$\max_{p_i \in P_i, q_i \in Q_i} U_i(\mathbf{p},\mathbf{q}), \quad \forall i \in \mathcal{N}. \tag{3}$$

3.1. The Nash equilibrium

Considering rationality of service providers, the Nash equilibrium concept is the natural concept solution of the NPQG game. We first will investigate the Nash equilibrium solution for the induced game as defined in the previous section. We will show that a Nash equilibrium solution exists and is unique by using the theory of concave games, [20]. We recall that a non-cooperative game **G** is called concave if all players' utility functions are strictly concave with respect to their corresponding strategies, [20].

According to, [20], a Nash equilibrium exists in a concave game if the joint strategy space is compact and convex, and the utility function that any given player seeks to maximize is concave in its own strategy and continuous at every point in the product strategy space. Formally, if the weighted sum of the utility functions with nonnegative weights:

$$\varphi = \sum_{i=1} x_i U_i, \;\; x_i > 0 \;\; \forall i. \tag{4}$$

is diagonally strictly concave, this implies that the Nash equilibrium point is unique. The notion of diagonal strict concavity means that an individual user has more control over its utility function than the other users have on it, and is proven using the pseudo-gradient of the weighted sum of utility functions, [20].

Fixed-Price Game : Considering some fixed price policy, a Nash equilibrium in QoS is formally defined as:

Definition 1. *A QoS vector* $\boldsymbol{q}^* = (q_1^*, .., q_N^*)$ *is a Nash equilibrium of the NPQG :* $\mathbf{G} = [\mathcal{N}, \{P_i, Q_i\}, \{U_i(.)\}]$ *if, for every* $i \in \mathcal{N}$, $U_i(q_i^*, \boldsymbol{q}_{-i}^*) \geq U_i(q_i', \boldsymbol{q}_{-i}^*)$ *for all* $q_i' \in Q_i$.

Theorem 1. *A Nash equilibrium in terms of QoS for game* $\mathbf{G} = [\mathcal{N}, \{P_i, Q_i\}, \{U_i(.)\}]$ *exists and is unique.*

Proof. To prove existence, we note that each SP's strategy space Q_i is defined by all QoSs in the closed interval bounded by the minimum and maximum QoSs. Thus, the joint strategy space Q is a nonempty, convex, and compact subset of the Euclidean space $\mathbb{R}^N$. In addition, the utility functions are concave with respect to QoSs as can be seen from the second derivative test:

$$\frac{\partial^2 U_i(\mathbf{p}, \mathbf{q})}{\partial q_i^2} = -2\vartheta_i < 0, \;\; \forall i \in \mathcal{N}, \tag{5}$$

which ensures existence of a Nash equilibrium.

In order to prove uniqueness, we follow, [20], and define the weighted sum of user utility functions.

$$\varphi(\mathbf{q}, \mathbf{x}) = \sum_{i=1}^{N} x_i U_i(q_i, \mathbf{q}_{-i}), \tag{6}$$

The pseudo-gradient of (6) is given by :

$$g(\mathbf{q},\mathbf{x}) = \left[x_1 \nabla U_1(q_1,\mathbf{q}_{-1}),...,\ x_N \nabla U_N(q_N,\mathbf{q}_{-N})\right]^T \tag{7}$$

The Jacobian matrix $\mathbf{J}$ of the pseudo-gradient (w.r.t. $\mathbf{q}$) is written

$$\mathbf{J} = \begin{pmatrix} x_1 \frac{\partial^2 U_1}{\partial q_1^2} & x_1 \frac{\partial^2 U_1}{\partial q_1 \partial q_2} & \cdots & x_1 \frac{\partial^2 U_1}{\partial q_1 \partial q_N} \\ x_2 \frac{\partial^2 U_2}{\partial q_2 \partial q_1} & x_2 \frac{\partial^2 U_2}{\partial q_2^2} & \cdots & x_2 \frac{\partial^2 U_2}{\partial q_2 \partial q_N} \\ \vdots & \vdots & \ddots & \vdots \\ x_N \frac{\partial^2 U_N}{\partial q_N \partial q_1} & x_N \frac{\partial^2 U_N}{\partial q_N \partial q_2} & \cdots & x_N \frac{\partial^2 U_N}{\partial q_N^2} \end{pmatrix}$$

$$= \begin{pmatrix} -2x_1\vartheta_1 & 0 & \cdots & 0 \\ 0 & -2x_2\vartheta_2 & \cdots & 0 \\ \vdots & \vdots & \ddots & \vdots \\ 0 & 0 & \cdots & -2x_N\vartheta_N \end{pmatrix}.$$

□

Thus, $\mathbf{J}$ is a diagonal matrix with negative diagonal elements. This implies that $\mathbf{J}$ is negative definite. Henceforth $[\mathbf{J} + \mathbf{J}^T]$ is also negative definite, and according to Theorem (6) in, [20], the weighted sum of the utility functions $\varphi(\mathbf{q},\mathbf{x})$ is diagonally strictly concave. Thus the fixed-price Nash equilibrium point

$$q_i^* \in \underset{q_i \in Q_i}{\operatorname{argmax}}\, U_i(q_i,\mathbf{q}_{-i}^*),\ \ \forall i \in \mathcal{N}. \tag{8}$$

is unique.

Fixed-QoS Game : When fixing the QoS, a Nash equilibrium in terms of price is formally defined as :

Definition 2. *A price vector* $\mathbf{p}^* = (p_1^*,..,p_N^*)$ *is a Nash equilibrium of the NPQG :* $\mathbf{G} = [\mathcal{N}, \{P_i, Q_i\}, \{U_i(.)\}]$ *if, for every* $i \in \mathcal{N}$, $U_i(p_i^*,\mathbf{p}_{-i}^*) \geq U_i(p_i',\mathbf{p}_{-i}^*)$ *for all* $p_i' \in P_i$.

Theorem 2. *A Nash equilibrium in terms of price for the game* $\mathbf{G} = [\mathcal{N}, \{P_i, Q_i\}, \{U_i(.)\}]$ *exists and is unique.*

Proof. To prove existence, we note that each SP's strategy space P_i is defined by all prices in the closed interval bounded by the minimum and maximum prices. Thus, the joint strategy space P is a nonempty, convex, and compact subset of the Euclidean space $\mathbb{R}^N$. In addition, the utility functions are concave with respect to prices as can be seen from the second derivative test:

$$\frac{\partial^2 U_i(\mathbf{p},\mathbf{q})}{\partial p_i^2} = -2\alpha_i^i < 0,\ \ \forall i \in \mathcal{N}, \tag{9}$$

which ensures existence of a Nash equilibrium.

To prove uniqueness we define now the weighted sum of user utility functions

$$\phi(\mathbf{p},\mathbf{x}) = \sum_{i=1}^{N} x_i U_i(p_i, \mathbf{p}_{-i}), \tag{10}$$

the pseudo-gradient of this later is given by

$$g(\mathbf{p},\mathbf{x}) = \left[x_1 \nabla U_1(p_1, \mathbf{p}_{-1}), ..., \; x_N \nabla U_N(p_N, \mathbf{p}_{-N})\right]^T. \tag{11}$$

In order to show that $\phi(\mathbf{p},\mathbf{x})$ is diagonally strictly concave in this case we use the following lemma proved in, [11].

Lemma 1. *If each $U_i(\boldsymbol{p})$ is a strictly concave function in p_i, each $U_i(\boldsymbol{p})$ is convex in p_{-i} and there is some $\boldsymbol{x} > 0$ such that $\phi(\boldsymbol{p},\boldsymbol{x})$ is concave in $\boldsymbol{p}$, then $[\boldsymbol{J}(\boldsymbol{p},\boldsymbol{x}) + \boldsymbol{J}^T(\boldsymbol{p},\boldsymbol{x})]$ is negative definite, where $\boldsymbol{J}(\boldsymbol{p},\boldsymbol{x})$ is the Jacobian of $g(\boldsymbol{p},\boldsymbol{x})$.*

From equation (9), we know that $U_i(\mathbf{p})$ is strictly concave in p_i. Further

$$\frac{\partial^2 U_i}{\partial p_j^2} = 0, \;\; \forall i \neq j,$$

which implies that $U_i(\mathbf{p})$ is convex in p_{-i} as well. Also, we have that

$$\begin{aligned}\frac{\partial^2 \phi(\mathbf{p},\mathbf{x})}{\partial p_i^2} &= x_i \frac{\partial^2 U_i(p_i, \mathbf{p}_{-i})}{\partial p_i^2} + \sum_{j\neq i}^{N} x_j \frac{\partial^2 U_j(p_i, \mathbf{p}_{-i})}{\partial p_i^2} \\ &= -2x_i \alpha_i^i < 0, \;\; \forall i,\end{aligned}$$

then $\phi(\mathbf{p},\mathbf{x})$ is concave in p_i and from *Lemma* 1 we have that $[\mathbf{J}(\mathbf{p},\mathbf{x}) + \mathbf{J}^T(\mathbf{p},\mathbf{x})]$ is negative definite. Thus the weighted sum of utility functions $\phi(\mathbf{p},\mathbf{x})$ is diagonally strictly concave. The fixed-QoS Nash equilibrium point is then unique and is given by

$$p_i^* \in \underset{p_i \in P_i}{\operatorname{argmax}} \, U_i(p_i, \mathbf{p}_{-i}^*), \quad \forall i \in \mathcal{N}. \tag{12}$$

□

3.2. The joint price and QoS game

As shown in equations (5) and (9), the utility functions $U_i(\mathbf{p},\mathbf{q}), \;\; \forall i \in \mathcal{N}$, are concave respectively w.r.t. q_i and p_i. So, for all, $i \in \mathcal{N}$, the QoS and price conditions which maximizes the utility given in equation (2) are respectively :

$$\begin{cases} \frac{\partial U_i(\mathbf{p},\mathbf{q})}{\partial q_i} = 0 \\ \frac{\partial U_i(\mathbf{p},\mathbf{q})}{\partial p_i} = 0 \end{cases}$$

Thus, the computation of Nash Equilibrium can be performed by solving latter system.

Now, we turn to develop a fully distributed algorithm to learn the two-parameter equilibrium. Designing distributed algorithms that converge quickly to equilibrium is one of the foremost research goals in algorithmic game theory, and convex programs have played a crucial role in the design of algorithms for markets. Assuming that Providers are selfish and choose dynamically each one the best price and QoS that maximize his profiles, the distributed algorithms can be thought of as protocols that players are programmed to follow. The design and analysis of distributed algorithms converging to equilibria in the context of games has also received considerable attention, most commonly convergence of best response dynamics.

Solutions of equations induces by vanishing the partial derivatives correspond respectively to the best response in terms of QoS $BR_q^i(.)$, and best response Price $BR_p^i(.)$, of each SP-i as a function of the strategies of its opponents. Since Nash equilibrium point is unique, then a best response-based dynamics would converge to the joint Price-QoS NE. The two-parameters best response dynamics is detailed in Algorithm 1.

Algorithm 1 Best response dynamics

1: Initialize price and QoS vectors $\mathbf{p}$ and $\mathbf{q}$ randomly;
2: For each service provider $i \in \mathcal{N}$ at iteration t :

a) $p_i^{t+1} = BR_p^i(\mathbf{p}^t, \mathbf{q}^t)$;

b) $q_i^{t+1} = BR_q^i(\mathbf{p}^t, \mathbf{q}^t)$.

3.3. Social welfare and price of anarchy

The concept of social welfare [17] or total surplus [21], is defined as the sum of the utilities of all agents in the systems (i.e. Providers). It is well known in game theory that agent selfishness, such as in a Nash equilibrium, does not lead in general to a socially efficient situation. As a measure of the loss of efficiency due to the divergence of user interests, we use the Price of Anarchy (PoA) [19], this latter is a measure of the loss of efficiency due to actors' selfishness. This loss has been defined in [19] as the worst-case ratio comparing the global efficiency measure (that has to be chosen) at an outcome of the noncooperative game played among actors, to the optimal value of that efficiency measure. A PoA close to 1 indicates that the equilibrium is approximately socially optimal, and thus the consequences of selfish behavior are relatively benign. The term Price of Anarchy was first used by Koutsoupias and Papadimitriou [19] but the idea of measuring inefficiency of equilibrium is older. The concept in its current form was designed to be the analogue of the "approximation ratio" in Approximation Algorithms or the "competitive ratio" in Online Algorithms. As in [12], we measure the loss of efficiency due to actors' selfishness as the quotient between the social welfare obtained at the Nash equilibrium and the maximum value of the social welfare:

$$PoA = \frac{\min_{\mathbf{p},\mathbf{q}} W_{NE}(\mathbf{p},\mathbf{q})}{\max_{\mathbf{p},\mathbf{q}} W(\mathbf{p},\mathbf{q})} \tag{13}$$

where $W(\mathbf{p},\mathbf{q}) = \sum_{i=1}^{N} U_i(\mathbf{p},\mathbf{q})$ is a welfare function and $W_{NE}(\mathbf{p},\mathbf{q}) = \sum_{i=1}^{N} U_i(\mathbf{p}^*,\mathbf{q}^*)$ is a sum of utilities of all actors at Nash Equilibrium.

4. Numerical investigations

To clarify and show how to take advantage from our theoretical study, we suggest to study numerically the market share game while considering the best response dynamics and expressions of demand as well as utility functions of SPs. Hence, we consider a system with two SPs seeking to maximize their respective revenues. Table 1 represents the system parameter values considered in this numerical study.

$\alpha_1^1 = \alpha_2^2$	$\alpha_2^1 = \alpha_1^2$	$\beta_1^1 = \beta_2^2$	$\beta_2^1 = \beta_1^2$	D_0^1	D_0^2
0.7	0.3	0.7	0.3	300	250
$\vartheta_1 = \vartheta_2$	$\underline{p_1} = \underline{p_2}$	$\overline{p_1}, \overline{p_2}$	$\underline{q_1} = \underline{q_2}$	$\overline{q_1}, \overline{q_2}$	
20	100	1000	0	10	

Table 1. System parameters used for numerical examples.

Figures 1 and 2 present respectively curves of the convergence to Nash Equilibrium Price and to Nash Equilibrium QoS. It is clear that the best response dynamics converges to the unique Nash equilibrium price and QoS. We also remark that the speed of convergence is relatively high (around 9 rounds are enough to converge to the joint price and QoS equilibrium).

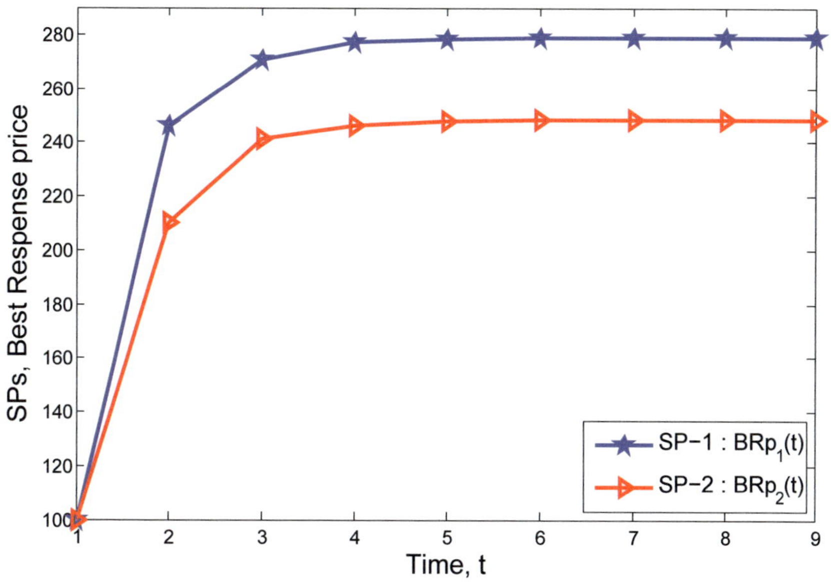

Figure 1. Price game : Convergence to the Price Nash equilibrium.

Next we plot in figures 3 and 4, respectively, the interplay of bandwidth cost (ϑ_i, $i \in \{1,2\}$ on the price and QoS at Nash equilibrium, for both SPs that we consider in this example. On one hand, we note that the equilibrium price for both SPs is increasing with respect to the

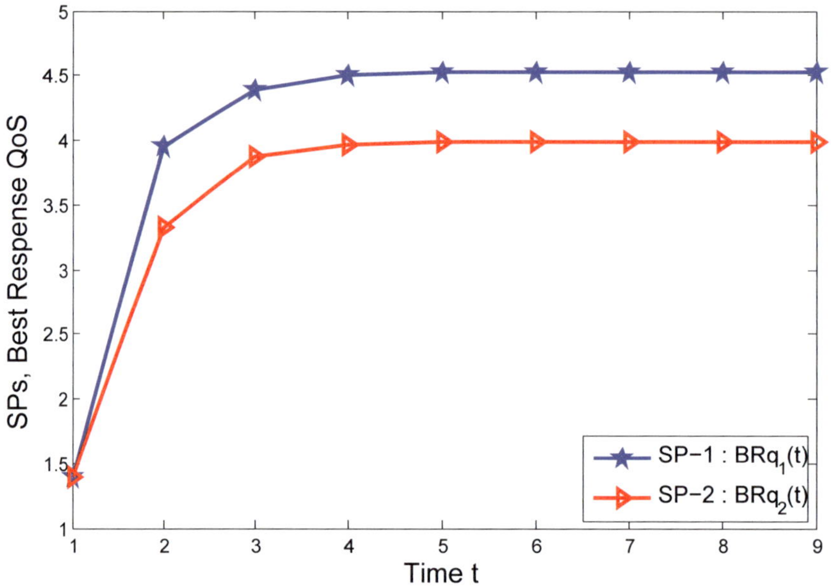

Figure 2. QoS game : Convergence to the QoS Nash equilibrium.

bandwidth cost. On the other hand, we note that the equilibrium QoS for all SPs is decreasing with the bandwidth cost. When the cost of bandwidth decided by the network owner is cheaper, the SPs invest for more bandwidth, so as to offer better QoS and an attractive price.

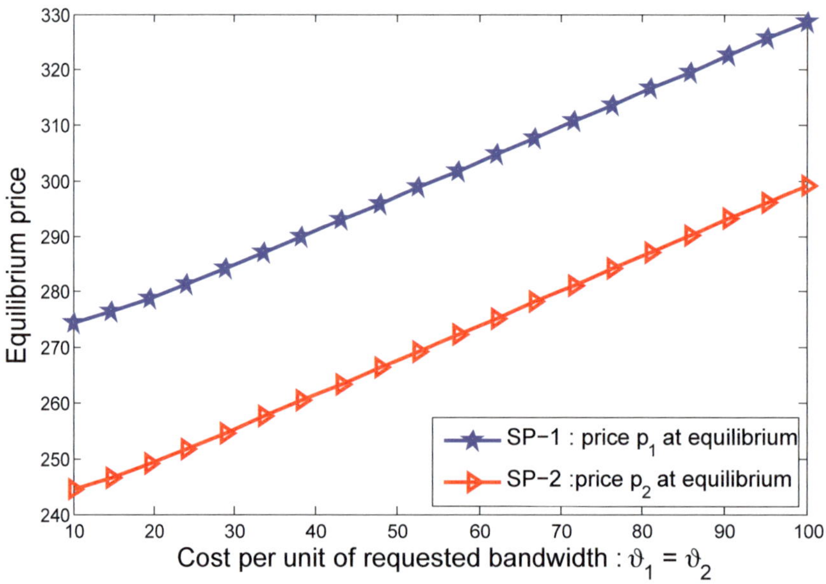

Figure 3. Equilibrium Prices w.r.t cost per unit of requested bandwidth ϑ_i.

In the following, we discuss the impact of the system parameters on the system efficiency in terms of Price of anarchy:

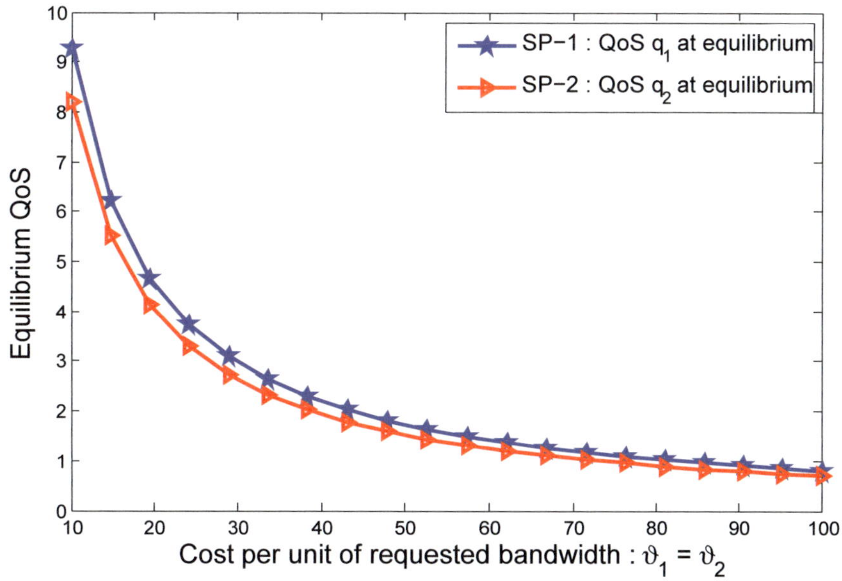

Figure 4. Equilibrium QoSs w.r.t cost per unit of requested bandwidth ϑ_i.

Influence of ϑ_i (cost per unit of requested bandwidth): Figure 5 shows the PoA variation curve as a function of the providers' bandwidth cost ϑ_i. Without loss of generality, we assume that $\vartheta_1 = \vartheta_2$. A special feature is that the Nash equilibrium performs well and the loss of efficiency is only around 8%. This result indicates that the Nash equilibrium of this game is fair and socially efficient. Henceforth, selfish players would not need the help of a third-part regulator (who recommends the players the best strategy profile to achieve their respective best outcomes) to get attracted by the optimum social welfare. However, the network owner can use the value of the bandwidth cost to control the selfishness/aggressiveness of the service providers, which will improve the whole network performance.

Influence of α (Sensitivity of SP-i to his price p_i) : Figure 6 plots the variation curve of price of anarchy with respect to α which represents the sensitivity of SP-i to his price p_i. In that figure, we first notice that the price of anarchy increases when α increases, the fact that the price of anarchy increases with α finds the simple intuition that increasing the sensitivity of SPs to their prices gives more and more freedom to SPs for optimizing the Nash equilibrium. On the other hand, when $\alpha = \alpha_1^1 = \alpha_2^2 = 1$, in the other word, when the sensitivity of an SP to the price of its competitor is zero ($\alpha_1^2 = \alpha_2^1 = 0$), price of anarchy converges to 1 and so the equilibrium is approximately socially optimal.

Influence of β (Sensitivity of SP-i to his QoS q_i) : Figure 7 illustrates variations of PoA as a function of, β, which is the sensitivity of SPs to their respective own QoS. We first notice that the loss of efficiency is around 8%. Moreover the curve of PoA is concave, this latter mean that there are some, $\beta^* < 1$, which optimizes the equilibrium, ($\beta^* = \beta_1^1 = \beta_2^2 = 0.76, \quad PoA^* = 0.925$). Surprisingly, the price of anarchy varies slightly (variation of almost 0.001). To explain this behaviour, Figures 8 and 9 depict, respectively, the curves of equilibrium Price and QoS of SP-1 and SP-2. We find that the induced variation of the price is much higher compared to that of QoS, and subsequently, β (Sensitivity of SPs to their QoS) has a smaller impact on the system.

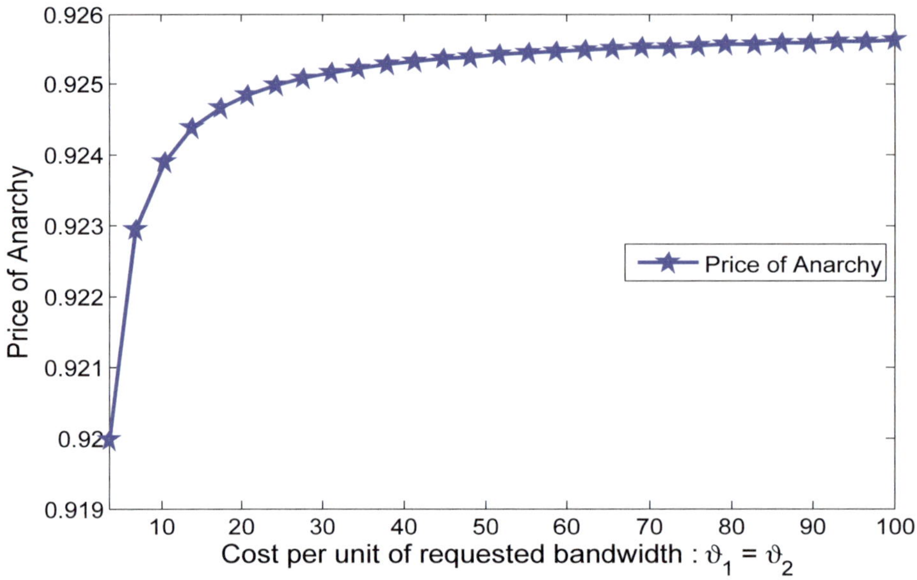

Figure 5. Price of Anarchy as a function of cost per unit of requested bandwidth ϑ_i.

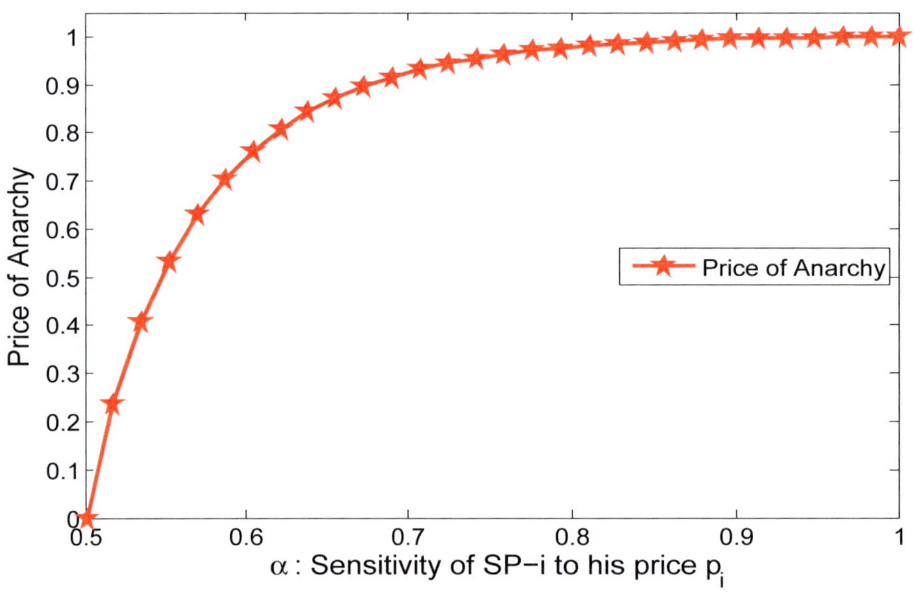

Figure 6. Price of Anarchy as a function of $\alpha = \alpha_1^1 = \alpha_2^2$ (Sensitivity of SP-i to his price p_i)

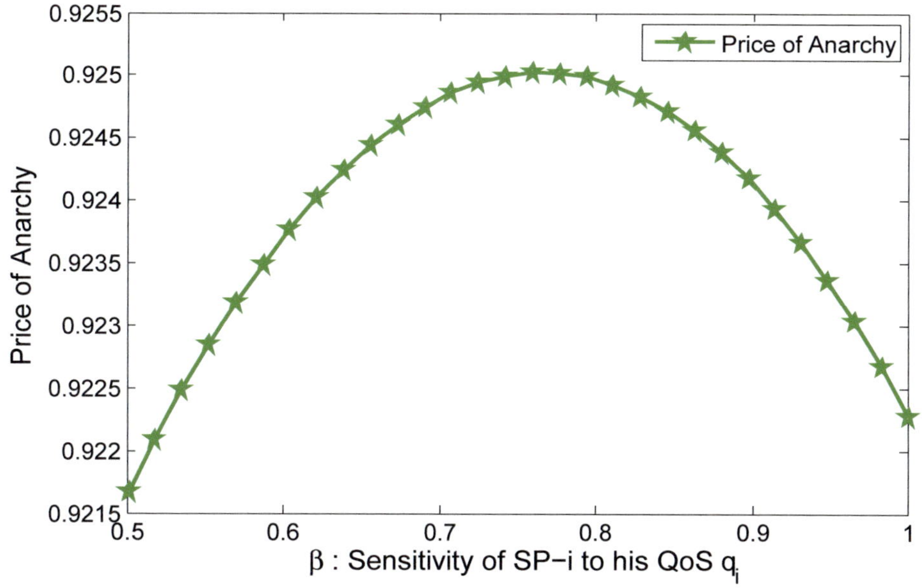

Figure 7. Price of Anarchy as a function of $\beta = \beta_1^1 = \beta_2^2$ (Sensitivity of SP-i to his QoS q_i)

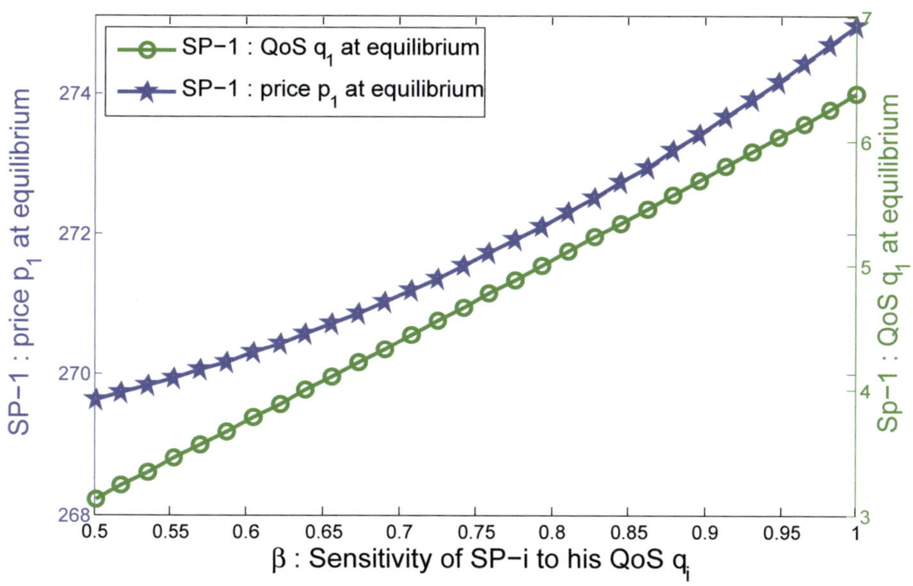

Figure 8. equilibrium Price and QoS of SP-1 as a function of $\beta = \beta_1^1 = \beta_2^2$ (Sensitivity of SP-i to his QoS q_i)

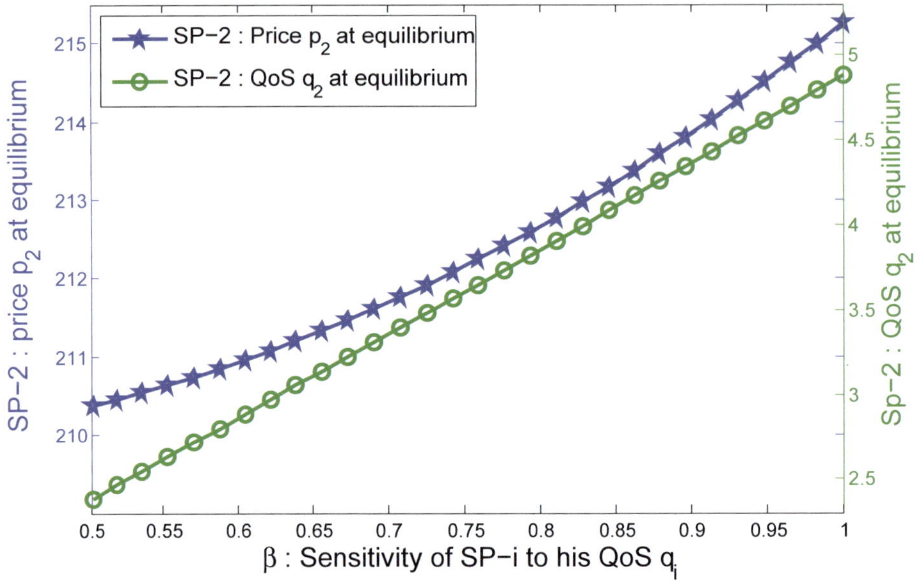

Figure 9. equilibrium Price and QoS of SP-2 as a function of $\beta = \beta_1^1 = \beta_2^2$ (Sensitivity of SP-i to his QoS q_i)

5. Conclusion

In this work, we presented and analyzed a framework to model the complex interactions among SPs as players through a class of two parameter Nash equilibrium models. The model is based on a simple linear demand functions which describe customer behaviour, take into account not only the characteristics of a current SP, (SP-i), but also of all other SPs, (SP-j, $j \neq i$), the presence of two parameters describing each SP's service price and QoS level. We established uniqueness of a Nash equilibrium point and developed a distributed algorithm to learn it. Then, our proposed algorithm finds very fast the equilibrium price and the equilibrium QoS to be chosen by each provider. Our scheme is different from previous approaches since it involves two varying parameters in a simple implementation and low complexity. Yet, we have obtained some insightful results such as the interplay of bandwidth cost. Results found in this work can be further extended to general network considerations, in particular under non-neutrality perspective or non-linear demand.

Author details

Mohamed Baslam and El-Houssine Bouyakhf
LIMIARF, University of Mohammed V, Faculty of Sciences, Rabat, Morocco

Rachid El-Azouzi
LIA-CERI, University of Avignon, Avignon, France

Essaid Sabir
RTSE Laboratory, GREENTIC/ENSEM, Hassan II University, Casablanca, Morocco

Loubna Echabbi
National Institute of Post and Telecommunication, Madinat Al-Irfane, Rabat, Morocco

6. References

[1] Altman, E. & Wynter, L. [2002]. Equilibrium, games, and pricing in transportation and telecommunications network, *submitted to the special issue of "Networks and Spacial Economics", on "Crossovers between Transportation Planning and Telecommunications* .
URL: *http://www-sop.inria.fr/mistral/personnel/Eitan.Altman/ntkgame.html*

[2] Altman, E. & Wynter, L. [2004]. Equilibrium, games and pricing in transportation and telecommunications networks, *Networks and Spacial Economics, special issue of on: Crossovers between Transportation Planning and Telecommunications* 4: 7–21.

[3] Anshelevich, E., Dasgupta, A., Tardos, E. & Wexler, T. [2008]. Near-optimal network design with selfsh agents, *Theory of Computing* 4: 77–109.

[4] Awerbuch, B., Azar, Y. & Epstein, A. [2005]. The price of routing unsplittable flow, *In Proc. 16th Symp. Theoretical Aspects of Computer Science (STACS)* pp. 331–337.

[5] Baslam, M., Azouzi, R. E., Sabir, E. & Echabbi, L. [2011]. Market share game with adversarial access providers : A neutral and a non-neutral network analysis, *Proc. of "NetGCOOP, Paris, France* .

[6] Baslam, M., Echabbi, L., Azouzi, R. E. & Sabir, E. [2011]. Joint price and qos market share game with adversarial service providers and migrating customers, *Proc. of GameNets, Shanghai, China* .

[7] Bernstein, F. & Federgruen, A. [n.d.]. general equilibrium model for decentralized supply chains with price and servicecompetition.
URL: *http://faculty.fuqua.duke.edu/fernando/bio/*

[8] Eidenbenz, S., Kumar, A. & Zust, S. [2006]. Equilibria in topology control games for ad-hoc networks, *Mobile Networks and Applications* 11: 143–159.

[9] El-Azouzi, R., Altman, E. & Wynter, L. [2003]. Telecommunications network equilibrium with price and quality-of-service characteristics, *Proc. of ITC, Berlin* .

[10] Friedman, J. [1996]. The rational choice controversy, *Yale University Press* .

[11] Goodman, J. C. [1980]. A note on existence and uniqueness of equilibriumpoints for concave n-person games, *In Econometrica* pp. 48–251.

[12] Guijarro, L., Pla, V., Vidal, J. & Martinez-Bauset, J. [2011]. Analysis of price competition under peering and transit agreements in internet service provision to peer-to-peer users, *IEEE Consumer Communications and Networking Conference (CCNC2011), Las Vegas, Nevada USA* pp. 9 – 12.

[13] Halldôrsson, M., Halpern, J., Li, L. & Mirrokni, V. [2004]. On spectrum sharing games, *In Proc. 22nd Symp. Principles of Distributed Computing (PODC)* pp. 107–114.

[14] Kelly, F. [1997]. Charging and rate control for elastic traffic, *European Trans. Telecommunications* 8: 33–37.
URL: *http://www.statslab.cam.ac.uk/frank/elastic.html*

[15] Low, S. & Lapsley, D. [1999]. Optimization flow control, i: Basic algorithm and convergence, *IEEE/ACM Transactions on Networking* 7: 961Ű874.

[16] Maillé, P., Naldi, M. & Tuffin, B. [2009]. Price war with migrating customers, *In: 17th Annual Meeting of the IEEE/ACM International Symposium on Modelling, Analysis and*

Simulation of Computer and Telecommunication Systems (MASCOTS 2009), IEEE Computer Society, London, UK .

[17] Maille, P. & Tuffin, B. [2008]. Analysis of price competition in a slotted resource allocation game, *in Proc. of IEEE INFOCOM* .

[18] Orda, A. [1993]. Competitive routing in multi-user environments, *IEEE/ACMTrans. on Netw.* pp. 510–521.

[19] Papadimitriou, K. & Koutsoupias, E. [1999]. Worst-case equilibria, *in STACS* pp. 404–413.

[20] Rosen, J. [1965]. Existence and uniqueness of equilibrium points for concave n-person games, *Econometrica* 33: 520–534.

[21] Varian, H. [1992]. Microeconomic analysis, *Norton New York* .

[22] Vetta, A. [2002]. Nash equilibria in competitive societies with application to facility location, *In Proc. 43th Symp. Foundations of Computer Science (FOCS)* p. 416.

[23] von Neumann, J. & Morgenstern, O. [1944]. Theory of games and economic behavior, *Princeton University Press* .

[24] Wynter, L. [2001]. Optimizing proportionally fair prices, *INRIA RR. 4311* .
URL: *http://www.inria.fr/rrrt/rr-4311.html*

Chapter 8

Cooperative Game Theory and Its Application in Localization Algorithms

Senka Hadzic, Shahid Mumtaz and Jonathan Rodriguez

Additional information is available at the end of the chapter

http://dx.doi.org/10.5772/53930

1. Introduction

Game theory is a field of applied mathematics for analyzing complex interactions among entities. It is basically a collection of analytic tools that enables distributed decision process. Game theory (GT) provides insights into any economic, political, or social situation that involves individuals with different preferences. GT is used in economics, political science and biology to model competition and cooperation among entities, and the role of threats/punishments in long term relations. Contemporary social science is based on game theory, economics, and psychology in which mathematical logic is applied. The formation of coalitions or alliances is omnipresent in many applications. For example, in political games, parties, or individuals can form coalitions for improving their voting power. Recently, computer science and engineering have been added to the list of scientific areas applying GT.

While in optimization theory the goal is to optimize a single objective over one decision variable, game theory studies multi-agent decision problems. In social sciences and economics, the focus of game is the design of right incentives/payoffs; in engineering it comes to efficiency – how to design efficient decentralized schemes that take into account incentives. However, there are still similarities when applying game theory to different disciplines. For example, a measurement allocation framework for localization in wireless networks, based on the idea to allocate more measurements to the nodes which contribute more, mimics a capitalist society where the gains are mostly reinvested where more profit is expected. It also replicates the concept of natural selection in population genetics.

In general, a game consists of a set of players (decision makers), while each player has its strategy, whereby utility (payoff) for each player measures its level of satisfaction. Each player's objective is to maximize the expected value of its own payoff (Myerson, 1997).

(Srivastava V., et all, 2005) proposed a mapping of network components to game components according to the following table:

Network component	Game component
Nodes	Players
Available adaptations	Action set
Performance metrics	Utility function

Table 1. Classification of coalitional games

Game theory can be applied to communication networks from several aspects: at the physical layer, link layer and network layer. However, there a certain challenges when applying game theory principles to wireless networks. For example, GT assumes that the players act rationally, which does not exactly reflect real systems. Furthermore, realistic scenarios necessitate complex models, yet the main challenge is to select the appropriate utility function, due to a lack of analytical models that would map each node's available actions to higher layer metrics.

1.1. Notation

A normal form representation of a game is given by $G = \langle N, S_i, \{u_i\} \rangle$, where $N = \{1,\ldots,n\}$ is the set of n of players. We indicate an individual player as $i \in N$ and each player i has an associated set $S_i = \{s_i^1,\ldots,s_i^m\}$ of possible strategies from which, in a pure strategy normal form game, it chooses a single strategy $s_i \in S_i$ to be realized. $\mathbf{s} = \{s_1,\ldots,s_N\}$ is the strategy profile of N players, i.e., the outcome of the game, while s_{-i} is the strategy profile of all players but the i-th, and $\{u_i\} = \{u_1,\ldots,u_N\}$ is the utility function of the i-th player. The utility function measures the preferences of each player to a given strategy, assuming the strategies of other players are known. If $\mathbf{s}$ is a strategy profile played in a game, then $u_i(\mathbf{s})$ denotes a payoff function defining i's payoff as an outcome of $\mathbf{s}$.

There are two main branches of game theory: cooperative and non-cooperative. Non-cooperative GT addresses interactions among individual players, each aiming to achieve their own goal, namely improving its utility, or reducing its costs. Specifically, in cooperative games the utility does not only depend on a single node's strategy, but also on the strategies of other nodes within a coalition. Hence, cooperative game theory is more elaborate. Especially in realistic situations where entities can participate in several coalitions, the potential structure of these coalition allocations is more complex; thus there is a need to for concepts that could reduce the complexity, without identifying and comparing all of the $2^n - 1$ possible coalitions.

One of the concepts for solving non-cooperative games is the Nash equilibrium. Nash equilibrium is a stable solution of the game such that no player has reason to unilaterally change its action, since it may not improve its utility function. More precisely, a strategy profile set $\mathbf{s}^* = \{s^*_1,\ldots,s^*_N\}$ is a NE if for $\forall s_i \in S_i$ and for $\forall i \in N$, $u(s^*_i, s^*_{-i}) \geq u(s_i, s^*_{-i})$. A strategy set that corresponds to the Nash equilibrium signifies a consistent prediction of the outcome of the game. In other words, if all players predict that Nash equilibrium will occur, there is no player in the game that has incentives to choose a different strategy. Any game allowing mixed strategies has at least one NE. However, some pure strategy normal form games may not have a NE solution at all. Therefore it is relevant to formulate the utility function in such a way that the game has at least one equilibrium point.

When efficiency is important, Pareto Optimality is used. The existence of Nash Equilibrium does not assure that the outcome of a game will be beneficial for all players. Mathematically formulated, a strategy set $\mathbf{s} = \{s_1,\ldots,s_N\}$ is Pareto optimal if and only if there exists no other strategy set $\mathbf{t} = \{t_1,\ldots,t_N\}$ such that $u_i(\mathbf{t}) \geq u_i(\mathbf{s})$ for $\forall i \in N$, and for some $k \in N$, $u_k(\mathbf{t}) > u_k(\mathbf{s})$

In other words, Pareto optimal outcome cannot be improved upon without hurting at least one player.

In this chapter we will focus on cooperative game theory and its application in localization algorithms.

2. Coalitional games in wireless communications

A coalition formation game is uniquely defined by the pair $(\mathcal{N}, v)$. $\mathcal{N} = \{1,2,\ldots,N\}$ denotes the set of players, e.g., network entities, pursuing to form sets in order to collaborate with each other. Any nonempty subset $S \in \mathcal{N}$ is called a coalition. Coalitions with cardinality $|S| = 1$, are called singleton coalitions and $\mathcal{N}$ is called the grand coalition. The set of all coalitions in a game is called coalition structure and is denoted by $\mathcal{P}$. v denotes the coalition value which quantifies the worth of a coalition in a game.

2.1. Coalitional games – background

Coalitional games in characteristic form are classified into two types based on the distributing of gains among users in a coalition:

i. A transferable utility (TU) game where the total gain achieved can be apportioned in any manner between the users in a coalition subject to feasibility constraints, and
ii. A non-transferable utility (NTU) game where the apportioning strategies have additional constraints that prevent arbitrary apportioning. Each payoff is dependent on joint actions within coalition.

In TU games, the cooperation possibilities of a game can be defined by a characteristic function v that assigns a value $v(S)$ to every coalition S. Here $v(S)$ is called the value of coalition S, and it characterizes the total amount of transferable utility that the members of S could gain without any help from the players outside of S. In general, we use the term coalition structure to refer to any mathematical structure that describes which coalitions (within the set of all $2^n - 1$ possible coalitions) can effectively negotiate in a coalitional game.

The overall goal is to find a coalition structure such that no group of players has the incentive to leave it – so called stable coalition structure. Superadditivity is defined in TU games as a property of the characteristic function:

$$v(S_1 U\, S_2) \geq v(S_1) + v(S_2);\ \forall S_1,\, S_2 \in N, S1 \cap S2 = \varnothing. \quad (1)$$

In other words, a TU game is superadditive if cooperation is always rewarding. Thus, grand coalition, i.e., the coalition comprising all sensors, is beneficial. The most notable solution concept for the coalition formation in superadditive games is the core; other solutions include Shapley value, kernel, and Nucleolus.

The superadditivity concept can be extended to NTU games, by:

$$\{\mathbf{x} \mid (x_i)_{i \in S1} \in v(S_1), (x_j)_{j \in S1} \in v(S_2)\} \subseteq v(S_1 \, U \, S_2) \tag{2}$$

In case of TU games, goal is to find a coalition structure that maximizes the total utility, while in NTU games it is the structure with Pareto optimal payoff distribution. A centralized approach can be used, but it is generally NP-complete. The reason is that finding an optimal partition requires iterating over all the partitions of the player set N. The number of partitions grows exponentially with the number of players in N. For example, for a game where N has 10 elements, the number of partitions that a centralized approach has to go through is 115,975 (easily computed through the Bell number (Saad W., et all, 2009c). Therefore, using a centralized approach for finding an optimal partition is, generally, computationally complex and not very practical. Nevertheless, many applications require the coalition formation process to take place in a distributed manner, so that the players have autonomy on the decision whether or not to join a coalition. Indeed, the complexity of the centralized approach has initiated a growth in the coalition formation literature, with the goal to find low complexity and distributed algorithms for establishing coalitions.

A novel classification of coalitional games has been proposed in (Saad W., et all, 2009c). Games are grouped into three types: canonical games, coalition formation games and coalitional graph games. Their properties are shown in the following table.

Canonical coalitional games	Coalition formation games	Coalitional graph games
Grand coalition is the optimal structure	Resulting coalitional structure depends on gains and costs	Interaction of players depends on communication graph structure
Goal: stabilize the grand coalition	Goal: form appropriate coalition structure	Goal: stabilize grand coalition or form network topology taking into account the communication graph

Table 2. Classification of coalitional games

In this chapter we will focus on coalition formation games. A generalized approach to coalition formation has been proposed in (Apt & Witzel, 2006). The notion of stable partition is used when there does not exist any other partition that would improve the total gain. In order to illustrate the coalition formation procedure, an abstract preference operator $\triangleright$ has been introduced, and coalitions are being transformed using merge and split rules.

2.2. Applications to communication networks

From the communication networks perspective, there is the need for developing distributed and flexible wireless networks, where the units make independent and rational strategic decisions. In addition, low complexity distributed algorithms are required, to capably represent collaborative scenarios between network entities. Non-cooperative games have been mainly applied for applications such as spectrum sharing, power control or resource allocation – mainly settings that can be seen as competitive scenarios. On the other hand, cooperative game theory provides analytical tools to study the behavior of rational players in cooperative scenarios. In particular, coalitional games show to be a very powerful tool for designing fair, efficient and robust cooperation strategies in communication networks. In order to highlight an expanding application field, in the following section we will give some examples on use of cooperative game theory for communication networks, and specifically for localization purposes.

Physical layer security has been studied via coalitional games in (Saad W., et all, 2009a), (Saad W., et all, 2009b). In a distributed way, wireless users organize themselves into coalitions (see Figure 1.) while maximizing their secrecy capacity - maximum rate of secret information sent from a wireless node to its destination in the presence of eavesdroppers (Saad W., et all, 2009a). This utility maximization is taking into consideration the costs occurring during information exchange. On the other hand, (Saad W., et all, 2009b) introduces a cooperation protocol for eavesdropper (attacker) cooperation. Here the utility function is formulated to capture the damage caused by the attackers, and the costs in terms of time spent for communication among the eavesdroppers. In both cases, independent disjoint coalitions will form in the network, as the grand coalition would involve various communication costs.

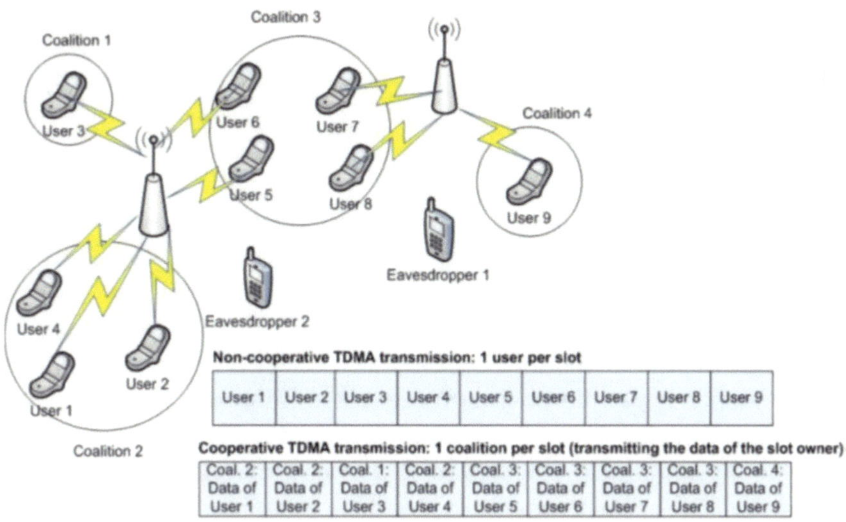

Figure 1. Wireless users organized into coalitions

(Mathur S., et all, 2006) and (Mathur S., et all, 2008) consider coalition structures in a wireless network where users are permitted to cooperate, while maximizing their own rates.

Here both transmitter and receiver cooperation in an interference channel is studied. Several models have been analyzed: a TU and an NTU model, and with perfect and partial cooperation. In (Mathur S., et all, 2006), the feasibility and stability of the grand coalition for all cases was evaluated, while the work in (Mathur S., et all, 2008) is focused on stable coalition structures. In (Saad W., et all, 2008) a game theoretical framework for virtual MIMO has been proposed, where single antenna transmitters self-organize into coalitions. The utility function denotes the total achieved capacity, and also includes the power constraint to account for the costs.

In (Hao X., et all, 2011) the multi-channel spectrum sensing problem is formulated as a coalitional game, where players are secondary users that cooperatively sense the licensed channels of primary users. The utility of each coalition reflects the sensing accuracy and energy efficiency. Distributed algorithms have been proposed to determine a stable coalition structure, maximizing the overall utility in the system. More game theory based solutions for spectrum sensing in cognitive radio have been proposed in (Khan Z., et all, 2010) and (Saad W., et all, 2009c).

A network-level study using coalition formation has been performed in (Singh C., et all, 2012), considering a scenario where service providers are cooperating in order to enhance the usage of the available resources. Particularly, different providers may serve each other's customers and thereby increase the throughput and reduce the overall energy consumption. The model supports multi-hop networks and is not limited to stationary users and fixed channel conditions. A game theory based framework is used to determine optimal decisions and a rational basis for sharing the aggregate utility among providers. The optimal coalition structure can be obtained by means of convex optimization.

Other applications of game theory include packet forwarding in ad hoc networks, distributed cooperative source coding, routing problems, and localization algorithms, which will be more elaborated in the next chapter.

3. Game theory for localization algorithms

The expansion and enhancement of wireless and mobile devices has aroused the demand of context-aware applications, in which location is often viewed as one of the most important contexts. Those applications include pervasive medical care, wireless sensor network surveillance, mobile peer-to-peer computing etc. The essential purpose of wireless sensor networks (WSN) is to provide information about observed events. Before the WSN can be exploited for various applications, knowledge about sensors' locations is crucial, as otherwise the data might become meaningless. Furthermore, location information can be used to improve the communication system itself. Geo-location information can serve as complementary data to estimate and predict critical parameters for improving wireless communication networks, such as setting up location dependent load balancing schemes (Yanmaz E. & Tonguz O.K.). Several studies have shown how the efficiency of available radio resources can be improved by the availability of position information to provide accurate scheduling and link adaptation (Tang S., et all, 2009), or even the prediction of required resources in a highly dynamic scenario. Additionally, localizing the nodes can help reduce power consumption in multi hop wireless networks.

Global navigation satellite systems (GNSS), such as Global Positioning system (GPS) or the European satellite navigation system Galileo, are providing positioning information. However their accuracy strongly depends on the scenario. Especially in dense urban or indoor environments, navigation based on GNSS becomes inaccurate or impossible, since the necessary amount of 4 directly visible satellites is not reached. In order to provide accurate MT position estimation, the MT position shall be estimated with alternative techniques focusing on radio signals which are provided by the terrestrial RANs itself. The rapid deployment of WLAN and WPAN technologies, especially in dense indoor environments, made it another compelling choice for localization, relying only on the existing network infrastructure.

Generally, the localization process assumes a number of location aware nodes, called anchors. In a typical two-stage positioning system, the first phase is the ranging phase, where nodes estimate the distances to their neighbors by observing time of arrival, received signal strength or some other distance dependent signal metric. In the second phase, nodes use the ranging information and the known anchor position for calculation of their coordinates.

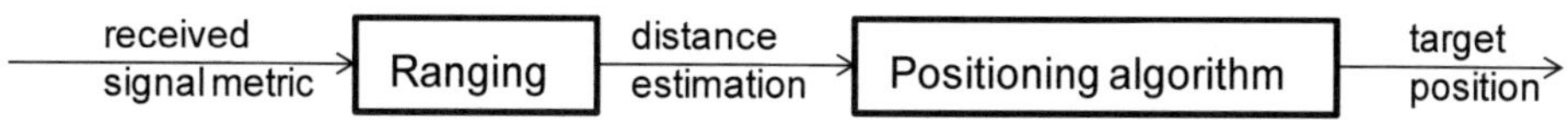

Figure 2. Two-stage positioning system

One simple way for position calculation is trilateration / triangulation, based on the least square algorithm. Trilateration uses distance estimates to anchor nodes as input, and estimates target's position based on geometric properties of triangles. Each estimated distance represents the radius of a circle centered at the corresponding reference node. For 2-D positioning, measurements from at least three reference nodes are required, and the location is obtained as the intersection of circles. This method is also used for GPS. Having in mind the errors in estimated distances to the anchors, the geometrical trilateration technique can only provide a region of uncertainty, instead of a single point. Therefore the solution is based on iterative algorithms to obtain the node position by formulating and solving a set of nonlinear equations.

The availability of positioning information depends on the existing infrastructure such as GPS satellites or base stations. Cooperative positioning techniques are used in scenarios where non-cooperative solutions are not feasible, or do not perform well in terms of accuracy, cost and complexity. The challenge is to allow nodes which are not in range of a sufficient number of anchors to be located, and hereby increase localization performance in terms of both accuracy and coverage. This can be achieved by means of iterative multilateration, among other solutions. Iterative multilateration is a way to expand localization coverage throughout the network in a step-by-step fashion, allowing also nodes which are not in range of a sufficient number of references to be localized. In this sense, coverage is the fraction of nodes that have an accurate position estimate. It follows an iterative scheme: once an unknown node estimates its position, it becomes an anchor and broadcasts its position estimate to all neighboring nodes. The process is repeated until all nodes that can have three or more reference nodes obtain a position estimate. As a newly localized node is becoming new anchor for its neighbors, the estimation error of the first node can propagate to other nodes and eventually get amplified. Over iterations the error could spread throughout the network, leading to abundant error in large topologies.

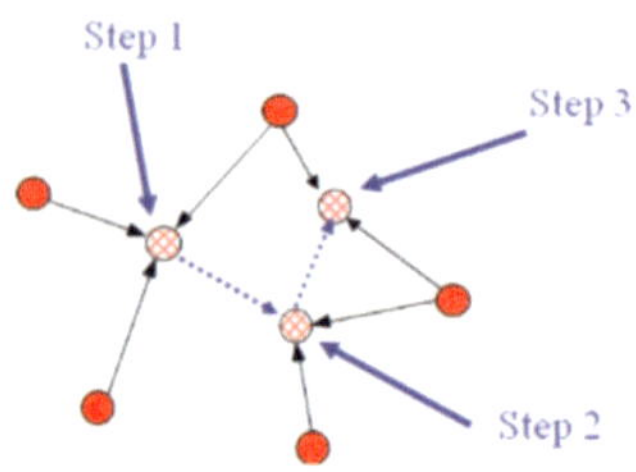

Figure 3. Iterative multilateration

The number of actively participating nodes should be kept to a minimum, and therefore an appropriate cooperation subset has to be chosen, while the other nodes can be ignored. Such a restrictive and selective use of references is crucial in networks with limited resources. A frequently used method is to select the nearest k anchor nodes. However, this method does not take into account node geometry. Therefore other metrics such as geometric dilution of precision, Cramer Rao lower bound or stochastic observability are more appropriate.

The geometric conditioning on localization accuracy is derived in the GDOP (geometric dilution of precision) metric (Spirito M.A., 2001). In brief, when reference nodes are well separated around the target, the GDOP is lower.

Localization can be defined as an estimation problem where measurements like wireless signal strength, angle or time of arrival are provided to an estimator (i.e. the localization algorithm) to obtain the most likely position in the assumed coordinate system. The Cramer Rao Lower bound (CRLB) provides a lower bound on covariance of any unbiased estimator. In case of localization, the CRLB captures information about node geometry and ranging quality, i.e., quality of distance estimates obtained from noisy measurements of received signal strength (RSS), time of arrival (TOA) or angle of arrival (AOA) (Patwari N., et all, 2003). Since the variance of position estimates is associated to the mean error, the lower bound on variance can be seen as the upper bound on accuracy.

3.1. Use of game theory in localization algorithms

Recently game theory has been applied in localization algorithms, mainly for modeling the cost-performance trade-off and for selection of reference nodes. The work in (Ghassemi F. & Krishnamurthy V., 2008a) applies game theory for sensor network localization, namely for measurement allocation among reference nodes localizing the target. The localization process has been modeled as a game belonging to the class of weighted-graph games. For such a representation, the vertices correspond to the players, and the coalition value can be obtained by summing the weights of the edges that connect a pair of vertices in the coalition with self-loop edges only considered with half of their weights. A weighted-graph game can therefore be well represented by $\frac{N(N-1)}{2}+N$ weights, in contrast to 2N numbers which are usually required to represent a cooperative game. Basic idea is to allocate more measurements to nodes that contribute more to the localization process. The allocation algorithm has been integrated into a Bayesian estimator. In (Ghassemi F. & Krishnamurthy V., 2008b), utility is defined as information gain from a node, i.e. the mutual information between the prior density of target position and the measurement. Additionally, a price for transmission is included to account for the current energy level in the nodes, and the energy needed for data transmission.

The algorithm proposed in (Moragrega A., et all, 2011) assumes a number of static anchor nodes, strategically placed to guarantee coverage to all unknown nodes. Anchors transmitting with lower energy can provide coverage to a smaller number of nodes; aim is to minimize power consumption at the anchor nodes, while assuring desired localization accuracy. The metric for positioning quality is the GDOP. The problem has been formulated as a noncooperative game, using Nash equilibrium as solution concept

In (Bejar B., et all, 2010) the coalition formation within the set of neighboring anchors helps reduce communication costs. Using only a subset of available reference nodes does not necessarily degrade the accuracy, since some of them provide redundant information. In some situations it might be even useful to discard ranging information from some reference nodes, after they have been identified as unreliable due to biases in the measurements. This paper the localization problem has been defined as a coalitional NTU game, where coalitions are formed based on the merge and split procedure. The utility function is defined to account for both a quality and cost indicator. While the quality function accounts for inconsistencies between each node's measured distance and the final joint estimated distance within the coalition, the cost function is related to communication costs. The target tracking task based on coalition formation has been implemented using a Kalman filter. For the coalition formation approach a higher mean estimation error has been observed than for grand coalition, i.e., when all nodes contribute to the tracking process. Nevertheless, in terms of communication costs the proposed scheme provides significant savings.

(Ghareshiran O. N. & Krishnamurthy V., 2010) proposes a dynamic coalition formation algorithm used for energy saving in multiple target localization. Assuming that nodes in sleep mode do not record any measurements and thereby save energy in both sensing and transmitting data, the optimization problem is formulated to maximize the average sleep time of all nodes in the network, assuring that targets are localized with desired accuracy. An important contribution is exploitation of spatial correlation of sensor readings. Accuracy metric used is the determinant of the Bayesian Fisher information matrix (B-FIM). The characteristic function is formulated in a way that larger coalitions of sensors do not necessarily lead to longer sleep times. This is mainly due to the fact that the B-FIM, depending on both relative angles and distances of sensors to the target, does not automatically increase as the number of sensor nodes in a coalition goes up. The trade-off between performance and average sleep time allocated in the network is demonstrated via Monte Carlo simulations.

4. Scenario

We propose the use of cooperative non-superadditive games for modeling localization algorithms. As stated in the previous section, a typical localization process consists of the ranging phase, where nodes estimate the distances to their neighbors, and a second phase where nodes use the ranging information and the known anchor position to calculate their coordinates. In a dense network one can assume a large number of available anchor nodes. However, transmitting and processing all the obtainable information would consume immense power, without necessarily leading to better localization performance. This is due to the fact that not all the anchors provide reliable measurements, what leads to erroneous distance estimates. Furthermore, the geometry of selected reference nodes shows to have significant impact on localization accuracy, what will be extensively elaborated in our work. Assuming that at each time instant a target has several neighboring anchor nodes in near vicinity, and different coalitions can be formed, the considered scenario is illustrated in Fig.4.

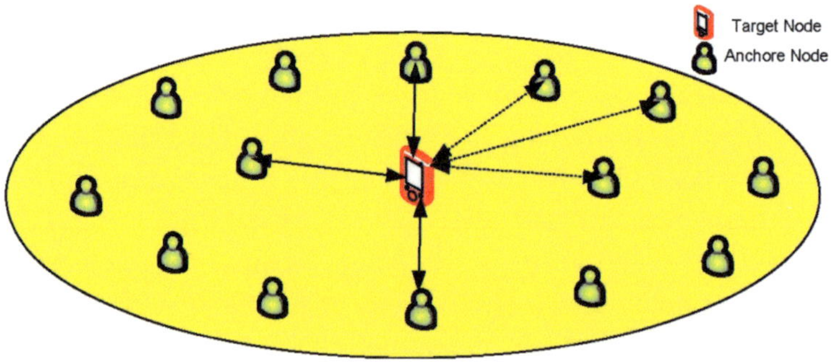

Figure 4. Scenario

We propose an algorithm for reference node selection based on coalitional games. We model the localization process as a cooperative game, and formulate the corresponding utility function. We define the node selection optimization as one that maximizes the accuracy subject to constraints given by nodes' limited processing capacity. Position estimates are obtained using the linearized least squares algorithm (trilateration).

4.1. Ranging error

We assume that the distance estimates between nodes are obtained using RSS measurements. We use the standard lognormal model for RSS with path loss parameter n_p and shadowing variance σ^2_{RSS}. Assuming that the received power $P_{i,j}$ between nodes i and j is lognormal, the random variable $P_{i,j}$ (dBm) $=10\log P_{i,j}$ is Gaussian. RSS based distance estimates are obtained using the lognormal model:

$$\tilde{d}_{i,j} = d_{i,j} * 10^{\frac{v_{i,j}}{10n_p}} \quad (3)$$

where $v_{i,j} \sim N(0, \sigma^2_{RSS})$ and n_p is the path loss exponent. We used values for indoor scenarios $n_p = 2.3$ and $\sigma^2_{RSS} = 3.92$ dB as in (Patwari N., et all, 2003).

4.2. Utility function

The following parameters are relevant for reference node selection: number of references, quality of range estimates and geometry. Therefore we propose a node selection mechanism based on the Cramer Rao Lower Bound. Since the CRLB gives the upper bound on accuracy, the utility function has to be inversely proportional to the CRLB. Besides the quality indicator, utility function also has to reflect the cost. Cost is related to the energy spent for message exchanges between nodes, and is proportional to the distances of target node to reference nodes. Having in mind the energy consumption if all reference nodes were used for localization, the grand coalition is not optimal. Therefore we define the problem as a nonsuperaditive cooperative game. Since least square localization is not possible for less than three reference nodes, we set the value of all coalitions containing less than three nodes to zero. For the remaining ones, the coalition value of each chosen subset of nodes S will be of the form:

$$v(S) = \frac{1}{CRLB_{i\in S}} - \sum_{i\in S} \frac{d_{i,t}}{R} \quad (4)$$

Where $CRLB_{i\in S}$ is the CRLB for the coalition S, $d_{i,t}$ is the distance from node i ∈ S to the target *t*, and *R* is the transmission range, used to normalize the cost function. In order to illustrate the performance of coalition formation based node selection, we will perform an exhaustive search over all possible coalition sets containing three nodes. The results are presented in the next section.

5. Results

In this section we show through simulations how localization performance can be improved using cooperative game theory. Performance metrics are accuracy, complexity and latency.

Accuracy is evaluated as the Euclidean distance between the estimated position, and the node's true location. Complexity is especially important in scenarios with low-power devices. In cellular scenarios computation is mainly performed in a central manner, e.g., at the base station with power supply, computational and processing complexity is not necessarily a limitation. In case of a moving target, its position needs to be updated with a frequency depending on the mobility model. Therefore it is important for the position calculation to be fast. We evaluate the localization accuracy as the root mean square error (RMSE) of location estimates. Complexity is assessed by means of amount of computation that has to be performed, while latency refers to the time needed to get a position estimate – particularly important for dynamic scenarios.

In our simulations we assume that the target node has a number of reference nodes in local vicinity, uniformly distributed within a 20 by 20 meters region. We show how appropriate selection of reference nodes outperforms the random selection, for cases of 10, 15 and 20 available references. We performed simulations for different node densities and compared them in terms of root-mean-square error (RMSE). We performed 1000 runs for each setup.

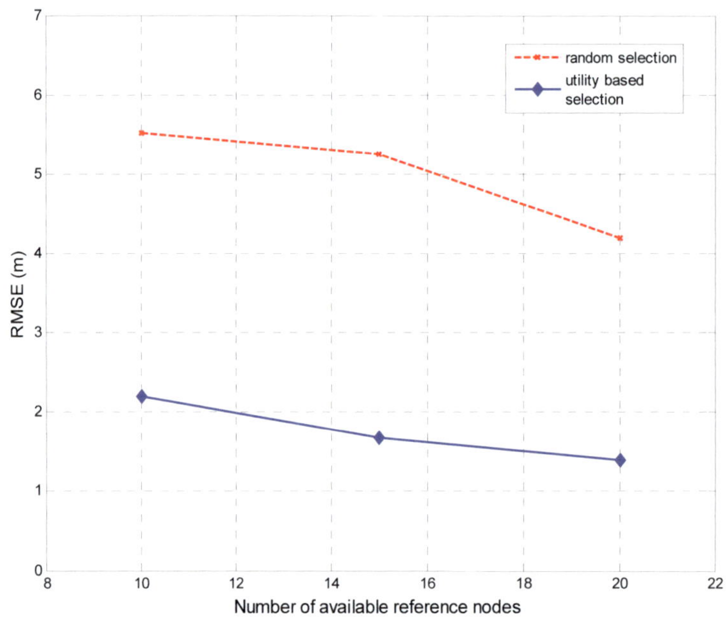

Figure 5. Localization accuracy for random anchor selection and utility based selection.

The following scatter plot illustrates how the coalition value reflects the localization error.

In Fig.6 we assumed 10 available reference nodes. Besides accuracy, we will assess the complexity of the algorithm depending on the number of available reference nodes, namely considering sets of 10, 15, 20, 25 and 30 nodes, respectively. From each of these sets, three anchors providing the best results are chosen. We define computational complexity as the amount of time spent on localization, in this case on a simulation run. The measurement of computation time is calculated using MATLAB functions *tic* and *toc*, which return the elapsed time in seconds. Knowing that combinatorial complexity increases with number of elements, Fig. 7 shows the expected result, namely significantly higher complexity as the number of references increases.

Since we consider a static scenario, the latency factor is not of particular significance. However, one can consider the computation time in Fig. 7 as a latency parameter as well.

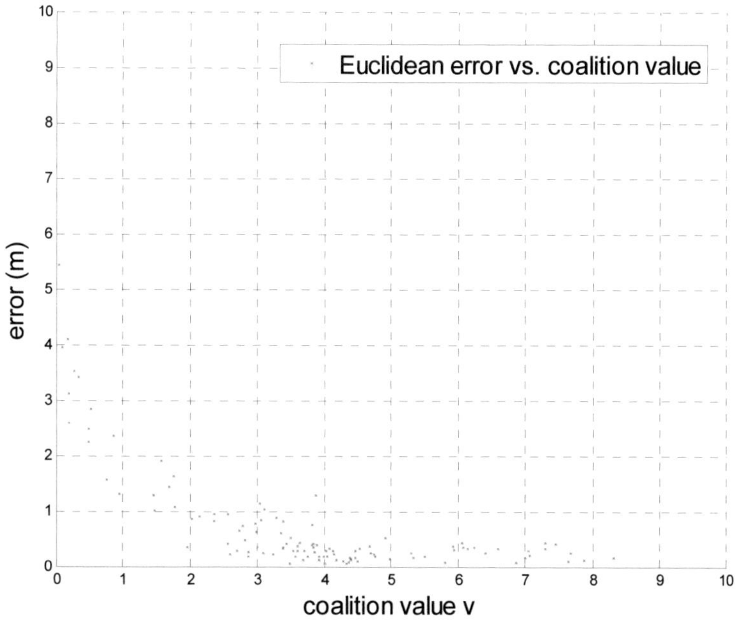

Figure 6. Localization error vs. coalition value

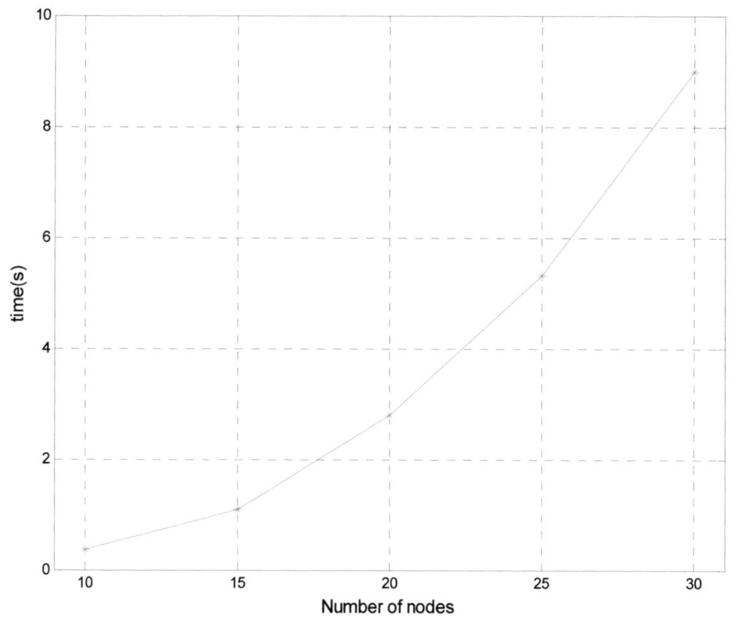

Figure 7. Computation time.

6. Conclusion

In this chapter we considered the application of coalitional games to communication networks, in particular to localization algorithms. Game theory proves to be a powerful tool for modelling various aspects of localization procedure, such as improved accuracy or energy saving. After giving an overview of the most significant contributions in the literature on this subject so far, we have proposed a localization procedure aiming to improve accuracy by selecting the references providing the best conditions in terms of channel conditions and node geometry. Besides providing better performance, choosing only a subset of available references contributes to resource saving. This is particularly important in wireless sensor networks, having in mind the nature of these networks, specifically the limited resources such as energy constraints, processing capacity and short transmission range.

The selection procedure is based on coalitional game theory. We proposed a utility function that reflects the contribution of each coalition to the localization accuracy, as well as a cost function related to energy consumption during the localization procedure. We compared the performance of utility based node selection vs. a random selection scheme. Even though the computational complexity significantly increases for a large number of available references, the achieved accuracy improvements make it a compelling concept for node selection.

Author details

Senka Hadzic, Shahid Mumtaz and Jonathan Rodriguez
University of Aveiro, Instituto de Telecomunicações, Portugal

Acknowledgement

This work has been performed in the framework of the ICT project ICT-248894 WHERE2, which is partly funded by the European Union. The work has been supported in part by the Portuguese Foundation for Science and Technology (Fundação para a Ciência e Tecnologia - FCT) under grant number SFRH / BD / 61023 / 2009.

7. References

Apt K. & Witzel A. (2006). A Generic Approach to Coalition Formation, *Proceedings of International Workshop Computational Social Choice (COMSOC)*, 2006.

Bejar B.; Belanovic P. & Zazo S. (2010). Cooperative localization in wireless sensor networks using coalitional game theory, *Proceedings of 18th European Signal Processing Conference EUSIPCO 2010*, Aalborg, Denmark, August 2010.

Ghareshiran O. N. & Krishnamurthy V. (2010). Coalition Formation for Bearings-Only Localization in Sensor Networks—A Cooperative Game Approach, *IEEE Transactions on Signal Processing*, Vol. 58, No. 8, August 2010, pp. 4322-4338, ISSN: 1053-587X.

Ghassemi F. & Krishnamurthy V. (2008a). A Cooperative Game-Theoretic Measurement Allocation Algorithm for Localization in Unattended Ground Sensor Networks, *Proceedings of 11th International Conference on Information Fusion*, pp. 1-7, ISBN: 978-3-8007-3092-6, Cologne, Germany, June 2008.

Ghassemi F. & Krishnamurthy V. (2008b). Decentralized Node Selection for Localization in Wireless Unattended Ground Sensor Networks, Proceedings of Second International Conference on Sensor Technologies and Applications (SENSORCOMM), pp. 294-299, ISBN: 978-0-7695-3330-8, Cap Esterel, France, August 2008.

Hao X.; Cheung M. H.; Wong V. W. S. & Leung V. C. M. (2011). A Coalition Formation Game for Energy-efficient Cooperative Spectrum Sensing in Cognitive Radio Networks with Multiple Channels, *Proceedings of GLOBECOM 2011*, Houston, TX, USA, December 2011.(to appear)

Khan Z.; Lehtomaki, J.; Latva-aho, M. & DaSilva, L.A. (2010). On Selfish and Altruistic Coalition Formation in Cognitive Radio Networks, *Proceedings of the Fifth International Conference on Cognitive Radio Oriented Wireless Networks & Communications (CROWNCOM)*, pp.1-5, ISBN: 978-1-4244-5885-1, Cannes, France, June 2010.

Mathur S., Sankar L. & Mandayam N. B. (2008). Coalitions in Cooperative Wireless Networks, *IEEE Journal on Selected Areas in Communications*, Vol. 26, No. 7, September 2008, pp. 1104-1115, ISSN: 0733-8716.

Mathur S.; Sankaranarayanan, L. & Mandayam, N.B. (2006). Coalitional Games in Cooperative Radio Networks, *Proceedings of Asilomar Conference on Signals, Systems and Computers (ACSSC '06)*, pp.1927-1931, ISBN: 1-4244-0784-2, Pacific Grove, CA, USA , October 2006.

Moragrega A.; Closas P. & Ibars C. (2011). Energy efficient positioning in sensor networks by a game theoretic approach, *Proceedings of 19th European Signal Processing Conference EUSIPCO 2011*, Barcelona, Spain, August 2011.

Myerson R. B. (1997). *Game theory, Analysis of conflict*, Harvard University Press, ISBN: 0-674- 34115-5, Cambridge, MA, USA.

Patwari N.; Hero A.O. III; Perkins M.; Correal N.S. & O'Dea R.J. (2003). Relative location estimation in wireless sensor networks, *IEEE Transactions on Signal Processing*, Vol. 51, No. 8, August 2003, pp. 2137-2148, ISSN: 1053-587X.

Saad W.; Han Z.; Basar T.; Debbah M.; Hjorungnes A. (2009a). Physical Layer Security: Coalitional Games for Distributed Cooperation, *Proceedings of 7th International Symposium on Modeling and Optimization in Mobile, Ad Hoc, and Wireless Networks (WiOPT)*, pp.1-8, ISBN: 978-1-4244-4919-4 , Seoul, South Korea, October 2009.

Saad W.; Han Z.; Basar T.; Debbah M.; Hjorungnes A. (2009b). Coalitional Games for Distributed Eavesdroppers Cooperation in Wireless Networks, *Proceedings of the Fourth International ICST Conference on Performance Evaluation Methodologies and Tools*, ISBN: 978-963-9799-70-7, Pisa, Italy, October 2009.

Saad W.; Han Z.; Debbah M.; Hjorungnes A & Basar T. (2009c). Coalitional Games for Distributed Collaborative Spectrum Sensing in Cognitive Radio Networks, *Proceedings of IEEE INFOCOM*, pp.2114-2122, ISBN: 978-1-4244-3512-8, Rio de Janeiro, Brazil, April 2009.

Saad W.; Han Z.; Debbah M.; Hjorungnes A & Basar T. (2009d). Coalitional game theory for communication networks, IEEE Signal Processing Magazine, Vol. 26, No. 5, September 2009, pp. 77-97, ISSN: 1053-5888.

Saad W.; Han Z.; Debbah M. & Hjorungnes (2008). A distributed merge and split algorithm for fair cooperation in wireless networks, *Proceedings of IEEE ICC Workshops*, pp. 311-315, ISBN: 978-1-4244-2052-0, Beijing,China, May 2008.

Singh, C.; Sarkar, S.; Aram, A. & Kumar, A. (2012). Cooperative Profit Sharing in Coalition Based Resource Allocation in Wireless Networks, *IEEE/ACM Transactions on Networking*, Vol.20., No. 1, February 2012, pp. 69-83, ISSN: 1063-6692.

Spirito M. A. (2001). On the accuracy of cellular mobile station location estimation, IEEE Transactions on Vehicular Technology, Vol. 50, No. 3, May 2001, pp. 674-685, ISSN: 0018-9545.

Srivastava, V.; Neel, J.; Mackenzie, A.B.; Menon, R.; Dasilva, L.A.; Hicks, J.E.; Reed, J.H.; Gilles, R.P. (2005). Using Game Theory to Analyze Wireless Ad Hoc Networks, *IEEE Communications Surveys & Tutorials*, Vol. 7, No. 4, February 2006, pp. 46-56, ISSN: 1553-877X.

Tang S.; Wu X.; Mao X.; Wu Y.W.; Xu P.; Chen G. & Li X. (2009). Low complexity stable link scheduling for maximizing throughput in wireless networks, *Proceedings of IEEE Communication Society Conference on Sensor, Mesh and Ad Hoc Communications and Networks 2009*, pp. 1-9, ISBN: 978-1-4244-2907-3, Rome, Italy, June 2009.

Yanmaz E. & Tonguz O.K. (2005). Location dependent dynamic load balancing, *Proceedings of GLOBECOM 2005*, vol.1, pp. 587-591, ISBN: 0-7803-9414-3 New York, USA, November 2005.

Section 3

Game-Theoretic Mathematics

Chapter 9

Models of Paradoxical Coincident Cost Degradation in Noncooperative Networks

Hisao Kameda

Additional information is available at the end of the chapter

http://dx.doi.org/10.5772/54377

1. Introduction

[**Networks**] As large-scale networks, one can think of transportation networks, which consist of very many roads, through which lots of vehicles run through. It is inappropriate if the utilization factors of the networks are so low that the amounts of traffic through the networks are too small, or if the sojourn times of vehicles are too long due to the congestion. In relation to the notion of transportation networks, we have the term 'information highway,' which means information networks that are represented by the Internet. The Internet is an information network consisting of a great many nodes (or routers) and the communication lines interconnecting them, through which packets (in place of vehicles) run. Commonly as transportation networks, it is inappropriate if the utilization factors of the networks are so low that the amounts of traffic through the networks are too small or if the sojourn times of packets are too long due to the congestion.

It is appropriate if each portion of the networks has a suitable amount of traffic and if vehicles or packets (later we call both of them 'users' or 'players') pass through it within suitable time lengths. We cannot be sure, however, that such good situations are always kept. In order to keep such good situations, we need to control the networks in one way or others. More concretely, we need to select adequately the paths for users to run through (*i.e.*, routing) or to decide adequately the rates of users to run through the networks (*i.e.*, flow control).

In distributed computer systems including recently highlighted Grids (a method of sharing computer resources by using communication networks), we have the problems of load balancing in order to have the high efficiency of utilizing computing resources (for example, [26, 34, 53]) that are equivalent to routing problems in the networks. In this chapter, we present some seemingly paradoxical results on the routing problems and on the equivalent load balancing problems. There have been found similar seemingly paradoxical results on flow control [19, 20], but we do not discuss them here.

[Distributed and independent decision making] On one hand, in general, the above-mentioned networks are of large scales and overall detailed control of them may be difficult. On the other hand, they are usually shared by users or organizations that make independent decisions; *e.g.*, in transportation networks, there are independent vehicle owners and enterprises that run buses and trucks; *e.g.*, in the Internet, there are Internet service providers that are private enterprises and universities/research organizations, each of which is considered to make independent decisions. These independent decision makers are equivalent to what are called 'players' in the framework of game theory.

Lots of people may believe that in a network or in a system, if each independent decision maker coincidently pursues the reduction of the cost relevant to itself, the overall utility of the network or of the system may increase. As to economic behavior, it appears to be generally believed that, if each decision maker seeks its profit independently, selfishly, and noncooperatively from others, the overall social system may achieve the most economical state by the guidance of so-called 'Invisible Hand' as mentioned by Adam Smith [50]. In the situations where the overall detailed control seems to be difficult in such as large-scale networks, the believes in the effectiveness of independent decision making, may lead to the expectations that if each user (or player) makes decisions only for its own objective, the entire problem of obtaining the best system state is divided into the collection of dispersed smaller-scale problems each of which is more easily solved than the entire problem.

[Paradox] There have been found phenomena seemingly quite paradoxical to the above-mentioned belief, for example, what is called the 'Braess paradox,' which we discuss later in this chapter. This chapter present some research results on the possible magnitudes of the harms brought by such paradoxical phenomena. In fact, it looks that no big harms induced by such paradoxical phenomena like large coincident performance deterioration of user costs, have been reported thus far. We have already the situations where the Internet is shared by a great number of independent Internet service providers and where online POS (Point of Sales) systems are shared by mutually independent chains of convenience stores. In spite of that, no big problems such as the above-mentioned severe performance degradation have been revealed.

We expect, however, that each independent organization will pursue its cost decrease much more seriously in the future and that the scales of the shared networks and the numbers of users sharing the networks will become much larger. We will thus need to investigate the above-mentioned problems and to gain much powerful insight into the problems.

1.1. Different degrees in the dispersion of decision making

We can think of different degrees in the dispersion of decision making.

(A)**[Completely centralized decision making]**: All users are regarded to belong to one group that has only one decision maker (Only one player in the game). The decision maker seeks to optimize a single performance measure such as the total cost over all users (for example, the expected sojourn time over all users). In the literature, the corresponding solution concept is referred to as a system optimum, overall optimum, cooperative optimum or social optimum.

In this chapter, we shall refer to it as the *overall optimum*. This may reflect the situation where the entire system is controlled by a single unified organization (*i.e.*, only one player).

(B)[**Completely dispersed decision scheme**]: Each of infinitely many infinitesimal users optimizes its own cost (for example, its own expected sojourn time), independently and selfishly of the others. In this optimized situation, each job cannot expect any further benefit by changing unilaterally its own decision. In this setting, the number of such users is so many that the impact of any one such user is infinitesimally small on the costs experienced by all other users. It is then assumed that the decision of a single job has a negligible impact on the performance of the entire system.

In these types of noncooperative networks, each infinitesimal, *i.e.*, "non-atomic" (as some game-theorists say), user makes its own routing decision so as to minimize its expected delay from its origin to its destination given the routing decisions of other users. In this case, the situation where every infinitesimal user has optimized its decision, given the decisions of other users, and would not unilaterally deviate from that choice, is called an *individual equilibrium*. The name given to this form of equilibrium is *Wardrop equilibrium*, *i.e.*, a Nash equilibrium with infinitesimal players (nonatomic users) ([18, 41, 54] *etc.*).

(C)[[**Intermediately dispersed decision scheme**]: Infinitely many jobs (users) are classified into a finite number ($N(>1)$) of classes or groups, each of which has its own decision maker and is regarded as one player or user. Each decision maker optimizes non-cooperatively its own cost (e.g., the expected sojourn time) over only the jobs of its own class. The decision of a single decision maker of a group has a non-negligible impact on the performance of other groups. In this optimized situation, each of a finite number of classes or players cannot receive any further benefit by changing unilaterally its decision. In the literature, the corresponding solution concept is referred to as a class optimum, Nash equilibrium, or user optimum. In this chapter, we shall refer to it as the *group optimum*. This may reflect the situation where the system is shared by a finite number of mutually independent organizations each of which is totally unified. We may have different levels in intermediately dispersed optimization.

In this situation, the users are referred to as "atomic" (as some game-theorists say) in that each user's decision has an impact on the costs experienced by the other users. The situation where, in such a scheme, every user has optimized his decision, given the decisions of other users, and furthermore, would not unilaterally deviate from this decision is called a *Nash equilibrium*, since it is, in this respect, "stable" [18, 24, 27].

Note that (C) is reduced to (A) when the number of players reduces to 1 ($N = 1$) and approaches (B) when the number of players becomes infinitely many ($N \to \infty$) [18]. In the cases of (B) and (C), there are plural decision makers and they can be regarded as 'games' (in particular, congestion games (for example, [42, 46])). In the terms of economics, (A), (B), and (C), respectively, present monopoly, perfect competition, and oligopoly.

1.2. Pareto inefficiency and paradox

We think that the cost (or utility) of each user is determined for each state of the system. For example, if the route through which each user runs through and the amount of traffic of each path is determined, the total cost (or utility) of each user (decision maker, or player) will be determined. In engineering fields, as the measure of evaluating the system status, a single measure such as the sum or the weighted means of the costs (or utilities) of all users (or players) has been used in general so far. It is questionable or problematic, however, to determine the superiority between the two states of the system if, in one system state, the utility of one user is *better* than that of another user and if, in the other system state, the the utility of the former user is *worse* than that of the latter user.

The exact definition of superiority among system states is given in terms of Pareto notions. The notions of Pareto optimality, superiority, and inefficiency have already been established. In the next section, we first confirm the notions and their definitions. Then, we discuss a definition of the measure of Pareto superiority.

1.3. Pareto optimality and superiority

We consider a system consisting of a number of users or players, numbered $1, 2, \cdots, n$ (Denote by $\mathbf{n}$ the set $\{1, 2, \cdots, n\}$). For each state of the system, each user has its own utility. Denote a combination of utilities of all users in a system state S by $\mathbf{U}(S) = (U_1(S), U_2(S), \ldots, U_n(S))$. We consider only the cases where $U_i(S) > 0$ for all i.

[**Pareto optimality and efficiency**]: There may exist a state of the system where we cannot improve the utility of each user without decreasing the utility of some other user. This is called a *Pareto optimum* or *efficient* state. In general, there may be infinitely many Pareto optimal states for a system. Consider the space of the combinations of utilities $\mathbf{U}(S)$ for all system state S. That is, each point $\mathbf{U}(S)$ in the utility space corresponds to a combination of utilities of the users of a system state. Then, each axis of the utility space shows the utility of a user given the system state. The set of points corresponding to Pareto optimal states forms the border (Pareto border) separating the set of achievable combinations of utilities from the set of unachievable ones in the utility space.

[**Pareto superiority and inferiority**]: Consider an arbitrary pair of two (achievable) states of the system, S^a and S^b: If $U_i(S^a) \leq U_i(S^b)$ for all i and $U_i(S^a) < U_i(S^b)$ for some i, then S^a is Pareto inferior to S^b and S^b is Pareto superior to S^a. Define $k_i \triangleq U_i(S^b)/U_i(S^a)$. Then, S^b is *Pareto superior* to S^a if and only if $k_i > 1$ for some i and $k_j \geq 1$ for all other j. S^b is *Pareto inferior* to S^a if and only if $k_i < 1$ for some i and $k_j \leq 1$ for all other j.

We define *strong Pareto superiority and inferiority*. That is, S^b is *strongly Pareto superior* to S^a iff $k_i > 1$ for all i. S^b is *strongly Pareto inferior* to S^a iff $k_i < 1$ for all i. A state to which some other state is (strongly) Pareto superior is (*strongly*) Pareto inefficient.

An overall optimum is evidently Pareto optimal. Individual optima or group optima (Nash equilibria) may be Pareto optimal (for example, [1, 7]), but may not be Pareto optimal as in the game called the prisoners' dilemma (for example, [38]). It has been shown that if the utility of each player is continuous, Nash equilibria are generally Pareto inefficient (See [13, 49]).

[**A measure of Pareto superiority/inferiority**]: As we see in the above, the definition of Pareto superiority/inferiority has already given and well accepted. It seems, however, that the measure of the *degree of Pareto superiority/inferiority* has not been generally accepted. The measure is necessary, *e.g.*, for defining the degree of paradoxical coincident cost degradation.

The Pareto superiority depends on the vector $(k_1, k_2, \ldots, k_n)$. It would, however, be convenient to express the degree of Pareto superiority by a single scalar measure. The primary concern must be the requirement that the value of the measure clearly distinguish cases of Pareto inferiority and, thus, paradoxes, from other cases, simply and almost clearly. Define $k_{\min} \triangleq \min_i k_i$ and $k_{\max} \triangleq \max_i k_i$. If $k_{\min} > 1$, the state S^b is (strongly) Pareto superior to S^a, and if $k_{\min} < 1$, the state S^b is Pareto indifferent or inferior to S^a. If $k_{\max} < 1$, the state S^b is (strongly) Pareto inferior to S^a, and if $k_{\max} > 1$, the state S^b is Pareto indifferent or superior to S^a. Thus, *the measures $k_{\min}$ and $k_{\max}$ may be used as primary measures of the degree of Pareto superiority and inferiority, respectively*. We note that if $k_{\min} < 1$ and $k_{\max} > 1$, states S^a and S^b are mutually Pareto indifferent to each other. On the other hand, for example, a measure X based on a certain average of all of k_i should be rejected, since it can hold that $X > 1$ even if some $k_i < 1$ for some i but if $k_j \gg 1$ for all other j's. Such a measure may be used as a secondary measure. (In many practical situations, the variables may have continuous values and truly exact equalities occur rarely. Or, the tie-breaking of the case that $k_{\min} = 1$ may depend on such a secondary measure.)

We propose that *$k_{\min}(>1)$ and $k_{\max}(< 1)$ are used as primary measures showing the degrees of Pareto superiority and inferiority, respectively*. The tie-breaking of the case that $k_{\min} = 1$ and that $k_{\max} = 1$ may depend on some other secondary measure.

Note, in passing, that a measure similar to the above has been used to discuss the effects of symmetry on globalizing separated monopolies to a Nash-Cournot oligopoly [28].

[Pareto inefficiency and paradox] Braess paradox presents an example of the case where for an equilibrium system state there exists a non-equilibrium system state that is Pareto superior to it as shown later. This chapter presents the research results on the cases where (paradoxical) coincident cost degradation of each user occurs and on the possible sizes of such coincident cost degradation. On the other hand please note that there exist cases where coincident cost improvement for each user is unlimitedly large [22].

The price of anarchy The idea of a measure, the price of anarchy, was mentioned by Koutsoupias and Papadimitriou [32], and its name 'the price of anarchy' appeared in Papadimitriou [39]. The term 'anarchy' is considered to mean the state of a Nash or Wardrop equilibrium which is reached by the situation where every player behaves selfishly or freely under no constraints imposed by a central controller to optimize its own cost or utility. The measure looks to show the degree how bad is the state of the worst-case Nash/Wardrop equilibrium against the best state. The proposer of the measure uses, as the best state, the state with the optimal social cost. Then, the price of anarchy is equal to the ratio of the social cost of the worst-case Nash/Wardrop equilibrium to the minimum social cost. A number of results have been obtained based on this measure, many of which are described by Roughgarden [43, 45]. In fact, before the idea of the measure, price of anarchy, was proposed, anomalous behaviors of Wardrop and Nash equilibria compared with social optima, like those expressed in terms of the price of anarchy, had already been discovered and investigated in the context of load balancing in distributed computer systems that was identical to routing in networks of particular types [25, 26, 55].

On the other hand, the measure of the Pareto inefficiency of a Nash equilibrium has to reflect the comparison with all the Pareto optima, whereas the state with the optimal social cost is only one Pareto optimum. Therefore, the price of anarchy cannot be a good measure of the Pareto inefficiency of a Nash equilibrium [33]. Following the spirit of the price of anarchy, we may think of the ratio of the social cost of state A to that of state B for comparing two states A and B. According to the discussion given above on the measure of Pareto superiority/inferiority and paradoxes, we do not use the above-mentioned measure that has the spirit of the price of anarchy as the primary measure of Pareto superiority/inferiority and paradoxes but may use it as a secondary measure. We note that the above-mentioned anomalous behavior of the Wardrop/Nash equilibrium necessarily occurs when the Braess paradox occurs, but not vice versa.

2. Braess paradox

[**Braess network**] Braess [5] considered a network consisting of 4 nodes, 1 origin (0), 2 relay nodes (1,2), and 1 destination (3) (Fig. 1). As shown in Fig. 1 left, before adding a link, the network has two paths, 0-1-3 (Path 1) and 0-2-3 (Path 2), each of which contains two links, Link 1 (from Node 0 to Node 1) and Link 2 (from Node 1 to Node 3) for the first path, and Link 3 (from Node 0 to Node 2) and Link 4 (from Node 2 to Node 3) for the second: See Fig. 1 left. After adding new link (Link 5), *i.e.*, a one-way link connecting the two relay nodes 1 and 2, the network has three paths including the new, third, path 0-1-2-3 (Path 3): See Fig. 1 right. In the original Braess network, the cost of each link is a linear function of the amount of the flow through the link ([5]).

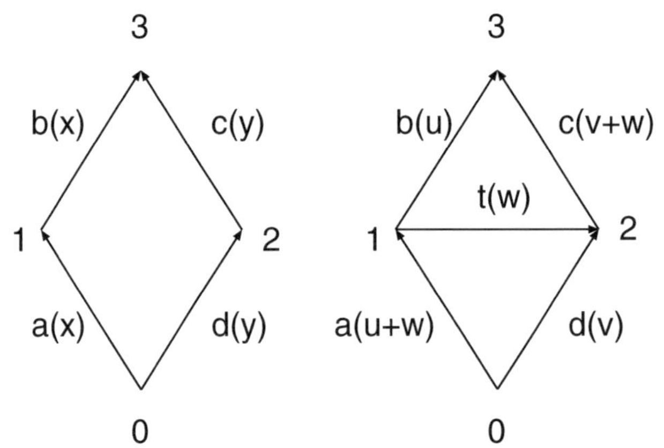

Figure 1. Braess network.

Lots of users pass through the network from the origin (0) to destination (3). The passage time through each link is determined by the rate of users that pass through the link. Each user strives to pass through the path of the minimum sojourn time. Each user cannot decrease its sojourn time by unilaterally changing the path it chooses to pass through in the equilibrium state, that is, the Wardrop equilibrium or individual optimum. Since the network has only one origin and one destination, any paths used have the same cost (the identical sojourn time). As shown in Fig. 1, if the rates of users that pass through links 01, 13, 23, 02, and 12, respectively, are $\eta_1, \eta_2, \eta_3, \eta_4$ and η_5, the passage times through links 01, 13, 23, 02, and 12, are $a(\eta_1)$, $b(\eta_2)$, $c(\eta_3)$, $d(\eta_4)$, and $t(\eta_5)$.

Denote by X the total rates of users that pass through network. Before adding a link, the two paths 0-1-3 and 0-2-3, respectively, have the rates of users passing through, x and y ($x + y = X$). After adding a link, the three paths, 0-1-3, 0-2-3, and 0-1-2-3, respectively, have the rates of users passing through, u, v, and w ($u + v + w = X$). Denote by C_o and C_c, respectively, the costs of the paths before and after adding a link. The ratio of the cost before adding a link and that after adding a link is denoted by $k = C_c/C_o$.

[**Braess paradox**] Braess [5] presents such an example that, if $a(\eta) = c(\eta) = 10\eta$, $b(\eta) = d(\eta) = \eta + 50$, $t(\eta) = \eta + 10$, $X = 6$, then $C_o = 83$, $C_c = 92$, and $k = C_c/C_o = 1.1084\ldots$. That is, by adding a link, the cost of all users (sojourn time) increases by about 10 percent. Adding a link leads to the augmentation of the degree of the freedom of decision making. In spite of it, it looks paradoxical that adding a link brings about the cost degradation to all users or decision makers. Thus the above-mentioned phenomenon is regarded as 'paradox.' In fact, it has been observed that similar paradox occurs in the real world [29]. Thus, we see that the existence of a state that is Pareto superior to a Wardrop equilibrium has been shown.

[**Cohen-Kelly paradox**] In the Braess network, linear functions are considered as the link costs. The link cost functions considered in the networks of queues are nonlinear in general. Cohen and Kelly [10] considered the following network with nonlinear link costs. λ and ϕ are system parameters. $X = 2\lambda$. Link flows $\eta_1, \eta_2, \eta_3, \eta_4$ and η_5, respectively, give $b(\eta_2) = d(\eta_4) = 2$, $t(\eta_5) = 1$, $a(\eta_1) = 1/(\phi - \eta_1)$ (if $0 \leq \eta_1 < \phi$, otherwise $a(\eta_1)$ is ∞), $c(\eta_3) = 1/(\phi - \eta_3)$ (if $0 \leq \eta_3 < \phi$, otherwise $c(\eta_3)$ is ∞). Then, assuming $2\lambda > \phi - 1 > \lambda > 0$, it is shown that $C_o = 1/(\phi - \lambda) + 2 < 3 = C_c$, that is, $1 < k < 3/2$. This also is regarded as a paradox. Thus, the ratio of degradation, k, is less than 1.5.

[**The researches related to Braess paradox**] Later, Braess paradox gradually caused attention of many scholars including economist Samuelson [48] and related studies have been accumulated including the above-mentioned study by Cohen and Kelly [10] (For example, [6, 9, 11, 12, 16, 17, 35, 36, 40, 51, 52]). In addition, it has been shown that for mechanical and electrical systems that have the topology similar to the Braess network, there may occur phenomena similar to Braess paradox, and the results were presented in a scientific journal, *Nature* [8]. A list of references on the Braess paradox is kept in Braess's home page (http://homepage.ruhr-uni-bochum.de/Dietrich.Braess/#paradox).

Almost all of the related results have been obtained as to Wardrop equilibria, and most of them have discussed the networks that have the same topology as that of Braess's or its generalized versions. Furthermore, some have handled only *weak Paradox* as explained later. Moreover, it has been shown that there exists a case of similar paradox as to a group optimum (Nash equilibrium of a finite number of users) in a network whose topology is similar to the Braess network [30]. Korilis et al. [31] have obtained a sufficient condition whereby no paradox occurs in a network with one origin and one destination and with plural groups of users of the same kind.

[**The bounds of the degrees of the paradox in networks of Wardrop equilibrium**] There seem to have been only a few studies that have provided an estimation of how harmful the paradox can be, *i.e.*, the worst-case degree of coincident cost degradation for all users by adding connections to a noncooperative system [21, 44, 47]. As to a generalized Braess network as shown in Fig. 1, it has been shown that, if functions a and c are increasing and if functions b, d, and t are non-decreasing, $k < 2$, *i.e.*, the degree of paradox cannot be over 2 [21]. Furthermore, if a generalized Braess network is embedded in a larger network, the degree of paradox with respect to the embedded network cannot be over 2 [21]. As a more general result, it has been shown that as to networks consisting of one origin, one destination, n nodes and links of nondecreasing costs, the degree of paradox cannot be over $\lfloor n/2 \rfloor$ [44]. An extreme case of it is shown in Fig. 2. The left-hand part of the figure shows the network before adding 3 links (depicted by vertical thin dotted lines) and the right-hand part of the figure presents the network after adding 3 links (depicted by vertical upward thick solid lines). In the figure, nodes means the point to which 3 lines are connected. Thus, the network has 8 nodes, A, B, C, D, E, F, G, and H. The total flow rate of users are 4. The extremely thin lines show zero flow rates and thick solid lines show positive flow rates. In the left-hand part of the figure, there are 4 paths, $ABH, ACDH, AEFH$ and AGH, each has flow rate 1, and in the right-hand part of the figure, there are 3 paths, $ACBH, AEDH$, and $AGFH$, each of which has flow rates $4/3$. Each line with 0 in the left part and with 1 in the right part has cost 0, if the flow rate through it is not larger than 1, and cost 1, if the flow rate through it is larger than 1. Each line with 1 both in the left part and in the right part has cost 1 regardless of the flow rate. Each of added link has cost 0. In this figure, for 8 nodes, $C_o = 1$, and $C_c = 4$, and it shows the worst-case paradox, $k = \lfloor 8/2 \rfloor = 4$.

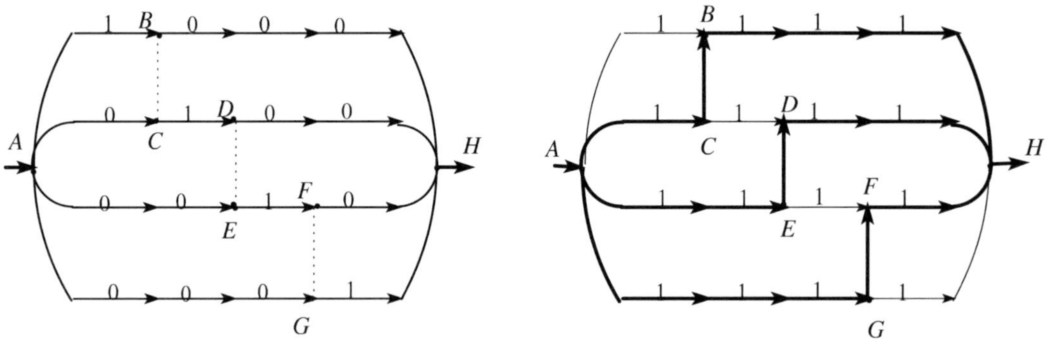

Figure 2. The network with 8 nodes that shows the worst case paradox $k = 4$.

In fact, it is likely that there has not been found any system in Wardrop equilibrium (with infinitesimal users) for which the degree of coincident cost degradation can increase without bound if the number of nodes in the network is bounded. On the other hand, we have also seen that the benefit brought by the addition of connections to a noncooperative network can increase without bound [22]. In contrast, it has been shown that there exists a system in a Nash equilibrium (with a finite number of users) for any size of the degree of the paradox as shown in the later section 4.1.1 and by [27].

3. Coincident cost degradation (Paradox) for all users by adding connections to networks

[The degree of coincident cost degradation (Paradox)] As to the above-mentioned networks in Wardrop equibria that has only one origin and one destination, the cost of all users are identical, and the comparison between the costs of before and after adding connections to the networks may be rather easy. In other networks, however, the costs of users are not necessarily identical, and we define here the degree of paradoxes for those networks. It seems that such definitions have not received attention until rather recently.

Following the section 1.2, we consider the concept of strong Pareto superiority. Denote by S^b and S^a, respectively, the states before and after adding connections. Assume that the costs of uses are positive. In fact, in the examples presented in this chapter, they are so.

Consider, for user $i \in \mathbf{n}$, k_i given by $k_i = C_i^a / C_i^b$. If $k_i > 1$ for all $i \in \mathbf{n}$, S^b is Pareto superior to S^a, which means a paradox. Define k_{min} such that $k_{min} = \min_p k_p$. That $k_i > 1$ for all $i \in \mathbf{n}$ is equivalent to that $k_{min} > 1$. In contrast, if $k_i \leq 1$ for some $i \in \mathbf{n}$, *i.e.*, $k_{min} \leq 1$, S^b is not Pareto superior to S^a, which implies no paradox. Thus, k_{min} shows whether a paradox occurs. We consider furthermore, that k_{min} shows the degree of a paradox. We thus consider k_{min} the measure of the degree of a paradox [22]. Note that the networks mentioned in Section 2 have only one origin and one destination, and that, in each Wardrop equilibrium, the utility (cost, *e.g.*, sojourn time) of every user is the same, and that k_{min} degenerates to k. Thus, as to the Braess-like paradoxes, only in this special cases, the price of anarchy can be a good measure of paradoxes.

[**Weak paradox**] In the paradoxes discussed in this chapter, the state S^b before adding connections is strongly Pareto superior to the state S^a after adding connections. On the other hand, even if S^b is not strongly Pareto superior to S^a, it is possible that the social or overall cost of users (for example, the overall average sojourn time or passage time for all users or packets) for S^b is better than S^a, which looks anomalous [26, 55]. In such a case, however, adding connections does not lead to the coincident cost degradation for all users. We call such an anomalous case a *weak paradox*. In the cases presented in Section 2, it holds that $k_i = k$, but it does not necessarily holds in general. In particular, the degree of cost degradation may be different for each user ($i \in \mathbf{n}$). The results that depend only on the price of anarchy may show only a weak paradox and not a (strong) paradox, and this chapter does not touch on such results.

4. The systems where the cost of each user may not necessarily be identical —— Paradoxes in the models of distributed computer systems

During about 30 years since the Braess paradox had started to attract attention of many authors, a lot of papers related to it were published, but in almost all of them, the networks discussed by them looked to have either the topology similar to the Baress's, or to have each user non-distinct with only one pair of the origin and the destination. As networks that have each user with distinct cost, we show the networks of distributed computer systems. That is, as to the group optima in a model of distributed computer systems, an example of (strong) paradox has been found. This type of paradox may occur for symmetric systems, and in some cases, the degree of paradox (coincident cost degradation) can increase without bound.

4.1. Analytic results on symmetric distributed computer systems

As to the paradoxes for symmetric distributed computer systems, there have been obtained very general and complicated analytical results [27], which is presented in the appendix to

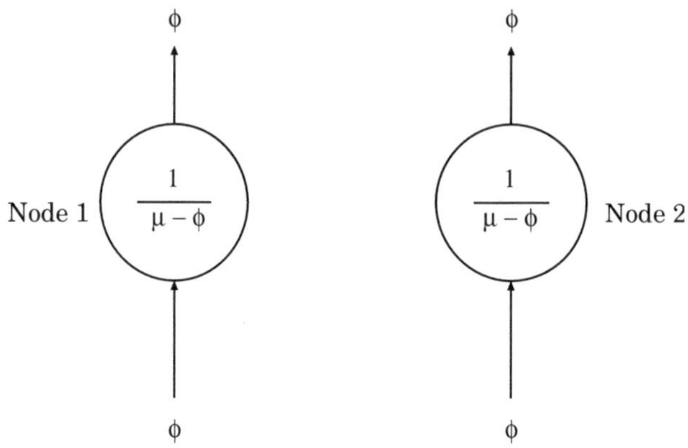

Figure 3. A model of distributed systems – Independent.

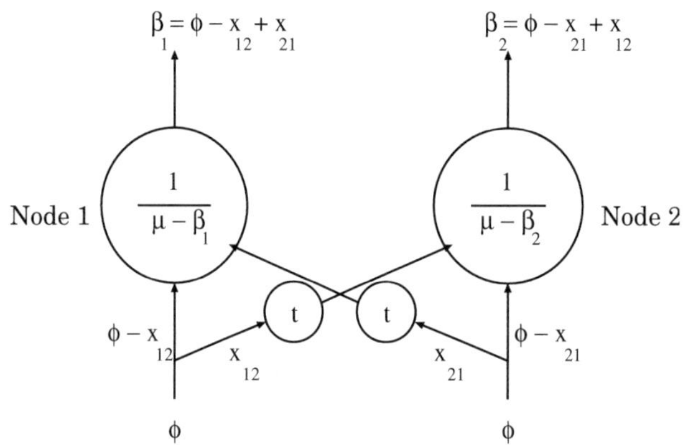

Figure 4. A model of distributed systems – Interconnected.

this chapter. In the next Section 4.1.1, by using a special case of the above-mentioned systems, some basic results that give intuitions into the nature of the paradox are presented.

4.1.1. A simple example of the paradox for symmetric distributed systems

Consider a model of distributed systems consisting of two nodes, as shown in Figs. 3 and 4. μ, ϕ, and T_i, respectively, denote the processing capacity of each node (computer), the job arrival rate to each node, and the expected sojourn time of the job that arrives at node i. Each node i is associated with one decision maker who decides the rate x_{ij} $(i \neq j)$ of jobs to forward to the other node j in order to minimize selfishly only the cost T_i for the jobs that arrive at node i. The equilibrium state is a group optimum (a Nash equilibrium). When there is no network connection between the two nodes as shown in Fig. 3, each decision maker cannot forward jobs to the other node, and, in the group equilibrium state,

$$T_1 = T_2 = \frac{1}{\mu - \phi} \quad \text{(M/M/1)}.$$

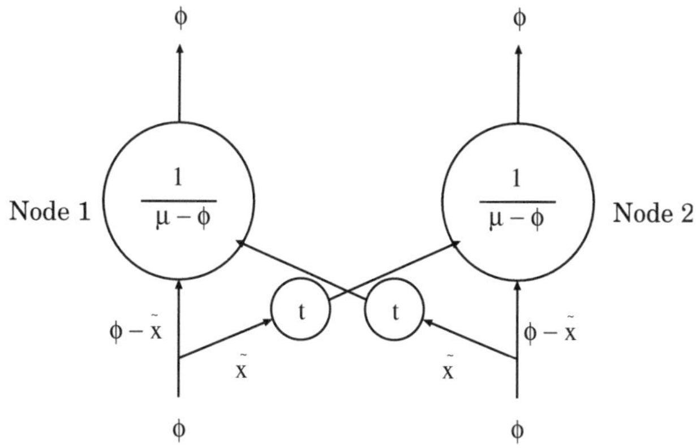

Figure 5. A model of distributed systems – A Nash equilibrium. $\tilde{x}_{ij} = \tilde{x}_{ji} = \tilde{x} \geq 0$.

After two nodes are inter-connected as shown in Fig. 4, denote by x_{ij} the rate of jobs forwarded from node i to node j. Then,

$$0 \leq x_{ij} \leq \phi,\ i,j = 1,2\ (i \neq j) \tag{1}$$

Define $\mathbf{x} = (x_{12}, x_{21})$. Denote $\mathbf{C}$ by the set of the vectors that satisfy relation (1). Assume that job forwarding takes the time length t irrespective of the value of $\mathbf{x}$. Then,

$$T_i(\mathbf{x}) = \frac{1}{\phi} \sum_k x_{ik} T_{ik}(\mathbf{x}), \tag{2}$$

$$\text{where } T_{ii}(\mathbf{x}) = \frac{1}{\mu - \beta_i},\ \ T_{ij}(\mathbf{x}) = \frac{1}{\mu - \beta_j} + t\ \ (j \neq i). \tag{3}$$

β_i denotes the workload on node i, and is obtained as follows:

$$\beta_i = \phi_i - x_i + x_j\ \ (i \neq j). \tag{4}$$

$T_{ik}(\mathbf{x})$ denotes the expected sojourn time of a job that arrives at node i and is processed by node k. After the network connection is added, denote by $\tilde{\mathbf{x}}$ the solution (the Nash equilibrium) of the group optimum (Fig. 5). Then,

$$T_i(\tilde{\mathbf{x}}) = \min_{x_{ij}} T_i(x_{ij}, \tilde{x}_{ji}),\ (i \neq j).$$

$\tilde{\mathbf{x}}$ is obtained as follows:

(i) The case where $t > \phi/(\mu - \phi)^2$: $\tilde{x} = \tilde{x}_{ij} = 0,\ \ T_i(\tilde{\mathbf{x}}) = \dfrac{1}{\mu - \phi}$.

The value of T_i is identical to the value before adding the connection, which implies no paradox occurring.

(ii) The case where $0 < t \leq \phi/(\mu - \phi)^2$:

$\tilde{x} = \tilde{x}_{ij} = \{\phi - t(\mu - \phi)^2\}/2 \geq 0, T_i(\tilde{\mathbf{x}}) = \dfrac{1}{\mu - \phi} + E$, where, $E = \dfrac{t}{2\phi}\{\phi - t(\mu - \phi)^2\} \geq 0$. E presents the difference of the value of T_i minus the value before adding the connection, and $E > 0$ shows the paradox occurring. Therefore, $0 < t < \phi/(\mu - \phi)^2$ is the necessary and sufficient condition for the occurrence of the paradox for this model.

[A derivation of the above-mentioned $\tilde{\mathbf{x}}$ and $T_i(\tilde{\mathbf{x}})$] From (2) and (4),

$$\frac{\partial T_i}{\partial x_i} = -\frac{\mu - x_j}{(\mu - \phi + x_i - x_j)^2} + \frac{\mu - \phi + x_j}{(\mu - \phi - x_i + x_j)^2} + t \quad (i \neq j). \tag{5}$$

From (5), we see that $\dfrac{\partial T_i}{\partial x_i}$ is increasing in x_i such that $\mathbf{x} \in \mathbf{C}$. If such $\tilde{\mathbf{x}}$ as satisfies the following:

$$\left.\frac{\partial T_i}{\partial x_i}\right|_{\mathbf{x}=\tilde{\mathbf{x}}} = 0, \quad i = 1, 2, \tag{6}$$

the value of $\tilde{\mathbf{x}}$ is the solution of the group optimum (the Nash equilibrium). From (5) and by defining $d = x_1 - x_2$,

$$\frac{\partial T_1}{\partial x_1} - \frac{\partial T_2}{\partial x_2} = \frac{2\mu - (\phi + d)}{(\mu - \phi - d)^2} - \frac{2\mu - (\phi - d)}{(\mu - \phi + d)^2} = \frac{2d}{(\mu - \phi)^2 - d^2}\left\{\frac{2\mu(\mu - \phi)}{(\mu - \phi)^2 - d^2} + 1\right\}, \tag{7}$$

If (6) hold, then, from (7), $d = 0$. Then, from (5),

$$\frac{\partial T_i}{\partial x_i} = \frac{2x_i - \phi}{(\mu - \phi)^2} + t = 0, \quad i = 1, 2. \tag{8}$$

Therefore,

$$x_i = \frac{1}{2}\{\phi - t(\mu - \phi)^2\}, \quad i = 1, 2, \quad \text{where } t \leq \frac{\phi}{(\mu - \phi)^2}. \tag{9}$$

From the above, we see that this is a unique solution (case (ii)). From this and (2), we can obtain $T_i(\tilde{\mathbf{x}})$.

For the case where $t > \dfrac{\phi}{(\mu - \phi)^2}$ (case (i)), from (8), we have for $x_i = 0$ $(i = 1, 2)$,

$$\frac{\partial T_i}{\partial x_i} = t - \frac{\phi}{(\mu - \phi)^2} > 0, \quad i = 1, 2. \tag{10}$$

Since $\dfrac{\partial T_i}{\partial x_i}$ is increasing in x_i, $\tilde{x}_i = 0$, $i = 1, 2$, is the solution of the group optimum (the Nash equilibrium). The uniqueness of the solution in this case is shown as outlined as follows: Assume $\tilde{x}_1 > 0$. From the definition of d, (5), and the condition (ii) with respect to t, we must have $d < 0$, and thus $\tilde{x}_2$ is not to be zero. If we have a similar argument as above with respect to $\tilde{x}_2$ we must have $d > 0$, which is a contradiction. Thus, we see that $\tilde{\mathbf{x}} = \mathbf{0}$ is a unique group optimum. For the existence and uniqueness of more general cases, see [2, 23, 37] and [27]. □

For any values of system parameters, the solutions of the overall optimum and the individual optimum are the same as the solution given above for case (i). Therefore, differently from the Braess networks, no paradox occurs for the individual optimum (the Wardrop equilibrium) in this model.

As in the above, we consider that the degree of the paradox is the ratio of the expected sojourn time after adding the connection to that before doing so. Denote it by $k(\mu, \phi, t)$ here.

$$k(\mu, \phi, t) = 1 + \frac{t}{2\phi}\{\phi - t(\mu - \phi)^2\}(\mu - \phi).$$

With ϕ and μ being fixed, the degree of the paradox is largest when $t = \phi/[2(\mu - \phi)^2]$, and thus

$$\max_t k(\mu, \phi, t) = 1 + \frac{\phi}{8(\mu - \phi)}.$$

Thus, as the arrival rate ϕ approaches to the processing capacity μ, the degree of the paradox can increase without bound. Denote $\Delta(\mu, \phi) = \max_t k(\mu, \phi, t) - 1$. Then, for example, $\Delta(1.00001, 1) = 12500$ (1250000% performance degradation), *etc.*

Therefore, we see that for any system including asymmetric one there exists a symmetric distributed system that has the larger degree of paradox than it. Thus, symmetric systems bring about the worst-case paradoxes. As to any group of systems that have finite degrees of paradox (the characteristics of the groups can be expressed in natural ways), for each group a symmetric system presents the worst-case paradox among the group as discussed in the later section 5.

4.1.2. Gereral results on the paradoxes for symmetric distributed systems

Analytic results have been obtained for the models with the number of nodes, the characteristics of job types, the processing capacities of nodes, and the characteristics of job-transfer capacities being much more general than those of the model of the above section. Details are given in the appendix.

5. Paradoxes for asymmetric distributed computer systems

Consider an extension of the models of symmetric distributed systems presented in Section 4.1 to those of asymmetric distributed systems. In the cases where each value of the parameter describing the system is not identical as to every user, the cost of each user is considered distinct in a Nash equilibrium. In this section, we consider the system consisting of m nodes, $1, 2, \ldots, m$ [26, 53]. Jobs are classified into groups, $i = 1, 2, \ldots, m$, depending on the node at which they arrive, and they arrive at node i according to Poisson distribution with the arrival rate ϕ_i. Out of them, the rate x_{ii} of jobs are processed at node i, and the rate x_{ij} $(i \neq j)$ of jobs are forwarded through the transfer facility to node $j (j \neq i)$ and processed there. Thus $\sum_p x_{ip} = \phi_i$, $x_{ij} \geq 0$, $i, j = 1, 2, \ldots, m$ are the constraints. Denote by $\mathbf{x}_i$ the vector $(x_{i1}, x_{i2}, \ldots, x_{im})$, and by $\mathbf{x}$ the vector $(\mathbf{x}_1, \mathbf{x}_2, \ldots, \mathbf{x}_m)$. We thus have $\mathbf{x} = (x_{11}, x_{12}, \ldots, x_{1m}, x_{21}, x_{22}, \ldots, x_{2m}, \ldots, x_{mm})$. Denote by $\mathbf{C}$ the set of $\mathbf{x}$'s that satisfy the constraints. Denote $\Phi = \sum_p \phi_p$. For each node i, there is a decision maker (or a player) i $(i = 1, 2, \ldots, m)$ who determines the values of x_{ij} $(j =$

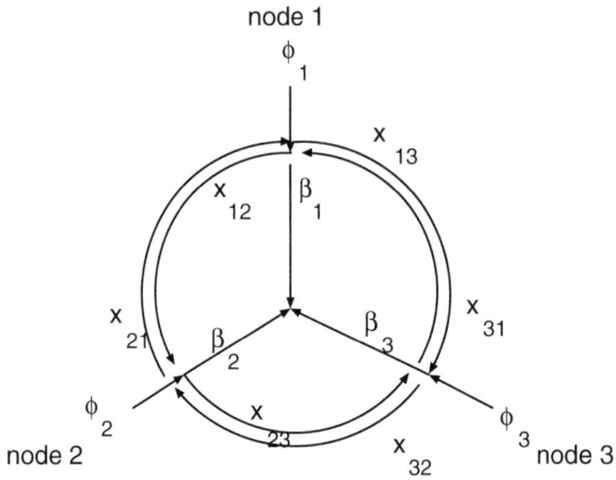

Figure 6. The model of a distributed computer system with asymmetric parameter values ($m = 3$).

$1, 2, \ldots, m$) unilaterally within the constraints so that the cost at node i may be minimum. As the result, node i has the workload $\beta_i = \sum_q x_{qi}$. This system is equivalent to the network consisting of m origin-destination pairs that have a common destination (See Fig. 6 for the case of $m = 3$. In the figure, the variable shown closely to each arrow presents the rate of jobs passing through the arrow.) Denote by $D_i(\beta_i)$ the expected sojourn time (including the waiting time) of a job that arrives at node i that has the workload β_i. It is assumed that $D_i(\beta_i)$ is convex and increasing in β_i. Denote by $G_{ij}(\mathbf{x})$ the expected transfer time of forwarding a job from node i to node $j (j \neq i)$. Denote by $T_i(\mathbf{x})$ the expected sojourn time that a job arrives at node i and finally goes out of the network. Then we have,

$$T_i(\mathbf{x}) = \sum_k x_{ik} T_{ik}(\mathbf{x}), \text{ where } T_{ii}(\mathbf{x}) = D_i(\beta_i), \text{ and, for } j \neq i, T_{ij}(\mathbf{x}) = D_j(\beta_j) + G_{ij}(\mathbf{x}).$$

The group optimal state (the Nash euilibrium) $\tilde{\mathbf{x}}$ satisfies the following (for example, [24]).

$$\tilde{T}_i = T_i(\tilde{\mathbf{x}}) = \min_{\mathbf{x}_i} T_i(\tilde{\mathbf{x}}_{-(i)}; \mathbf{x}_i), \text{ Constraints: } (\tilde{\mathbf{x}}_{-(i)}; \mathbf{x}_i) \in \mathbf{C}. \tag{11}$$

where $(\tilde{\mathbf{x}}_{-(i)}; \mathbf{x}_i)$ is $m \times m$ dimensional vector that is made from $\tilde{\mathbf{x}}$ by replacing its elements corresponding to $\tilde{\mathbf{x}}_i$ by $\mathbf{x}_i$. As to the existence and uniqueness of the group optimum (the Nash equilibrium), see [2, 3].

In the next section, we present some results on the asymmetric and symmetric distributed systems wherein the degrees of paradoxes are finite. The results show that, in the group of distributed systems characterized in natural ways, the degrees of paradoxes have the worst-case values for symmetric systems.

5.1. The models with multiple nodes with nonlinear costs

Consider the model of distributed computer systems consisting of $m (\geq 2)$ nodes. With the workload β_i, the expected node passage time (the cost) is: $D_i(\beta_i) = \dfrac{1}{\mu_i - \beta_i}$ (if $\beta_i < \mu_i$, otherwise ∞). The transfer cost $G_{ij}(\mathbf{x})$ has the following two cases (A) and (B):

(A) $G_{ij}(\mathbf{x}) = \dfrac{t}{1-\lambda t}, j\ (\neq i)$, if $\lambda t < 1$, otherwise ∞.
where, $\lambda = \sum_p \sum_{q,(q\neq p)} x_{pq}$ denotes the rate of jobs being transferred through the interconnection. One job-transfer channel is shared by the entire system.

(B) $G_{ij}(\mathbf{x}) = \dfrac{t}{1-x_{ij}t},\ j\ (\neq i)$, if $x_{ij}t < 1$, otherwise ∞. The entire system has $m(m-1)$ channels.

In either case, t denotes the expected transfer time when there is no waiting on the transfer line. Thus, $1/t$ is the transfer capacity. Furtheremore, denote $\boldsymbol{\phi} = (\phi_1, \phi_2, \ldots, \phi_m)$ and $\boldsymbol{\mu} = (\mu_1, \mu_2, \ldots, \mu_m)$.

5.1.1. *Numerical experiments*

The following algorithm is used to obtain the solution of a group optimum (a Nash equlibrium).

Given the values of $\boldsymbol{\phi}, \boldsymbol{\mu}$, and t,

- Initialization $\mathbf{x}^0 = (\mathbf{x}_1^0, \mathbf{x}_2^0, \ldots, \mathbf{x}_m^0) \in \mathbf{C}$.
- Obtain $\mathbf{x}^n$ $(n = (k\text{-}1)m + i,\ i = 1, 2, \ldots, m,\ k = 1, 2, \ldots)$
  as $(\mathbf{x}_1^{n-1}, \ldots, \mathbf{x}_{i-1}^{n-1}, \mathbf{x}_i^n, \mathbf{x}_{i+1}^{n-1}, \ldots, \mathbf{x}_m^{n-1})$,
  where $\mathbf{x}_i^n = \arg\min_{\mathbf{x}_i} T_i(\mathbf{x}_1^{n-1}, \ldots, \mathbf{x}_{i-1}^{n-1}, \mathbf{x}_i, \mathbf{x}_{i+1}^{n-1}, \ldots, \mathbf{x}_m^{n-1})$.
  Repeat this step until the conversion.

If the above algorithm converges, a group optimum is obtained. Then, we have the group optimum cost $\tilde{T}_i(\boldsymbol{\phi}, \boldsymbol{\mu}, t)$ of decision maker i for the values of $\boldsymbol{\phi}, \boldsymbol{\mu}, t$. It has been shown that, in the cases where the costs of nodes, *etc.* are linear, the algorithm converges [4]. In fact, for all the cases we examined, the algorithm converged. For a given combination of the values of $\boldsymbol{\mu}_i$ and $\boldsymbol{\phi}$, there exists the value of t^∞, such that, in the group optimum, if $t \geq t^\infty$, the transfer facility is not used.

If the following holds, a paradox occurs.

$$k_i(\boldsymbol{\phi}, \boldsymbol{\mu}, t) > 1,\ i = 1, 2, \ldots, m, \text{ where } 0 < t < t^\infty. \tag{12}$$

Here, $k_i(\boldsymbol{\phi}, \boldsymbol{\mu}, t) = \dfrac{\tilde{T}_i(\boldsymbol{\phi}, \boldsymbol{\mu}, t)}{\tilde{T}_i(\boldsymbol{\phi}, \boldsymbol{\mu}, t^\infty)}$. $\tilde{T}_i(\boldsymbol{\phi}, \boldsymbol{\mu}, t)$ gives the cost for the decision maker at node i, given the parameter values $\boldsymbol{\phi}, \boldsymbol{\mu}$, and t. Therefore, (12) shows that, if the capacity of the job transfer increases such that the transfer parameter decreases from t^∞ to t, the costs of decision makers at all nodes increases, which looks paradoxical. The maximum value, $\Gamma(\mu, \phi)$, of the degree of paradox $k_{\min}(\boldsymbol{\phi}, \boldsymbol{\mu}, t)$ is presented as follows:

$$\Gamma(\mu, \phi) = \max_t\{\min_i\{k_i(\phi, \mu, t)\}\}. \tag{13}$$

In the next section, the values of Γ for the combinations of the values of $\boldsymbol{\mu}$ and $\boldsymbol{\phi}$ are given.

5.1.2. The results of numerical experiments

We examined the cases of various combinations of the values $\boldsymbol{\mu}$ and $\boldsymbol{\phi}$ for some $m(\geq 2)$. The left and right parts of Fig. 7 presents, respectively, the results of examples of the job transfer facility of types (A) and (B). We can assume $\phi_1 = 1$ without losing the generality. The horizontal and vertical axes of each part of the figure show, respectively, the values of μ_1 and the worst-case degrees of paradoxes, Γ. The solid curve shows the values of Γ for the corresponding values of μ_1 for the completely symmetric cases ($\mu_i = \mu_1$ and $\phi_i = 1$ for all $i \in \mathbf{n}$). The dotted curve shows, for the corresponding values of μ_1, the maximum values of Γ among those with all combinations of values of μ_i and ϕ_i. The left and right parts of Fig. 7 show, respectively, the cases of 2 and 4 nodes ($m = 2$ and $m = 4$), with the transfer facilities (A) and (B). In both cases, as μ_1 increases, the worst-case paradox Γ increases up to some limit. It seems to be clear that the symmetric systems approaches to the limit. We can observe that there exists no asymmetric system whose values of worst-case paradox Γ depicted by the dotted curve are over the limit. That is, we can see that for any asymmetric system within each group, there exists a completely symmetric system whose degree of paradox is equal to or larger than that of the asymmetric system. In other experiments than those presented here, we found the same tendencies as shown here [14, 15]. These numerical examinations may imply that, for any group that can be expressed in natural (not excentric) ways and that has the finite value of the worst-case paradox, the worst-case paradox can be achieved by a completely symmetric system within the group.

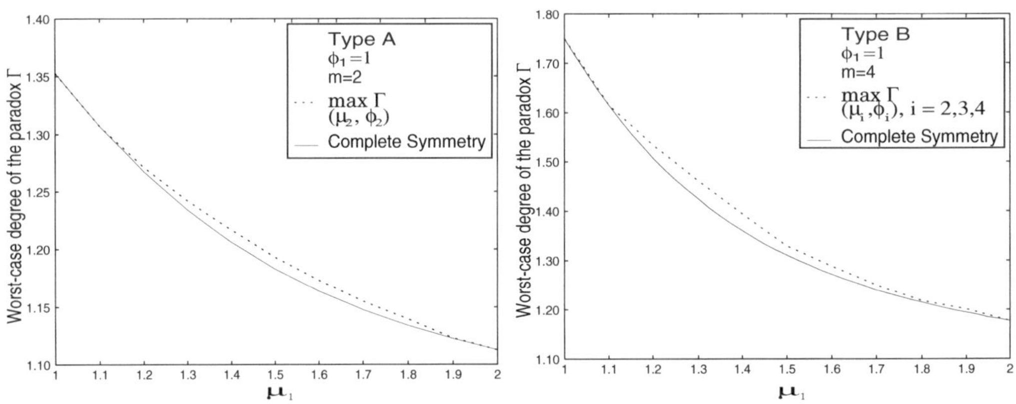

Figure 7. Comparison of the values of the worst-case paradoxes for completely symmetric and asymmetric systems

6. Concluding remarks

In information or transportation networks or distributed computer systems wherein a plural number of users make independent decisions to minimize their own costs, it may be possible that all the users have coincident cost degradation similarly as the prisoners' dilemma or the Braess paradox. Motivated by the above, in this chapter, we have presented an overview of the analytical and numerical results on the worst-case degree of paradoxes such that in selfish routing in networks or in noncooperative load balancing in distributed systems, as the degree of freedom in making decisions increases, the costs of all users degrade coincidently. Furthermore, we have considered the relation between the paradox and the Pareto superiority, and have shown a measure that gives the degree of paradox. As to the networks in an individual optimum (an Wardrop equilibrium) wherein decision making is completely dispersed, the degree of the worst-case paradox is limited for a finite number of nodes in the networks. On the other hand, as to the networks such as distributed computer systems like GRID, in an individual optimum, no paradox may occur whereas in a group optimum (a Nash equilibrium) wherein decision making is intermediately dispersed, paradoxes may occur and the degree of the paradoxes can increase without bound. Furthermore, for a group of systems, the worst-case paradox for the group may be achieved by a symmetric system in the group. For symmetric distributed systems, analytic results have been obtained for general systems. For asymmetric systems, however, it may not be so easy to obtain analytical results, and so we have to rely on numerical investigation in order to understand the problems in question.

Even in the direction of research (such as routing in networks and load balancing in distributed systems), there may remain to be solved many explicit and implicit problems. We have discussed here static or quasi-static controls, but it may be far more difficult to obtain rich results on dynamic controls. Paradoxes in flow controls that we have not discussed here may be pursued later.

Appendix A: The results on the paradoxes for general symmetric distributed computer systems

The following shows the results on the models generalized with respect to the number of nodes, the characteristics of job types, the node processing times, and the job transfer facilities from the models presented in Section 4.1 [27].

A.1. The model and assumptions

The model described in Section 4.1.1 is generalized as in the following. We consider a system with m (≥ 2) nodes (host computers or processors) connected with a job-transfer means. Jobs that arrive at each node i, $i = 1, 2, \cdots, m$, are classified into n types k, $k = 1, 2, \cdots, n$. Consequently, we have mn different job classes R_{ik}. Each class R_{ik} is distinguished by the node i at which its jobs arrive and by the type k of the jobs. We call such a class *local class*, or simply *class*. We assume that each node has an identical arrival process and identical processing capacity. Jobs of type k arrive at each node with node-independent rate ϕ_k. We denote the total arrival rate to the node by ϕ ($= \sum_k \phi_k$), and without loss of generality, we assume a time scale such that $\phi = 1$. We also consider what we call *global class* J_k that consists of the collection of local class R_{ik}, *i.e.*, $J_k = \bigcup_i R_{ik}$. J_k thus consists of all jobs of type k. Whereas, for local class R_{ik}, all the jobs arrive at the same node i, the arrivals of the jobs of global class J_k are equally distributed over all nodes i.

The average processing (service) time (without queueing delays) of a type-k job at any node is $1/\mu_k$ and is, in particular, node-independent. We denote ϕ_k/μ_k by ρ_k and $\rho = \sum_k \rho_k$. Out of type-k jobs arriving at node i, the rate x_{ijk} of jobs is forwarded upon arrival through the job-transfer means to another node j ($\neq i$) to be processed there. The remaining rate $x_{iik} = \phi_k - \sum_{j(\neq i)} x_{ijk}$ is processed at node i. Thus $\sum_q x_{iqk} = \phi_k$. That is, the rate x_{ijk} of type-k jobs that arrive at node i is forwarded through the job-transfer means to node j, while the rate x_{iik} of local-class R_{ik} jobs is processed at the arrival node i. We have $0 \leq x_{ijk} \leq \phi_k$, for all i, j, k. Within these constraints, a set of values for $\mathbf{x}_{ik}$ $(i = 1, 2, \cdots, m, k = 1, 2, \cdots, n)$ are chosen to achieve optimization, where $\mathbf{x}_{ik} = (x_{i1k}, \cdots, x_{imk})$ is an m-dimensional vector and called 'local-class R_{ik} strategy'. We define a global-class J_k strategy as the mm-dimensional vector $\mathbf{x}_k = (\mathbf{x}_{1k}, \mathbf{x}_{2k}, \cdots, \mathbf{x}_{mk})$. We will also denote, by an mmn-dimensional vector $\mathbf{x}$, the vector of strategies concerning all local classes, $\mathbf{x} = (\mathbf{x}_1, \mathbf{x}_2, \cdots, \mathbf{x}_n)$.We call $\mathbf{x}$ the *strategy profile*.

For a strategy profile $\mathbf{x}$, the load β_i on node i is

$$\beta_i = \beta_i(\mathbf{x}) = \sum_{j,k} \mu_k^{-1} x_{jik}. \tag{14}$$

The contribution $\beta_i^{(k)}$ on the load of node i by type-k jobs is

$$\beta_i^{(k)} = \beta_i^{(k)}(\mathbf{x}) = \sum_{j} \mu_k^{-1} x_{jik}, \tag{15}$$

and clearly $\beta_i = \beta_i^{(1)} + \beta_i^{(2)} + \cdots + \beta_i^{(n)}$.

Clearly from (14)

$$\sum_l \beta_l = \sum_{q,l,k} \mu_k^{-1} x_{qlk} = \sum_k \sum_l \sum_q \mu_k^{-1} x_{qlk} = \sum_k \rho_k m = m\rho. \tag{16}$$

We denote the set of $\mathbf{x}$'s that satisfy the constraints (*i.e.*, $\sum_l x_{ilk} = \phi_k, x_{ijk} \geq 0$, for all i, j, k) by $\mathbf{C}$. Note that $\mathbf{C}$ is a compact set.

We have the following assumptions:

Assumption $\Pi 1$ *The expected processing (including queueing) time of a type-k job that is processed at node i (or the cost function at node i), is a strictly increasing, convex and continuously differentiable function of β_i, denoted by $\mu_k^{-1} D(\beta_i)$ for all i, k.*

Assumption $\Pi 2$ *The expected job-transfer delay (including queueing delay) or the cost for forwarding type-k jobs arriving at node i to node j ($i \neq j$), denoted by $G_{ijk}(\mathbf{x})$, is a positive, nondecreasing, convex and continuously differentiable function of $\mathbf{x}$. $G_{iik}(\mathbf{x}) = 0$. Each job is forwarded at most once.*

We refer to the length of time between the instant when a job arrives at a node and the instant when it leaves one of the nodes, after all processing and job-transferring, if any, are over as *the sojourn time* for the job. The expected sojourn time of a local-class R_{ik} job that arrives at node i, $T_{ik}(\mathbf{x})$, is expressed as,

$$T_{ik}(\mathbf{x}) = \sum_j x_{ijk} T_{ijk}(\mathbf{x}), \tag{17}$$

where

$$T_{iik}(\mathbf{x}) = \mu_k^{-1} D(\beta_i(\mathbf{x})), \quad \text{and} \tag{18}$$

$$T_{ijk}(\mathbf{x}) = \mu_k^{-1} D(\beta_j(\mathbf{x})) + G_{ijk}(\mathbf{x}) = T_{iik}(\mathbf{x}) + G_{ijk}(\mathbf{x}), \text{ for } j \neq i. \tag{19}$$

Using the fact that all nodes have the same arrival process, the expected sojourn time of a global-class k job is

$$T_k(\mathbf{x}) = \frac{1}{m} \sum_i T_{ik}(\mathbf{x}). \tag{20}$$

The overall expected sojourn time of a job that arrives at the system is

$$T(\mathbf{x}) = \sum_k \phi_k T_k(\mathbf{x}) = \frac{1}{m} \sum_{i,k} \phi_k T_{ik}(\mathbf{x}), \tag{21}$$

A.2. Results

(A) [Completely centralized decision scheme: Overall optimization] The overall optimum $\bar{\mathbf{x}}$ is unique and given as follows: For all $i, j(\neq i), k$, $\bar{x}_{ijk} = 0$ and $\bar{x}_{iik} = \phi_k$ (No transfer facility is used). The expected sojourn time is, for all i, k,

$$T_k(\bar{\mathbf{x}}) = T_{ik}(\bar{\mathbf{x}}) = \mu_k^{-1} D(\rho), \ T(\bar{\mathbf{x}}) = \rho D(\rho).$$

(B) [Completely dispersed decision scheme: Individual optimization] The individual optimum $\hat{\mathbf{x}}$ is unique and equal to the overall optimum $\bar{\mathbf{x}}$. Therefore, in contrast to the Braess network, no paradox occurs in the individual optima.

(C) [Intermediately dispersed decision scheme: Group optimization] Furthermore, we have the following assumption on the job transfer facility.

[Assumption Π3] Define the following function $G_{ijk}(\mathbf{x})$:

Type G-I: $G_{ijk}(\mathbf{x}) = \omega_k^{-1} \underline{G}(\omega_k^{-1} x_{ijk})$

(One dedicated line for each combination of a pair of origin and destination nodes, and a local class: *i.e.*, $m(m-1)n$ lines in total).

Type G-II(a): $G_{ijk}(\mathbf{x}) = \omega_k^{-1} \underline{G}(\sum_{p,q \neq p} \omega_k^{-1} x_{pqk})$

(One bus line for each global class: *i.e.*, n bus lines in total),

Type G-II(b): $G_{ijk}(\mathbf{x}) = \omega_k^{-1} \underline{G}(\sum_{p,q(\neq p),r} \omega_r^{-1} x_{pqr})$

(One common bus line for the entire system: *i.e.*, 1 bus line.)

where ω_k is a constant, $\underline{G}(0) = 1$, and $\underline{G}(x)$ is a nondecreasing, convex, and differentiable function of x.

Remark A.1 ω_k^{-1} can be regarded as the expected job transfer time (without queueing delays) for forwarding a Type-k job from the arrival node to another processing node. □

The group optimum $\tilde{\mathbf{x}}$ satisfies, for all i and k, the following:

$$T_{ik}(\tilde{\mathbf{x}}) = \min_{\mathbf{x}_{ik}} T_{ik}(\tilde{\mathbf{x}}_{-(ik)}; \mathbf{x}_{ik}), \quad \text{with constraints: } (\tilde{\mathbf{x}}_{-(ik)}; \mathbf{x}_{ik}) \in \mathbf{C},$$

where $(\tilde{\mathbf{x}}_{-(ik)}; \mathbf{x}_{ik})$ is the mmn-dimensional vector with the elements corresponding to $\tilde{\mathbf{x}}_{ik}$ being replaced by $\mathbf{x}_{ik}$.

Define $\tilde{g}_{ijk}(\cdot)$ as follows:

$$\tilde{g}_{ijk}(\mathbf{x}) = \frac{\partial}{\partial x_{ijk}} \{\phi_k \sum_{p \neq i} x_{ipk} G_{ipk}(\mathbf{x})\}. \tag{22}$$

In the case where the assumption II3 holds, for all $i, j(\neq i), k$ that satisfies $x_{ijk} = x_k$, denote as follows:

$$G_k(\mathbf{x}) = G_{ijk}(\mathbf{x}) \text{ and } \tilde{g}_k(\mathbf{x}) = \tilde{g}_{ijk}(\mathbf{x}).$$

Group optimum: Denote $\Gamma_k = \rho_k^2 \sigma_k^{-1}$ and $\sigma_k = \phi_k / \omega_k$. The group optimum $\tilde{\mathbf{x}}$ is unique and is given as follows:

The cases of G-I and G-II(a)

(a) As to group R_{ik} such that $\Gamma_k D'(\rho) \leq \underline{G}(0)$: For all $i, j(\neq i)$ $\tilde{x}_{ijk} = 0$, and $\tilde{x}_{iik} = \phi_k$. This is identical to the overall optimum $\bar{\mathbf{x}}$. Similarly, the expected sojourn time is, for all i, k,

$$T_k(\tilde{\mathbf{x}}) = T_{ik}(\tilde{\mathbf{x}}) = \mu_k^{-1} D(\rho), \;\; T(\tilde{\mathbf{x}}) = \rho D(\rho).$$

(b) As to group R_{ik} such that $\Gamma_k D'(\rho) > \underline{G}(0)$: For all $i, j(\neq i), k,$ $\tilde{x}_{ijk} = \tilde{x}_k$, where $\tilde{x}_k$ is the unique solution of the following:

$$\begin{aligned} &\rho_k^2 \phi_k^{-1}(\phi_k - m\tilde{x}_k) D'(\rho) = \tilde{g}_k(\tilde{x}_k) \\ &= \sigma_k[\underline{G}(m(m-1)\omega_k^{-1}\tilde{x}_k) + \omega_k^{-1}(m-1)\tilde{x}_k \underline{G}'(m(m-1)\omega_k^{-1}\tilde{x}_k)]. \end{aligned}$$

The expected sojourn time is, for all i, k,

$$T_k(\tilde{\mathbf{x}}) = T_{ik}(\tilde{\mathbf{x}}) = \mu_k^{-1} D(\rho) + \phi_k^{-1}(m-1)\tilde{x}_k G_k(\tilde{\mathbf{x}}). \tag{23}$$

The case of G-II(b)

The group optimum is obtained by the following steps: First, reorder k as follows:

$$\Gamma_1 \geq \Gamma_2 \geq \cdots \geq \Gamma_k \geq \cdots \geq \Gamma_n.$$

The we have the following 3 cases as to K,

$$\Gamma_K D'(\rho) > \underline{G}(0),\; \Gamma_{K+1} D'(\rho) \leq \underline{G}(0) \tag{24}$$

$$\text{or } \Gamma_n D'(\rho) > \underline{G}(0) \text{ (that is, } K = n) \tag{25}$$

$$\text{or } \Gamma_1 D'(\rho) \leq \underline{G}(0). \tag{26}$$

In the case (26), the unique solution $\tilde{x}_k = 0$, for all k, is obtained. In the case (24) or (25), the unique solution is obtained in the following way. Define $F_k(X)$ as follows:

$$F_k(X) = \left\{ \sum_{l=1}^{k} \frac{\sigma_l[\Gamma_l D'(\rho) - \underline{G}(X)]}{m\Gamma_k D'(\rho) + (m-1)\omega_l^{-1}\underline{G}'(X)} \right\} - \frac{X}{m(m-1)}.$$

Obtain the largest k and $X = \tilde{X}_{\tilde{k}} (> 0)$ that satisfy $F_{\tilde{k}}(\tilde{X}_{\tilde{k}}) = 0$ and $[\Gamma_{\tilde{k}} D'(\rho) - \underline{G}(\tilde{X}_{\tilde{k}})] > 0$. Then, by using the next equation (27), we obtain $\tilde{x}_k$ for $k = 1, 2, \cdots, \tilde{k}$,

$$\sigma_k[\Gamma_k D'(\rho) - \underline{G}(\tilde{X}_{\tilde{k}})] = \omega_k^{-1} \tilde{x}_k [m\Gamma_k D'(\rho) + (m-1)\omega_k^{-1}\underline{G}'(\tilde{X}_{\tilde{k}})]. \tag{27}$$

We can thus obtain the unique set of values such that $\tilde{x}_k > 0, k = 1, 2, \cdots, \tilde{k}$ and that $\tilde{x}_{\tilde{k}+1} = \tilde{x}_{\tilde{k}+2} = \cdots = \tilde{x}_n = 0$. This is the unique solution. The expected sojourn time is, for all i, k,

$$T_k(\tilde{\mathbf{x}}) = T_{ik}(\tilde{\mathbf{x}}) = \mu_k^{-1} D(\rho) + \phi_k^{-1}(m-1)\tilde{x}_k G_k(\tilde{\mathbf{x}}). \tag{28}$$

Therefore, we have the following conclusion: In symmetric distributed systems, the necessary and sufficient condition for the occurrence of paradoxes is that there exists a job type k such that $\Gamma_k D'(\rho) > 1$.

Remark A.2 Thus, the possibility of the coincident cost degradations depends on the value Γ_k $(= \rho_k^2/\sigma_k = \phi_k\omega_k/\mu_k^2)$, and different for each group of jobs. The probability of paradox is higher for groups with the larger arrival rate (ϕ_k), with the longer processing time (μ_k^{-1}), and with the smaller job-transfer capacity (ω_k^{-1}). Furthermore, in the case of higher utilization factors of each node (ρ $(= \sum_k \rho_k)$), paradoxes may occur more easily. By noting that $\sum_k \phi_k = \phi = 1$, as the number n of job types is greater, each ϕ_k becomes smaller, and the possibility of paradox may be smaller. On the other hand, the number of nodes m may not have big influence on the occurrence of paradoxes. □

Author details

Hisao Kameda
University of Tsukuba, Japan

7. References

[1] Altman, E., Başar, T., Jiménez, T. & Shimkin, N. [2002]. Competitive routing in networks with polynomial cost, *IEEE Trans. Automatic Control* 47(1): 92–96.

[2] Altman, E. & Kameda, H. [2005]. Equilibria for multiclass routing in multi-agent networks, *in* A. Nowak & K. Szajowski (eds), *Advances in Dynamic Games: Annals of International Society of Dynamic Games Vol. 7*, Birkhäuser, Boston, pp. 343–367. (An extended version of the paper that appeared in *Proc. 40th IEEE Conference on Decision and Control* (*IEEE CDC'01*), Orlando, U.S.A., pp. 604–609, Dec. 2001).

[3] Altman, E., Kameda, H. & Hosokawa, Y. [2002]. Nash equilibria in load balancing in distributed computer systems, *International Game Theory Review* 4(2): 91–100.

[4] Boulogne, T., Altman, E. & Pourtallier, O. [2002]. On the convergence to Nash equilibrium in problems of distributed computing, *Annals of Operations Research* 109: 279–291.
[5] Braess, D. [1968]. Über ein Paradoxon aus der Verkehrsplanung, *Unternehmensforschung* 12: 258–268.
[6] Calvert, B., Solomon, W. & Ziedins, I. [1997]. Braess's paradox in a queueing network with state-dependent routing, *J. Appl. Prob.* 34: 134–154.
[7] Cohen, J. E. [1998]. Cooperation and self-interest: Pareto-inefficiency of Nash equilibria in finite random games, *Proc. Natl. Acad. Sci. USA* 95: 9724–9731.
[8] Cohen, J. E. & Horowitz, P. [1991]. Paradoxial behaviour of mechanical and electrical networks, *Nature* 352: 699–701.
[9] Cohen, J. E. & Jeffries, C. [1997]. Congestion resulting from increased capacity in single-server queueing networks, *IEEE/ACM Trans. Networking* 5(2): 305–310.
[10] Cohen, J. E. & Kelly, F. P. [1990]. A paradox of congestion in a queuing network, *J. Appl. Prob.* 27: 730–734.
[11] Dafermos, S. & Nagurney, A. [1984a]. On some traffic equilibrium theory paradoxes, *Transpn. Res. B* 18: 101–110.
[12] Dafermos, S. & Nagurney, A. [1984b]. Sensitivity analysis for the asymmetric network equilibrium problem, *Mathematical Programming* 28: 174–184.
[13] Dubey, P. [1986]. Inefficiency of Nash equilibria, *Mathematics of Operations Research* 11(1): 1–8.
[14] El-Zoghdy, S. F., Kameda, H. & Li, J. [2003]. Numerical studies on paradoxes in non-cooperative distributed computer systems, *Game Theory and Application*, Vol. 9, pp. 1–16.
[15] El-Zoghdy, S. F., Kameda, H. & Li, J. [2006]. Numerical studies on a paradox for non-cooperative static load balancing in distributed computer systems, *Computers and Operations Research* 33: 345–355.
[16] Frank, M. [1981]. The Braess paradox, *Mathematical Programming* 20: 283–302.
[17] Frank, M. [1984]. Cost effective links of ladder networks, *Methods of Operations Research* 45: 75–86.
[18] Haurie, A. & Marcotte, P. [1985]. On the relationship between Nash-Cournot and Wardrop equilibria, *Networks* 15: 295–308.
[19] Inoie, A., Kameda, H. & Touati, C. [2004]. A paradox in flow control of $M/M/m$ queues, *Proceedings of the 43rd IEEE Conference on Decision and Control*, Paradise Island, The Bahamas, pp. 2768–2773.
[20] Inoie, A., Kameda, H. & Touati, C. [2006]. A paradox in flow control of $M/M/n$ queues, *Computers and Operations Research* 33: 356–368.
[21] Kameda, H. [2002]. How harmful the paradox can be in the Braess/Cohen-Kelly-Jeffries networks, *Proc. IEEE INFOCOM 2002*, New York, pp. 437–445.
[22] Kameda, H. [2009]. Coincident cost improvement vs. degradation by adding connections to noncooperative networks and distributed systems, *Networks and Spatial Economics* 9(2): 269–287.
[23] Kameda, H., Altman, E. & Kozawa, T. [1998]. Braess-like paradoxes of Nash equilibria for load balancing in distributed computer systems, *Technical Report ISE-TR-98-157*, Institute of Information Sciences and Electronics, University of Tsukuba.

[24] Kameda, H., Altman, E., Kozawa, T. & Hosokawa, Y. [2000]. Braess-like paradoxes in distributed computer systems, *IEEE Trans. Automatic Control* 45(9): 1687–1691.
[25] Kameda, H., Kozawa, T. & Li, J. [1997]. Anomalous relations among various performance objectives in distributed computer systems, *Proc. 1st World Congress on Systems Simulation*, IEEE, pp. 459–465.
[26] Kameda, H., Li, J., Kim, C. & Zhang, Y. [1997]. *Optimal Load Balancing in Distributed Computer Systems*, Springer.
[27] Kameda, H. & Pourtallier, O. [2002]. Paradoxes in distributed decisions on optimal load balancing for networks of homogeneous computers, *J. ACM* 49(3): 407–433.
[28] Kameda, H. & Ui, T. [2012]. Effects of symmetry on globalizing separated monopolies to a Nash-Cournot oligopoly, *International Game Theory Review* (to appear) .
[29] Knödel, W. [1969]. *Graphentheoretishe Methoden und ihre Anwendungen*, Springer-Verlag, Berlin.
[30] Korilis, Y. A., Lazar, A. A. & Orda, A. [1995]. Architecting noncooperative networks, *IEEE J. Selected Areas in Communications* 13: 1241–1251.
[31] Korilis, Y. A., Lazar, A. A. & Orda, A. [1999]. Avoiding the Braess paradox in noncooperative networks, *J. Appl. Prob.* 36: 211–222.
[32] Koutsoupias, E. & Papadimitriou, C. [1999]. Worst-case equilibria, *Proceedings of the 16th Annual Symposium on Theoretical Aspects of Computer Science*, pp. 404–413.
[33] Legrand, A. & Touati, C. [2007]. How to measure efficiency?, *Proceedings of the 1st International Workshop on Game theory for Communication networks (Game-Comm'07)*.
[34] Li, J. & Kameda, H. [1998]. Load balancing problems for multiclass jobs in distributed/parallel computer systems, *IEEE Trans. Comput.* 47(3): 322–332.
[35] Masuda, Y. & Whang, S. [2002]. Capacity management in decentralized networks, *Management Science* 48: 1628–1634.
[36] Murchland, J. D. [1970]. Braess's paradox of traffic flow, *Transpn. Res.* 4: 391–394.
[37] Orda, A., Rom, R. & Shimkin, N. [1993]. Competitive routing in multiuser communication networks, *IEEE/ACM Trans. Networking* 1: 614–627.
[38] Osborne, M. J. & Rubinstein, A. [1994]. *A Course in Game Theory*, The MIT Press, Cambridge, Mass.
[39] Papadimitriou, C. H. [2001]. Algorithms, games and the internet, *STOC*, ACM Press, pp. 749–753.
[40] Pas, E. I. & Principio, S. L. [1997]. Braess's paradox: Some new insights, *Transpn. Res. B* 31: 265–276.
[41] Patriksson, M. [1994]. *The Traffic Assignment Problem – Models and Methods*, VSP, Utrecht.
[42] Rosenthal, R. W. [1973]. A class of games processing pure-strategy Nash equilibria, *International Journal of Game Theory* 2: 65–67.
[43] Roughgarden, T. [2005]. *Selfish Routing and the Price of Anarchy*, The MIT Press.
[44] Roughgarden, T. [2006a]. On the severity of Braess's paradox: Designing networks for selfish users is hard, *Journal of Computer and System Sciences* 72(5): 922–953.
[45] Roughgarden, T. [2006b]. Selfish routing and the price of anarchy, *OPTIMA*.
[46] Roughgarden, T. & Tardos, É. [2004a]. Bounding the inefficiency of equilibria in nonatomic congestion games, *Games and Economic Behavior* 47(2): 389–403.
[47] Roughgarden, T. & Tardos, É. [2004b]. A stronger bound on Braess's paradox, *Proceedings of the 15th Annual ACM-SIAM Symposium on Discrete Algorithm*, pp. 333–334.

[48] Samuelson, P. A. [1992]. Tragedy of the open road: Avoiding paradox by use of regulated public utilities that charged corrected Knightian tolls, *J. Int. and Comparative Econ.* 1: 3–12.
[49] Smale, S. [1973]. Optimizing several functions, *Proc. of the Int. Conf. on Manifolds and Related Topics in Topology (Manifolds Tokyo 1973)*, Univ. Tokyo Press, Tokyo, Japan, pp. 69–75.
[50] Smith, A. [1776]. *The Wealth of Nations, Book IV, Ch. II*, Modern Library.
[51] Steinberg, R. & Zangwill, W. I. [1983]. The prevalence of Braess's paradox, *Transportation Science* 17(3): 301–318.
[52] Taguchi, A. [1982]. Braess's paradox in a two terminal transportation network, *J. Oper. Res. Soc. of Japan* 25(4): 376–388.
[53] Tantawi, A. N. & Towsley, D. [1985]. Optimal static load balancing in distributed computer systems, *J. ACM* 32(2): 445–465.
[54] Wardrop, J. G. [1952]. Some theoretic aspects of road traffic research, *Proc. Inst. Civ. Eng., Part 2* 1: 325–378.
[55] Zhang, Y., Kameda, H. & Shimizu, K. [1992]. Parametric analysis of optimal load balancing in distributed computer systems, *Journal of Information Processing* (Info. Proc. Soc. of Japan) 14(4): 433–441.

Chapter 10

On the Long-Run Equilibria of a Class of Large Supergames*

Anjiao Wang, Zhongxing Ye and Yucan Liu

Additional information is available at the end of the chapter

http://dx.doi.org/10.5772/54317

1. Introduction

This paper studies a broad class of large supergames, i.e.,infinitely repeated games played by (infinitely) many players. By relating each supergame to a relevant stochastic process of strategy configuration, we first investigate the existence, uniqueness and stability of invariant measures which represent the long-run equilibrium plays. Then we study the relationship between those invariant measures and the solution concepts evolved from the game theory literature.

In our stylized class of supergames, game players are located on the vertex set of a graph, typically the d-dimensional integer lattice, Z^d (see Fig.1-3). The spatial arrangement and the location of a particular player have no tangible restriction other than offering a convenient way to establish a neighborhood structure, since in the class of games which we study the individual players are assumed to be identical.

Each player plays a continent stage game, only with her neighbor, in each and every period of discrete times. Players may or may not have chances to change their strategy simultaneously at every period of time. We endow the graph with a pre-specified ordering according to which is the global updating rule. Supergames differ from one another possibly because of differences in the stage games or differences in the orderings.

When a player is on her turn to update her strategy, the following features of bounded rationality are assumed:

(1) Although all players may employ mixed strategy, each player can only observes an track of the history of pure strategies of her relevant neighbors. In a series of studies Kalai and

*This work was presented in part at the International Conference on Intelligent Computation Technology and Automation(ICICTA), May 11-12, 2010, Changsha, China, and at The International Conference on Computer and Management (CAMAN 2011), May 19-21, 2011, Wuhan, China. A slightly different version has been published in Journal of Mathematical Sciences: Advances and Applications, Vol.13, No.1, 21-46, 2012.

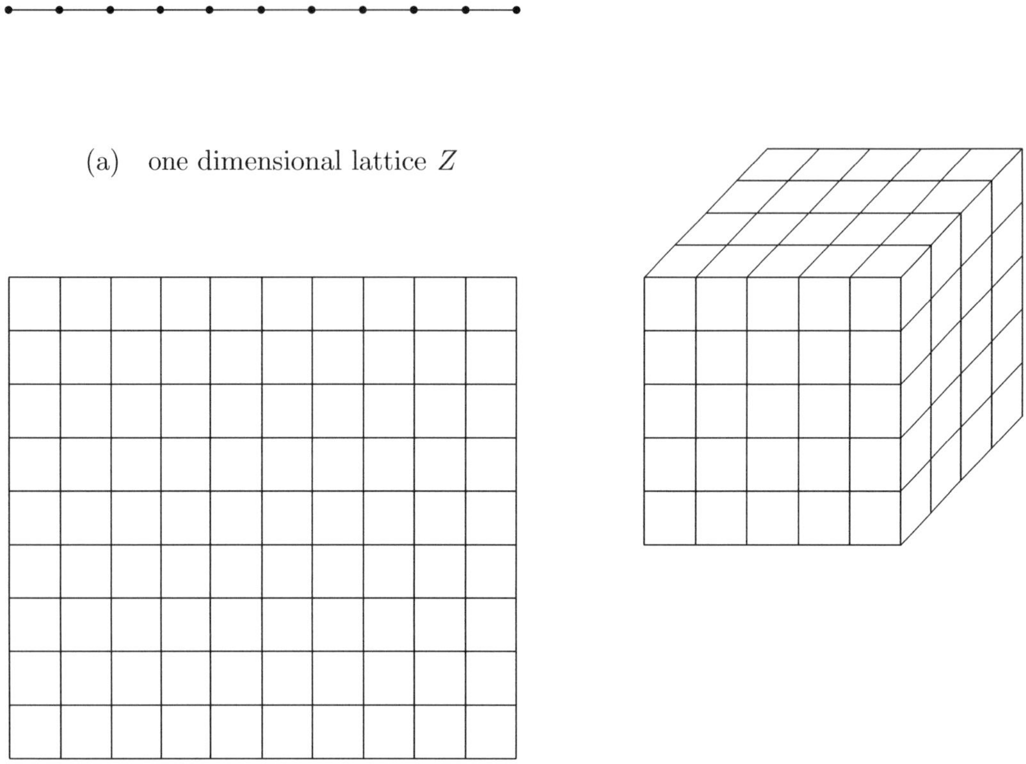

(a) one dimensional lattice Z

(b) 2-dimensional rectangle lattice Z^2

(c) 3-dimensional cubic lattice Z^3

Figure 1. Lattices Z^d

Labra(1990[6],1991a[7],1991b[8]) investigated, in a finite-player supergame, the possibility for the players to learn to play Nash equilibria by keeping track of the history players, engaging in Baysian learning and taking best response policy. In this paper we do not intend to extend their results to large supergames although it may be an interesting topic. Relevant problems have been investigated in the literature of Cellular Automata (for example,see [4,10,11]).

(2) To make a strategy choice, the player only takes into account her neighbors' plays in the previous period. She is not sensitive enough to instaneously make response to the change in their neighbors' plays and to play based on her inference of her neighbors' current and future plays.

(3) The player may or may not have full control over her choices in the sense that she may or may not be able to use the best response strategy. The possibility of choosing strategy a versus strategy b depends on the difference in average payoffs accrued by playing a against her neighbors versus by playing b.

In summary, the class of games which we study in this paper is a class of infinitely repeated games with infinite numbers of players. Each player is directly connected only with her finite neighbors. The information flow is very close to the traditional open-loop setting.

In Section 2 we describe the general formulation of the class of supergames by offering the ingredients needed. We relate each game setting to a stochastic process which represents the evolution of the strategy configuration of the game. This enables us to investigate the evolution of the supergame by looking at the invariant measure of the relevant stochastic process. For the whole class of games we first prove the existence of invariant measures and then study the relationship among the concepts of invariant measures, reversible measures, ergodic measures, Gibbs measures and Markovian random fields.

In Section 3, we derive invariant measures for some special supergames. We assume that each person located on the vertices of the rectangle lattice plays two-person two strategies games with her 4 neighbors simultaneously. The payoff function is symmetric. We prove some results on the existence of ergodic measures. In this Section, we also investigate another class of supergames. We assume that each person located on the vertices of the rectangle lattice plays four 4-person team games or 4-person 2-pairs team games simultaneously with her neighbors. The formula of invariant measure which represents the long-run equilibrium plays with symmetric payoffs is obtained.

Section 4 is the conclusion. Some speculations on possible future research of other type supergames on different lattices are discussed.

2. General formulation of a class of large supergames

This section is devoted to describe the general formulation of a class of supergames. Subsection 2.1 presents the ingredients. Subsection 2.2 introduces the strategy evolution process(SEP). Subsection 2.3 depicts all the subclass of games. Subsection 2.4 contains general results on the existence of invariant distribution, reversibility and ergodicity of the SEP.

2.1. Ingredients

The class of large supergames which we investigate in this paper has the following ingredients:

The players:

We assume that players are located on the sites of a graph $G = \{V, E\}$, where V is the vertex set and E the edge set of the graph. In this work we assume that V is a lattice, usually Z^d or its finite sublattice. We also assume that all the players are identical.

Neighborhood:

A neighborhood system $\mathcal{N} = \{N_i;\ i \in V\}$ is assigned with each specific model. $\mathcal{N}$ is a collection of nonempty subsets of the vertices of V such that

(i) i does not belong to N_i for all $i \in V$;

(ii) $i \in N_j$ if and only if $j \in N_i$, for all $i, j \in V$;

N_i is called the neighborhood of i. We define the set $W_i = N_i \cup \{i\}$. A clique C of V is either a single vertex or a nonempty subset of vertex set of V such that all vertices that belong to C are neighbors of each other.

To simplify the model we assume that the neighborhood structure is translation invariant. For example, in Z^d cases, the invariance of the neighborhood system means that $i \in N_j$ if and only if $i+k \in N_{j+k}$. Since $N_i = (i-j) + N_j$ for any $i, j \in Z^d$, from $N_i = i + N_0$, which 0 is the origin vertex of Z^d, we deduce that all the information of the neighborhood structure is contained in N_0. For $V \subset Z$, the neighborhood N_0 is usually taken as $N_0 = \{-s, -s+1, \cdots, -1, +1, \cdots, s\}$ for some $s > 0$. For $V \subset Z^2$, the two commonest choices for N_0 are the following

$N_0^N = \{(1,0), (-1,0), (0,1), (0,-1)\}$ the von Neumann neighborhood,

$N_0^M = N_0^N \cup \{(1,1), (1,-1), (-1,1), (-1,-1)\}$ the Moore Neumann neighborhood.

Stage games:

In our class of supergames, some stage games are played over discrete time $t \in Z^d$. At each discrete time every player plays a finite-strategy n-person game simultaneously with his neighbors. Mixed strategy is used in general. At the end of each stage game, every player obtains information about the pure strategies that his neighbors took in the finished game and then may revise his strategy, under some global ordering of updating, for the next game according to these information and the payoff that he received. Then the game is repeated again.

Strategy and payoff:

Let A_i be the finite set of all possible pure strategies that the player i can take and assume that $A_i = A$ for all $i \in V$. In a n-person game, let $Q(x_1, \cdots, x_n), x_1, \cdots, x_n \in A$ be the payoff to the player who plays $x_1 \in A$, when his relavent neighbors play $(x_2, \cdots, x_n) \in A^{n-1}$. We assume $Q(x_1, \cdots, x_n)$ that is invariant under any permutation of $(x_2, \cdots, x_n)$.

Ordering of strategy changes:

At each period of time, all players may or may not update their strategies simultaneously. Instead, associated with each game is an ordering according to which the players change their strategies. Such an ordering over V which is pre-specified as in extensive form games, is represented by the global updating rule.

The global updating rule will be called *synchronous* if all the players change their strategies simultaneously at the same time; *sequential*, if they change their strategies one by one under a fixed ordering; *group − sequential*, if the players within a group change their strategies simultaneously at the same time, but different groups change their strategies one group at a time under a fixed ordering; and *asynchronous* if at a given time only one player - selected by random with uniform probability - updates his strategy. The *sequential* and *asynchronous* updating rules are applicable only for the case with finite players(i.e., V is finite).

2.2. Strategy evolution process(SEP)

The dynamics of a supergame is characterized by a stochastic process which is called strategy evolution process(SEP) in this paper. Technically, the SEP for a large supergame is a Markov chain whose state at time t is denoted by $\mathbf{X_t} = \{X_{t,i}; i \in V\}$. It takes value over $\Omega_t = \Omega = A^V$ which is called configuration space of the SEP at time t. $\mathbf{x_t} = \{x_{t,i}; i \in V\}$ is the realization of $\mathbf{X_t}$. Equivalently we may model the state of SEP at time t by a probability distribution μ_t

on A^V. Suppose that the configuration $\mathbf{x_{t-1}}$ determines the strategy of player i at time t with probability(called local transition probability)

$$p_i(x_{t,i}|\mathbf{x_{t-1}}) = p_i(x_{t,i}|x_{t-1,j}; j \in W_i) \tag{2.1}$$

Note that

$$\sum_{x_{t,i}\in A} p_i(x_{t,i}|x_{t-1,j}; j \in W_i) = 1, \; for\ all\ i \in V. \tag{2.2}$$

Let $P(\mathbf{y}|\mathbf{x})$ be the global one-step transition probabilities from $\mathbf{x}$ to $\mathbf{y}$. They are defined for different global updating modes as follows, respectively.

(i)Synchronous: the global transition probabilities of the SEP are defined by

$$P(\mathbf{x}_t|\mathbf{x}_{t-1}) = \prod_{i\in V} p_i(x_{t,i}|x_{t-1,j};\ j \in W_i) \tag{2.3}$$

(ii)Group-sequential:

In this work we discuss a specific mode of group-sequential rules, say even-odd sequential rule for Z^d model. We partition the lattice Z^d into two disjoint equivalent sublattices, say V^E and V^O, arranged so that the nearest neighbor sites lie in different sublattices. The player $i \in V^E$ updates his strategy at even time t with the local transition probability given by (2.1). Then the global transition probabilities for the SEP is given by

$$P(\mathbf{x}_t|\mathbf{x}_{t-1}) = \begin{cases} \prod_{i\in V^E} p_i(x_{t,i}|x_{t-1,j}; j \in W_i), \ x_{t,i} = x_{t-1,i}, & \text{for any } i \in V^O \\ 0, & \text{otherwise} \end{cases} \tag{2.4}$$

If t is odd, the updating rule is obtained by reversing the rules of V^E and V^O.

(iii)Asynchronous:

In this case, we assume that V is finite, $\|V\| = M$. We denote by $\mathbf{x}(i, y)$ the configuration that is identical to $\mathbf{x}$, except the strategy of player i is $y \in A$. Then

$$\begin{aligned} P(\mathbf{x}|\mathbf{y}) &= P(\mathbf{X}_t = \mathbf{x}|\mathbf{X}_{t-1} = \mathbf{y}) \\ &= \begin{cases} \frac{1}{M}\sum_{i\in V} P_i(X_{t,i} = x_i|\mathbf{X}_{t-1} = \mathbf{x}), & \text{if } \mathbf{y} = \mathbf{x} \\ \frac{1}{M}P_i(X_{t,i} = x_i|\mathbf{X}_{t-1} = \mathbf{x}(i, y)), & \text{if } \mathbf{y} = \mathbf{x}(i, y) \neq \mathbf{x} \\ 0, & \text{otherwise} \end{cases} \end{aligned} \tag{2.5}$$

2.3. Subclasses

In the remaining of this section, we prove some results which apply to the whole class. The class of supergames that we study is rather broad. All the subclasses can be represented in the following chart:

The ordering pattern of strategy updating	Number of persons in a stage game two	four	four in two pairs
Synchronous	yes	yes	yes
Asynchronous	yes	yes	yes
Group-sequential	yes	yes	yes

Each cell in the chart can be further divided into four subcells according to homogeneity and symmetry of the payoff. For detail see the next section.

2.4. Invariant measures, ergodicity and reversibility

We are interested in the condition on the local transition probability for the existence and the uniqueness of the invariant measure, the ergodicity and reversibility of the SEP. For the same local transition rule given by (2.1), what are the differences between the invariant measures for the SEP with different type of global updating rules? In certain cases there may exist multiple invariant measures. This phenomena is called phase transition.

We are also interested in the inverse problem-for a given distribution π on A^V find all the SEP with π as their invariant measures, specifically when π is Gibbsian.

The answers of these problems vary for different types of transition and updating rules. Finite V or infinite V will imply different results too. We will discuss them separately in the next section.

The global transition probabilities (2.3), (2.4) or (2.5) defines a discrete-time Markov process on the configuration space A^V. Given a measure ρ_{t-1} on the configuration $\mathbf{x}_{t-1}$ (2.3), (2.4) or (2.5) defines a probability measure $\rho_t = \rho_{t-1}P$ on $\mathbf{x}_t$.

$$\rho_t(d\mathbf{x}_t) = \int \rho_{t-1}(d\mathbf{x}_{t-1})P(d\mathbf{x}_t|\mathbf{x}_{t-1})$$

We say that a measure ν is *stationary* or time *invariant* if $\nu = \nu P$. The following result is well known.

Lemma 2.1. *The invariant measures for the time evolution form a nonempty convex set.*

Proof: See [9].

For the SEP with certain type of updating rule we define the following. A SEP is *ergodic* if the chain is regular, i.e., it has a unique invariant measure which almost surely describes the limit behavior of the SEP. A SEP will be called *Gibbsian*, if its invariant measure corresponds to the probability distribution of a Markov random field(MRF) on A^V, i.e.,

$$\lim_{t \longrightarrow \infty} Pr(\mathbf{X}_t = \mathbf{x}) = \pi(\mathbf{x}) = \frac{1}{\Lambda} exp[-U(\mathbf{x})], \tag{2.6}$$

with $\pi(\mathbf{x}) > 0$ for all $\mathbf{x} \in \Omega$, Λ the normalization constant, and $U(\mathbf{x}) = \sum_C v_C(\mathbf{x})$, where the summation is taken over the cliques of V, and where the function $v_C(.)$ is called potential function. We call a SEP *reversible* if the corresponding chain is reversible. It is well known that the reversibility is equivalent to the detailed balance condition.

$$\pi(\mathbf{y})P(\mathbf{x}|\mathbf{y}) = \pi(\mathbf{x})P(\mathbf{y}|\mathbf{x}). \tag{2.7}$$

Any reversible probability distribution is invariant since (2.7) implies that

$$\pi(\mathbf{x}) = \sum_{\mathbf{y}} \pi(\mathbf{y})P(\mathbf{x}|\mathbf{y}). \tag{2.8}$$

Theorem 2.1. *A SEP is Gibbsian if and only if it is reversible.*

Proof: See [9].

3. Invariant measures for some special models

In this section we study the above-mentioned problems in detail and depth for some special game setting. The discussion is organized according to the number of players in a game.

3.1. Two-person game

In this subsection we discuss the case that the players are located on V which may be Z^d or its finite sublattice with the neighborhood structure of von Neuman type, i.e., $N_0 = \{\mathbf{j} = (j_1, \cdots, j_d); |j| = \sum_{k=1}^{d} j_k^2 = 1\}$. Every player plays a q-strategy two person game simultaneously with each of his nearest neighbors (Z^2 case see Fig. 2). We denote by $\mathbf{Q}_{ij} = \{Q_{ij}(x,y); x, y \in A\}$ the payoff matrix of player i playing with player j. $\mathbf{Q}_{ij}$ is called symmetric if $Q_{ij}(x,y) = Q_{ji}(y,x)$ for all $x, y \in A$.

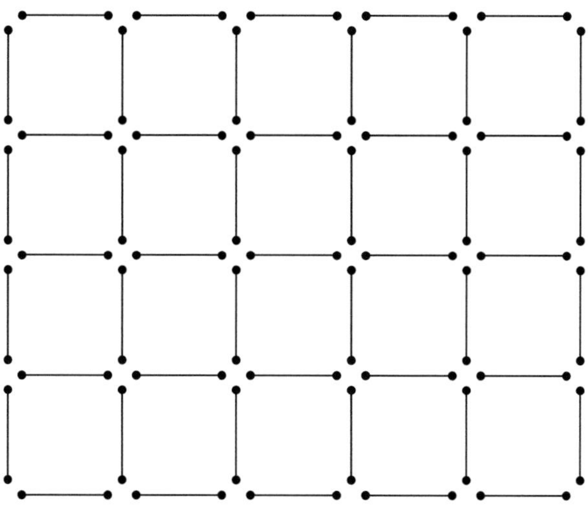

Figure 2. Supergame based on basic two-person games on Z^2

A game played by all nearest pairs on V is called homogeneous if the same payoff function is assigned for all pair players. Otherwise, it is called nonhomogeneous. We will discuss them separately.

At the end of each game, player i receives payoff $Q_{ij}(y,x)$ if he plays strategy y while his neighbor j plays strategy x; so his total payoff from playing strategy y is the sum of the payoffs received from playing y against each of his neighbors. Then player i may revises his strategy from y to z, under the given global updating rule, for the next game with probability

$$\begin{aligned} p_i(z|\mathbf{x}(i,y)) &= \frac{1}{\lambda} exp\{\beta \frac{1}{\|N_i\|} \sum_{j\in N_i} [Q_{ij}(z,x_j) - Q_{ij}(y,x_j)]\} \\ &= \frac{1}{\lambda'} exp\{\beta \frac{1}{\|N_i\|} \sum_{j\in N_i} Q_{ij}(z,x_j)\}. \end{aligned} \tag{3.1}$$

where

$$\lambda = \sum_{z\in A} exp\{\beta \frac{1}{\|N_i\|} \sum_{j\in N_i} [Q_{ij}(z,x_j) - Q_{ij}(y,x_j)]$$

$$\lambda' = \sum_{z\in A} exp\{\beta \frac{1}{\|N_i\|} \sum_{j\in N_i} Q_{ij}(z,x_j)\}$$

$\|N\|$ is the cardinality of set N. Note that both λ and λ' depend on $x^{N_i} = \{x_j : j \in N_i\}$. We may write them by $\lambda(x^{N_i})$ and $\lambda'(x^{N_i})$. Roughly speaking, the probability the player i switchs his strategy from y to z is proportional to the utility differences for these two strategies. We will discuss three cases.

3.1.1. Homogeneous game with symmetric payoff

In this case all the payoff functions equals to $\mathbf{Q}$ which is symmetric. The local transition probability will read as

$$\begin{aligned} p_i(z|\mathbf{x}(i,y)) &= \frac{1}{\lambda} \exp\{\beta \frac{1}{\|N_i\|} \sum_{j\in N_i} [Q_i(z,x_j) - Q_i(y,x_j)]\} \\ &= \frac{1}{\lambda'} \exp\{\beta \frac{1}{\|N_i\|} \sum_{j\in N_i} Q_i(z,x_j)\}. \end{aligned} \tag{3.2}$$

To discuss different global updating rule we denote $V_n = \{\mathbf{j} = (j_1,\cdots,j_d) \in Z^d; |j_k| \leq n \; for \; 1 \leq k \leq d\}$ and assume that V is some finite cubic with this type. $\|V\| = M$.

(i)Asynchronous and even-odd sequential cases

Theorem 3.1. *Consider a large homogeneous supergame with finite players located on a lattice V. The payoff matrix of a two-person game is symmetric. Then the SEP whose asynchronous global transition probability is given by (2.5) with the above local transition rule (3.2) and the SRP whose even-odd sequential global transition probability is given by (2.4) with the above local transition rule (3.2) have the following distribution on A^V as their reversible invariant measure*

$$\pi(\mathbf{x}) = \frac{1}{\Lambda} exp\{\beta \frac{1}{\|N_0\|} \sum_{<i,j>} Q(x_i,x_j)\} \tag{3.3}$$

where the summation is taken over all nearest neighboring pairs of players, and

$$\Lambda = \sum_{\mathbf{x}} \exp\{\beta \frac{1}{\|N_0\|} \sum_{<i,j>} Q(x_i, x_j)\}$$

Proof: We only need to check (2.7) for $\mathbf{y} = \mathbf{x}(i, y)$ for the asynchronous case and $\mathbf{y}$ whose even (or odd) coordinates are different from those of $\mathbf{x}$ for even-odd sequential case.

(a)Asynchronous cases

$$\begin{aligned}
&\pi(\mathbf{x})P(\mathbf{x}(i,y)|\mathbf{x}) \\
&= \frac{1}{\Lambda} \exp\{\frac{\beta}{\|N_0\|}[\sum_{<j,k>,j,k\neq i} Q(x_j, x_k) + \sum_{j\in N_i} Q(x_i, x_j)]\} \cdot \frac{1}{M\lambda'(x^{N_i})} \exp\{\frac{\beta}{\|N_0\|} \sum_{j\in N_i} Q(y, x_j)\} \\
&= \frac{1}{\Lambda} \exp\{\frac{\beta}{\|N_0\|}[\sum_{<j,k>,j,k\neq i} Q(x_j, x_k) + \sum_{j\in N_i} Q(y, x_j)]\} \cdot \frac{1}{M\lambda'(x^{N_i})} \exp\{\frac{\beta}{\|N_0\|} \sum_{j\in N_i} Q(x_i, x_j)\} \\
&= \pi(\mathbf{x}(i,y))P(\mathbf{x}|\mathbf{x}(i,y))
\end{aligned}$$

(b)Even-odd sequential cases

We denote by $\mathbf{x}(E, y^E)(\mathbf{x}(O, y^O))$ for the configuration whose even components equals $y^E = \{y_j, j \in V^E\}$, while the odd components equal $y^O = \{x_i, i \in V^O\}$(vice versa for $\mathbf{x}(O, y^O)$). We need to prove (2.7) for $\mathbf{y} = \mathbf{x}(E, y^E)$(or $\mathbf{x}(O, y^O)$). In fact

$$\begin{aligned}
&\pi(\mathbf{x})P(\mathbf{x}(E,y^E)|\mathbf{x}) \\
&= \frac{1}{\Lambda} \exp\{\frac{\beta}{\|N_0\|} \sum_{<j,k>} Q(x_j, x_k)\} \prod_{i\in V^E} \frac{1}{\lambda'(x^{N_i})} \exp\{\frac{\beta}{\|N_0\|} \sum_{j\in N_i} Q(y, x_j)\} \\
&= \frac{1}{\Lambda} \exp\{\frac{\beta}{\|N_0\|} \sum_{<i,j>,i\in V^E,j\in V^O} Q(y_i, x_j)\} \prod_{i\in V^E} \frac{1}{\lambda'(x^{N_i})} \exp\{\frac{\beta}{\|N_0\|} \sum_{j\in N_i} Q(x_i, x_j)\} \\
&= \pi(\mathbf{x}(E,y^E))P(\mathbf{x}|\mathbf{x}(E,y^E)).
\end{aligned}$$

(ii)Synchronous cases

The global transition probability is given by (2.3) with the above local transition rule (3.2). The above distribution $\pi(\cdot)$ in (3.3) is not the invariant distribution of this SEP.

For even-odd sequential and asynchronous models, it has been realized that there is an intimate relation between d-dimensional time evolution and equilibrium statistical model(ESM) in $(d+1)$-dimension, the extra dimension being the discrete time ([3], [9]). In fact, we can consider $\underline{\mathbf{x}} = \{\mathbf{x}_t\}_{t\in Z}$ as a configuration on the space-time lattice Z^{d+1}.

It is easy to see that if the transition probability $p_i(x_{t,i}|\mathbf{x}_{t-1})$ are all strictly positive, then μ_v is a Gibbsian measure on $A^{Z^{d+1}}$. When V is infinite there are various ways to define finite-volume Gibbs states in the thermodynamical limit, yielding the space-time measure μ_v of the time evolution as an infinite-volume Gibbs measure. From the theory of ESM, it is important to note that there may exist more than one Gibbs measure on $A^{Z^{d+1}}$ which indicates the existence of more than one stationary or periodic measure ν for the time evolution as a phase transition.

Example : binary strategy game

We assume that each player has only two choices of strategies which may be identified as {-1,+1} for simplicity and convenience. The payoff matrix is given by

$$Q = \begin{pmatrix} Q(+1,+1) & Q(+1,-1) \\ Q(-1,+1) & Q(-1,-1) \end{pmatrix} = \begin{pmatrix} a & b \\ b & d \end{pmatrix} \tag{3.4}$$

Or we may write $Q(x,y)$ in the following form

$$Q_i(x,y) = Q(x,y) = Jxy + K(x+y) + L \qquad \text{for all } i \in I \tag{3.5}$$

where J, K and L are uniquely determined by a, b and d; and vice versa.

$$Q(+1,+1) = a = J + 2K + L$$

$$Q(+1,-1) = Q(-1,+1) = b = -J + L$$

$$Q(-1,-1) = d = J - 2K + L;$$

or

$$J = \frac{1}{4}(a - 2b + d)$$

$$K = \frac{1}{4}(a - d)$$

$$L = \frac{1}{4}(a + 2b + d).$$

(i)Asynchronous case

The invariant probability measure can be written as the following form

$$\begin{aligned} \pi(\mathbf{x}) &= \frac{1}{\Lambda} \exp\{\beta \frac{1}{\|N_0\|} \sum_{<i,j>} Q(x_i, x_j)\} \\ &= \frac{1}{\Lambda} \exp\{\beta \frac{1}{\|N_0\|} \sum_{<i,j>} [Jx_i x_j + K(x_i + x_j) + L]\} \\ &= \frac{1}{\Lambda'} \exp\{\tilde{J} \sum_{<i,j>} x_i x_j + \tilde{K} \sum_{i \in V} x_i\} \end{aligned} \tag{3.6}$$

where $\tilde{J} = \frac{\beta J}{\|N_0\|}$ and $\tilde{K} = \frac{4\beta K}{\|N_0\|}$. This is the well known Ising model. From standard point of view $\tilde{J}$ represents local pair interaction, and $\tilde{K}$ represents the global interactions. It is interesting to notice that the payoff matrix $\mathbf{Q}$ contains information of both local and global interactions. This is not surprising because the game is homogeneous, i.e., the same payoff matrix is assigned for all two-person games.

(ii)Even-odd sequential case

The invariant measure has same form of (3.6). It is well known that this model exhibits the phenomena of phase transition in certain cases when V is infinite. To see this we first consider the SRP with even-odd sequential updating rule for each V_n and denote by $\pi_n(\cdot)$ the corresponding invariant measures. Then let $n \longrightarrow \infty$, V_n spreads to cover the whole lattice Z^d. It is well known that there is a unique limiting distribution of sequence $\{\pi_n\}$ when $d = 1$. While $d = 2$ the situation becomes complex but more interesting.

There is no phase transition whenever $\tilde{K} \neq 0$, i.e., there exists only one invariant measure. When $\tilde{K} = 0$(i.e., $a = d$) there is a value $\tilde{J}_c > 0$ called critical value such that, for $0 \leq \tilde{J} \leq \tilde{J}_c$, no phase transition occurs; but if $\tilde{J} > \tilde{J}_c$, phase transition does occur. The critical value is about $\tilde{J}_c = 0.44$(see [1],[9]). Notice that $a = d$ means $Q(+1,+1) = Q(-1,-1)$, and $J = \frac{1}{2}(a-b)$. So when a is sufficiently larger than b, phase transition occurs.

Furthermore, there exist only two extreme distributions π^+ and π^- which are all transition invariant in 2-dimension such that all other invariant measure π can be expressed as convex combination of them, i.e.,

$$\pi = p\pi^+ + (1-p)\pi^-$$

where $0 \leq p \leq 1$(see [1]). The marginal distributions of π^+ and π^-at single site satisfy

$$\pi^+(+1) = \pi^-(-1) > 2/3$$

(see [9]). This means that under $\pi^+(\pi^-)$, all players favor strategy +1(-1). Note that there are two Nash Equilibrium states for which all players play strategy +1 or -1.

For $d \geq 3$, the phase diagrams are more complex and not completely known.

(iii)Synchronous case

To find the invariant measure is, in general, a difficult task.

Remarks: For the SEP of a large homogeneous supergame with asymmetric payoff under different global updating rule with the local transition probability given by (3.1), the invariant Gibbsian measure may not exist, in general. But for some special payoff it exists. Also for binary strategy game it always exists.

3.1.2. Nonhomogeneous game with symmetric payoff

Now we consider the case of nonhomogeneous game with symmetric payoff. We assign different payoff function $Q_{ij}(x,y)$ for different pairs of players i and j, but assume that all payoff are symmetric, i.e., $Q_{ij}(x,y) = Q_{ji}(y,x)$. Then we have similar results. For the SEP with synchronous and even-odd sequential global updating rule associated with the local transition probability (3.1) the invariant measure is then given by

$$\pi(\mathbf{x}) = \frac{1}{\Lambda}\exp\{\frac{\beta}{\|N_0\|}\sum_{<i,j>} Q_{ij}(x_i,x_j)\}$$

3.2. Four-person team game

We return to the model on Z^2. This time we assume that the four players located on the vertices of a basic square $\square = \{(i_1, i_2), (i_1, i_2+1), (i_1+1, i_2+1), (i_1+1, i_2)\}$ form a team to play a four-person team game. Let $\square(i, j, k, l)$ to be the square with vertices i, j, k and l in clockwise order. Denote by S_i the set of basic squares which contain vertex i, and S the set of all basic square.

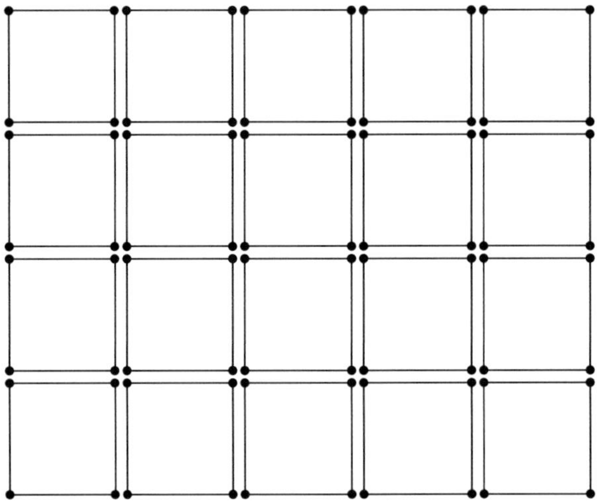

Figure 3. Supergame based on 4-person team games

Each player is a member of four four-person teams consisting of his neighbors. At each time every player plays finite-strategy four-person team games with four neighboring teams simultaneously(see Fig. 3).

Suppose that the payoff function $\mathbf{Q} = \{Q(x, y, z, u); x, y, z, u \in A\}$ is symmetric. The local transition probability for the SEP is given by

$$p_i(z|\mathbf{x}(i, y)) = \frac{1}{\lambda} \exp\{\frac{\beta}{4} \sum_{\square(i,j,k,l) \in S_i} [Q_i(z, x_j, x_k, x_l) - Q_i(y, x_j, x_k, x_l)]\}$$
$$= \frac{1}{\lambda'} \exp\{\frac{\beta}{4} \sum_{\square(i,j,k,l) \in S_i} Q_i(z, x_j, x_k, x_l)\} \tag{3.7}$$

where.

$$\lambda = \sum_{z \in A} \exp\{\frac{\beta}{4} \sum_{\square(i,j,k,l) \in S} [Q_i(z, x_j, x_k, x_l) - Q_i(y, x_j, x_k, x_l)]\}$$
$$\lambda' = \sum_{z \in A} \exp\{\frac{\beta}{4} \sum_{\square(i,j,k,l) \in S_i} Q_i(z, x_j, x_k, x_l)\}.$$

We discuss homogeneous game with symmetric payoff only and claim that for the SEP with asynchronous global updating rule associated with the local transition probability (3.7). The invariant measure is given by

$$\pi(\mathbf{x}) = \frac{1}{\Lambda}\exp\{\frac{\beta}{4}\sum_{\square(i,j,k,l)\in S} Q(x_i, x_j, x_k, x_l)\}. \quad (3.8)$$

Example: binary strategy game

We assume that each player has only two choices of strategies which may be identified as $\{+1, -1\}$. The payoff function is symmetric. We denote

$$\begin{aligned} Q(x,y,z,u) &= Hxyzu + I(xyz + yzu + zux + xyu) \\ &+ J(xy + xz + xu + yz + yu + zu) + K(x + y + z + u) + L \end{aligned}$$

where $H,\ I,\ J,\ K$ and L are uniquely determined by $a,\ b,\ c,\ d$ and e and vice versa.

$$\begin{aligned} Q(+1,+1,+1,+1) &= a = H + 4I + 6J + 4K + L \\ Q(+1,+1,+1,-1) &= b = -H - 2I + 2K + L \\ Q(+1,+1,-1,-1) &= c = H - 2J + L \\ Q(+1,-1,-1,-1) &= d = -H + 2I - 2K + L \\ Q(-1,-1,-1,-1) &= e = H - 4I + 6J - 4K + L \end{aligned}$$

or

$$\begin{aligned} H &= \frac{1}{16}(a - 4b + 6c - 4d + e) \\ I &= \frac{1}{16}(a - 2b + 2d - e) \\ J &= \frac{1}{16}(a - 2c + e) \\ K &= \frac{1}{16}(a + 2b - 2d - e) \\ L &= \frac{1}{16}(a + 4b + 6c + 4d + e) \end{aligned}$$

The invariant measure $\pi(\mathbf{x})$ is a Ising model with one-site, two-site,three-site and four-site intersections.

$$\pi(\mathbf{x}) = \frac{1}{\Lambda}\exp\{\tilde{H}\sum_{\square(i,j,k,l)\in S} x_i x_j x_k x_l + \tilde{I}\sum_{\Delta(i,j,k)\in T} x_i x_j x_k + \tilde{J}\sum_{<i,j>} x_i x_j + \tilde{K}\sum_{i\in I} x_i\} \quad (3.9)$$

where $\tilde{H} = \frac{\beta H}{4}$, $\tilde{I} = \frac{4\beta I}{4}$, $\tilde{J} = \frac{6\beta J}{4}$ and $\tilde{K} = \frac{4\beta K}{4}$; $\Delta(i,j,k)$ the triangle with the vertices i, j and k.

3.3. Four-person two-pair game

We consider another type of 4-person game for Z^2 model. This time we assume that four players within a basic square form two pairs along the diagonal lines (example includes bridge-a card play). Therefore each player plays four 4-person 2-pairs games simultaneously every time (see Fig. 4).

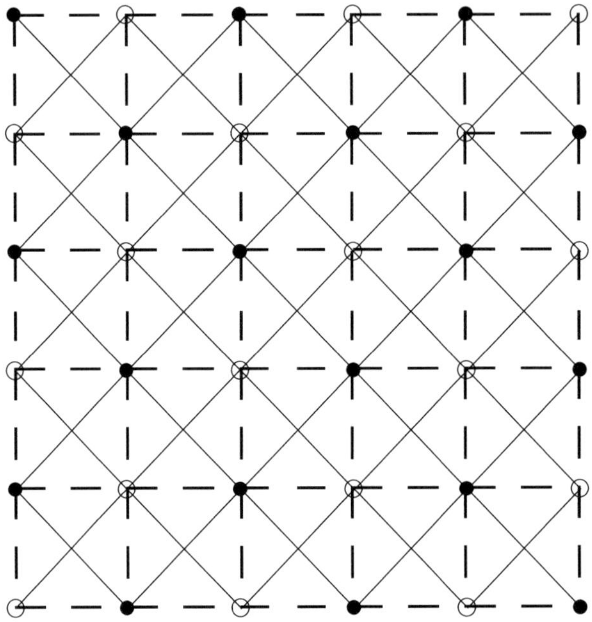

Figure 4. Supergame based on 4-person 2-pairs games on Z^2

If the strategies of four players located on the sites of a basic square $\square(i,j,k,l)$ are (x,y,z,u), then the payoff function is defined by $Q((x,z),(y,u)), x,y,z,u \in A$. Again we assume that each player has only two choices of strategies which may be identified as $A = \{+1,-1\}$, and the payoff function Q is symmetric between pairs (x,z) and (y,u) and within each pair, i.e.,

$$Q((x,z),(y,u)) = Q((z,x),(y,u)) = Q((y,u),(x,z))$$

It can be represented as the following:

$$Q(x,y,z,u) = Hxyzu + I(xyz + yzu + zux + xyu)$$

$$+ J_1(xz + yu) + J_2(xy + yz + zu + ux) + K(x + y + z + u) + L$$

where $H,\ I,\ J_1,\ J_2,\ K$ and L are uniquely determined by $a,\ b,\ c,\ d,\ e$ and f and vice versa.

$$Q((+1,+1),(+1,+1)) = a = H + 4I + 2J_2 + 4J_2 + 4K + L$$

$$Q((+1,+1),(+1,-1)) = b = -H - 2I + 2K + L$$

$$Q((+1,+1),(-1,-1)) = c = H + 2J_1 - 4J_2 + L$$

$$Q((+1,-1),(+1,-1)) = d = H - 2J_1 + L$$

$$Q((+1,-1),(-1,-1)) = e = -H + 2I - 2K + L$$

$$Q((-1,-1),(-1,-1)) = f = H - 4I + 2J_1 + 4J_2 - 4K + L$$

or

$$H = \frac{1}{16}(a - 4b + 2c + 4d - 4e + f)$$

$$I = \frac{1}{16}(a - 2b + 2e - f)$$

$$J_1 = \frac{1}{16}(a + 2c - 4d + f)$$

$$J_2 = \frac{1}{16}(a - 2c + f)$$

$$K = \frac{1}{16}(a + 2b - 2e - f)$$

$$L = \frac{1}{16}(a + 4b + 2c + 4d + 4e + f)$$

We define synchronous, even-odd sequential and asynchronous global updating rules as before. The local transition probability is defined by

$$P_i(z|\mathbf{x}(i,y)) = \frac{1}{\Lambda}\exp\{\frac{\beta}{4}\sum_{\square(i,j,k,l)\in S_i} Q((z,x_k),(x_j,x_l))\}. \tag{3.10}$$

where

$$\lambda = \sum_{z\in A}\exp\{\frac{\beta}{4}\sum_{\square(i,j,k,l)\in S_i} Q((z,x_k),(x_j,x_l))\}.$$

For the SEP with asynchronous and even-odd sequential global updating rules we can prove that the invariant measure is given by

$$\pi(\mathbf{x}) = \frac{1}{\Lambda}\exp\{\tilde{H}\sum_{\square(i,j,k,l)\in S} x_i x_j x_k x_l + \tilde{I}\sum_{\Delta(i,j,k)\in T} x_i x_j x_k + \tilde{J}_1\sum_{diagonal<i,k>} x_i x_k$$
$$+ \tilde{J}_2\sum_{horizonal\ or\ vertical<i,j>} x_i x_j + \tilde{K}\sum_{i\in I} x_i\}. \tag{3.11}$$

where

$$\Lambda = \sum_{\mathbf{x}}\exp\{\frac{\beta}{4}\sum_{\square(i,j,k,l)\in S} Q((x_i,x_k),(x_j,x_l))\}.$$

and $\tilde{H} = \frac{\beta H}{4}$, $\tilde{I} = \frac{\beta I}{4}$, $\tilde{J}_1 = \frac{\beta J_1}{4}$, $\tilde{J}_2 = \frac{2\beta J_2}{4}$, $\tilde{K} = \frac{4\beta K}{4}$. Notice that the pair interaction for diagonal pairs is different from that for horizontal and vertical pairs. When $c = d$, i.e., $J_1 = J_2, (\tilde{J}_1 = \tilde{J}_2)$ this reduces the model of 4-person game. The invariant measure is another new type of Ising model with one-site, two-site, three-site and four-site interactions. We also conjecture that for some set of parameters there exists phase transition. This is again an interesting problem needed to be pursued.

4. Further researches

We conclude with a few remarks about the possible problems for future research along this line.

(i) In Section 3 we have discussed some reversible SEP with certain types of global and local transition probability. We treated symmetric payoff. For asymmetric payoff, the situation becomes more complicated. A famous example is so-called Prisoner's Dilemma. The payoff of this two-person two-strategies game is given by

$$Q = \begin{pmatrix} Q(+1,+1) & Q(+1,-1) \\ Q(-1,+1) & Q(-1,-1) \end{pmatrix} = \begin{pmatrix} (a,a) & (b,c) \\ (c,b) & (d,d) \end{pmatrix} \tag{4.1}$$

The long run behavior exhibits complex dynamics.

(ii) For q-strategies ($q > 2$) with the general payoff function which is not necessarily symmetric, the process may not be reversible. A simple example is called Rock-Paper-Scissors game, in which there are three strategies: rock(R), paper(P) and scissor(S). The payoff is given by

$$Q = \begin{pmatrix} Q(R,R) & Q(R,P) & Q(R,S) \\ Q(P,R) & Q(P,P) & Q(P,S) \\ Q(S,R) & Q(S,P) & Q(S,S) \end{pmatrix} = \begin{pmatrix} (0,0) & (-1,1) & (1,-1) \\ (1,-1) & (0,0) & (-1,1) \\ (-1,1) & (1,-1) & (0,0) \end{pmatrix} \tag{4.2}$$

This is also an example of zero-sum game. It is interesting to find other or possible all asymmetric payoff functions with which the SEP could be reversible.

(iii) For synchronous global updating rule, it seems more difficult to find the invariant measure.

(iv) We may consider various types of team games. For example we may consider the 2-dimensional Union Jack lattice (see Figure 5). The player located on the vertex of the lattice plays three-person team game for players located on the triangle$((i_1,i_2),(i_1+1,i_2),(i_1,i_2+1)),((i_1,i_2),(i_1-1,i_2),(i_1,i_2-1)),((i_1,i_2),(i_1+1,i_2),(i_1,i_2-1)),((i_1,i_2),(i_1-1,i_2),(i_1,i_2+1))$ for Z^2 model, or five-person star-team game for site $((i_1,i_2),(i_1+1,i_2),(i_1,i_2+1),(i_1-1,i_2),(i_1,i_2-1))$. They may deduce different results, specifically, different behavior of phase transition are expected.

(v) We may consider models on other lattices. Besides the above-mensioned 2-dimensional Union Jack lattice, there is a rich theory on the lattices including tree, other two and three-dimensional lattice models, such as 2-dimensional triangle lattice, 2-dimensional honeycomb lattice, 2-dimensional Kagome lattice, 3-dimensional cubic lattice, 3-dimensional face-centered lattice, 3-dimensional body-centered cubic lattice. Also the supergame on small world network is worth to be pursued. Different behavior of phase transition of the invariant measures for these lattices has been found. we have treated the supergame on trees and the result is reported in [12].

(vi) It is also interesting to pursue the inverse problem-for a given Gibbsian invariant measure what is the sufficient and necessary conditions on the local transition probabilities(or on the payoff) for Gibbsian SEP with different global updating rules.

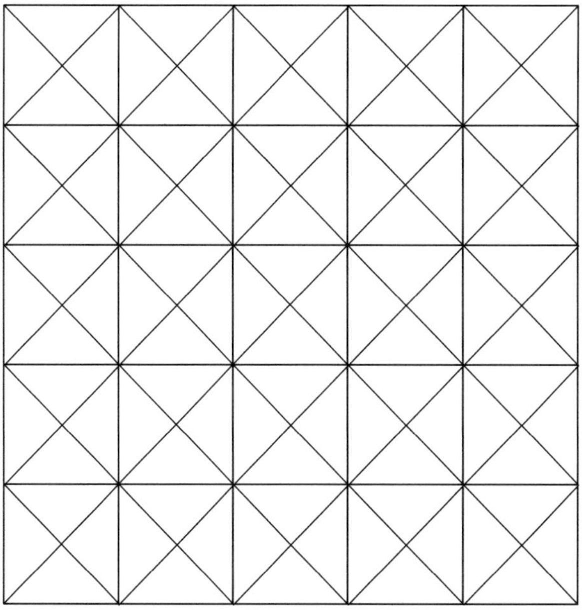

Figure 5. 2-dimensional Union Jack lattice

(vii) For those models for which the theoretical analysis is difficult, the numerical simulation could be of great help. We will report some numerical results elsewhere.

Overall this work is the first step to treat the SEP for large supergame over discrete time. More and deeper results are expected when we pursue the above problems.

Acknowledgment

The authors would like to thank the National Natural Science Foundation of China(no.11171215) and Shanghai 085 Project for financial supports.

Author details

Anjiao Wang
Department of Mathematics, Jiao Tong University, Shanghai 200240, China
School of Business Information Management, Shanghai Institute of Foreign Trade, Shanghai 201620, China

Zhongxing Ye
Department of Mathematics, Jiao Tong University, Shanghai 200240, China
School of Business Information Management, Shanghai Institute of Foreign Trade, Shanghai 201620, China

Yucan Liu
Department of Mathematics, Jiao Tong University, Shanghai 200240, China
School of Economics and Management, Nanjing University of Science and Technology, Nanjing 210094, China

5. References

[1] Aizenman M.,ąąTranslation Invaraince and Instability of Phase Coexistence in Two Dimensional Ising System, (Communications in Mathematical Physics)*Commun. Mth. Phys.* 73(1), (1980)83-94.

[2] Blouce L., The Statitical Methanics of Best-Response Strategy Revision Process, *Games and Economic Behavior*, Vol. 11(2), (1995)111-145, .

[3] Friedman J.W.,ąąGame Theory with Applications to Economics, Oxford University Press, New York(1986).

[4] Georges A. and Doussal P.L., From Equilibrium Spin Models to Probabilitic Cellular Automata, *J. Stst. Phys.*, 54 (1989)1011-1064.

[5] Georgi H-O , *Gibbs Measures and Phase Thransitions*, Walter de Gruyter, Berlin, New York(1988).

[6] Gilboa I., Kalai E. and Zemel E., On the Order of Eliminating Dominated Strategies, *Operations Research Letters*, 9(2), (1990)85-89.

[7] Kalai E., and Lehrer E., Rational Learning Leads to Nash Equilibrium, *Econometrica*, 61, No. 5, (1993)1019-1045.

[8] Kalai E. and Lehrer E., Subjective Equilibrium in Repeated Games, *Econometrica*, 61, No. 5, (1993)1231-1240.

[9] Kinderman R. and Snell J.L., *Markov Random Fields and Their Applicarions*, (American Mathematical Society) AMS Providence Rhode Island(1980).

[10] Lebowtiz J.L., Maes C. , and Speer E.R., Statistical Mechanics of Probabilistic Cellular Automata, *J. Stst. Phys.*, 39 (1990)117-170.

[11] Marroquin J.L. and Rumirez A., *Stochastic Celluar Automata with Gibbsian Invariant Measyres*, IEEE Trans Vol-It 37, No.3 (1991)541-551.

[12] Ou H. and Ye Z.,"Dynamic supergames on trees", Proceeding of 2010 International Conference on Progress in Informatics and Computing(PIC-2010), 340-344čňShanghai, December 10-12, 2010

Chapter 11

Cooperative Trust Games

Dariusz G. Mikulski

Additional information is available at the end of the chapter

http://dx.doi.org/10.5772/53928

1. Introduction

In certain multi-agent systems, the interactions between agents result in the formation of relationships, which can be leveraged for cooperative or collaborative activities. These relationships generally constrain individual-agent actions, since relationships imply that at least one contract (or mutual agreement) between the agents must exist. There is always some uncertainty as to whether or not either agent can or will satisfy some contract requirement – especially at the creation of a new contract. But in order to maintain the existence of a contract, each agent must overcome this uncertainty and assume that the other will do the same. The mechanism that facilitates this "act of faith" is generally regarded as "trust." In essence, each agent (whether a person or organization) in a relationship mutually trusts that the loss of some control will result in cooperative gains that neither agent could achieve alone.

In general, trust helps agents deal with uncertainty by reducing the complexity of expectations in arbitrary situations involving risk, vulnerability, or interdependence [1]. This is because agents rely on trust whenever they need to gauge something they cannot ever know precisely with reasonable time or effort. The benefits of trustworthy relationships include lower defensive monitoring of others, improved cooperation, improved information sharing, and lower levels of conflict [2]. But the reliance on trust also exposes people to vulnerabilities associated with betrayal, since the motivation for trust – the need to believe that things will behave consistently – exposes individuals to potentially undesirable outcomes. Thus, trust is a concept that must not only be managed, but also justified [3].

Since agents in an arbitrary system are always assumed to have selfish interests, the goal of each agent is to try to find the most fruitful relationships in a pool of potential agents [4]. That said, we cannot assume that agents do not already have pre-existing relationships with other agents. Furthermore, some agents may actually be within strongly-connected sub-system groups known as *coalitions*, where every agent in the coalition has a relationship with every other agent in the coalition. A coalition may contain a mixture of trustworthy and untrustworthy agents – but as a group, achieves cooperative gains that no sub-coalition could match. Thus, agents may be justified in forming relationships with coalition members who are not ideally trustworthy in order to acquire these cooperative gains as well.

As a simple example to illustrate this concept, consider two geographically-separated agents (who never physically met) who would like to conduct a financial transaction in exchange for some good. One agent must provide the good (through the mail) and the other must provide the payment (through the mail or electronically). If both agents follow their economic best interest, then neither agent should participate in the transaction since both agents are vulnerable to betrayal. This is because neither agent can truly verify the intent of the other agent before they act. Thus, if a transaction takes place, it can be entirely attributed to trust since both agents needed to overcome the uncertainty associated with the transaction. Let us suppose, however, that the value of the good and the size of the payment are sufficiently high such that no amount of mutual trust allows a direct transaction to take place. To handle this situation, both agents could form a coalition with a mutually trusted third party, such as an escrow agent. The escrow agent would receive the payment from one agent to verify that the good can be shipped, and then later disperse the payment to the other agent (minus the escrow fee) when the good has been verified as received. Here, each agent benefits from the cooperative gains of the transaction. These gains would not be possible if even one agent chose to disband from the coalition.

This chapter intends to show how one could mathematically describe these types of trust-based interactions via the *cooperative trust game* to predict coalition formation and disbanding. It presents a rigorous treatment of coalition formation using cooperative game theory as the underlying mathematical framework. It is important to highlight that cooperative game theory is significantly different than the more widely recognized competitive (non-cooperative) game theory. Cooperative game theory focuses on what groups of self-interested agents can achieve. It is not concerned with how agents make choices or coordinate in coalitions, and does not assume that agents will always agree to follow arbitrary instructions. Rather, cooperative game theory defines games that tell how well a coalition can do for itself. And while the coalition is the basic modeling unit for coalition game, the theory supports modeling individual agent preferences without concern for their possible actions. As such, it is an ideal framework for modeling trust-based coalition formation since it can show how each agent's trust preferences can influence a group's ability to reason about trustworthiness. We refer the reader to [5] for an excellent primer on cooperative game theory.

2. Classes of trust games

This section characterizes different classes of trust games within the context of cooperative game theory. Our characterizations provide the necessary conditions for a coalition trust game to be classified into a particular class. We start with additive and constant-sum trust games, which have limited value for cooperative applications, but are included for completeness. Then, we discuss superadditive and convex trust games, which show conditions for agents to form a grand coalition. In general, grand coalition solution concepts presented here can also be applied to smaller coalitions within a trust game through the use of a trust subgame.

2.1. Preliminaries

Let $\Gamma = (N, v)$ be a coalitional trust game with transferable utility where:

- N is a finite set of agents, indexed by i
- $v: 2^N \to \mathbb{R}$ associates with each coalition $S \subseteq N$ a real-valued payoff $v(S)$ that is distributed between the agents. Singleton coalitions, by definition, are assigned no value; i.e. $v(i) = 0\ \forall i \in N$.

The transferable utility assumption means that payoffs in a coalition may be freely distributed among its members. With regards to payoff value of trust between agents, this assumption can be interpreted as a universal means for agents to mutually share the value of their trustworthy relationships. Trust cultivation often requires reciprocity between two agents as a necessary behavior to develop trust, and a transferable utility is a convenient way to model the exchange for this notion.

In defining a transferable payoff value of trust, one aspect to consider are the "goods of trust". These refer to opportunities for cooperative activity, knowledge, and autonomy. In this chapter, we refer to these goods as *trust synergy* $s(S)$, which is a trust-based result that could not be obtained independently by two or more agents. We may also interpret trust synergy as the value obtained by agents in a coalition as a result of being able to work together due to their attitudes of trust for each other. In defining a set function for trust synergy, it is important to explicitly show how each agent's attitude of trustworthiness for every other agent in a coalition affects this synergy. In general, higher levels of trust in a coalition should produce higher levels of synergy.

The payoff value of trust, however, also includes an opposing force in the form of vulnerability exposure, which we refer to as *trust liability* $l(S)$. Trusting involves being optimistic that the trustee will do something for the truster; and this optimism is what causes the vulnerability, since it restricts the inferences a truster makes about the likely actions of the trustee. However, the refusal to be vulnerable tends to undermine trust since it does not allow others to prove their own trustworthiness, stifling growth in trust synergy. Thus, we see that agents in trust-based relationships with other agents must be aware of the balance between the values of the trust synergy and trust liability in addition to their relative magnitudes.

Let the characteristic payoff function of a trust game be the difference between the trust synergy and trust liability of a coalition S.

$$v(S) = s(S) - l(S) \tag{1}$$

2.2. Additive trust game

Additive games are considered inessential games in cooperative game theory since the value of the union of two disjoint coalitions ($S_1 \cap S_2 = \emptyset$) is equivalent to the sum of the values of each coalition.

$$v(S_1 \cup S_2) = v(S_1) + v(S_2)\ \forall S_1, S_2 \subset N \tag{2}$$

We see that the total value of the trust relationships between any two disjoint coalitions must always be zero. In other words, the trust synergy between any two disjoint coalitions must always result in a value that is equal to their trust liability. Thus, by expanding this definition for trust games and rearranging the terms, we can characterize an additive trust game as:

$$s(S_1 \cup S_2) - l(S_1 \cup S_2) = s(S_1) - l(S_1) + s(S_2) - l(S_2)$$
$$\{\forall S_1, S_2 \subset N: S_1 \cap S_2 = \emptyset\}$$

$$s(S_1 \cup S_2) - s(S_1) - s(S_2) = l(S_1 \cup S_2) - l(S_1) - l(S_2)$$
$$\{\forall S_1, S_2 \subset N: S_1 \cap S_2 = \emptyset\}) \tag{3}$$

2.3. Constant-sum trust game

In constant-sum games, the sum of all coalition values in N remains the same, regardless of any outcome.

$$v(N) = v(S) + v(N \backslash S) = k\ \forall S \subset N \tag{4}$$

By expanding this definition for trust games and rearranging the terms, we can see that the constant-sum trust game is a special case of a two-coalition additive trust game involving every agent in the game.

$$s(N) - l(N) = s(S) - l(S) + s(N \backslash S) - l(N \backslash S)\ \forall S \subset N$$
$$s(N) - s(S) - s(N \backslash S) = l(N) - l(S) - l(N \backslash S)\ \forall S \subset N \tag{5}$$

Definition: An agent is a *dummy agent* if the amount the agent contributes to any coalition is exactly the amount that it is able to achieve alone.

Theorem: Γ is a constant-sum trust game implies that Γ is a zero-sum trust game.

Proof: If Γ is a constant-sum game, the following constraint for singleton coalitions must always hold:

$$s(N) - s(i) - s(N \backslash i) = l(N) - l(i) - l(N \backslash i)\ \forall i \in N \tag{6}$$

By rearranging the terms, combining, and substituting, we get:

$$s(N) - l(N) = s(i) - l(i) + s(N \backslash i) - l(N \backslash i)\ \forall i \in N$$
$$v(N) = v(i) + v(N \backslash i)\ \forall i \in N$$

$$v(N) = v(N \backslash i) \; \forall i \in N \tag{7}$$

The last equation implies that every agent in N must behave like a dummy agent if Γ is a constant-sum trust game. Since all agents behave like dummy agents and $v(i) = 0$ for all $i \in N$, then any coalition that forms in Γ will have no value. Hence, the value of the grand coalition is zero (i.e. $v(N) = k = 0$). Therefore, the only possible constant-sum trust game is the zero-sum trust game. This completes the proof.

Corollary: Γ is a zero-sum trust game if $s(S) = l(S) \; \forall S \subset N$.

Proof: If $s(S) = l(S) \; \forall S \subset N$, then $v(S) = 0 \; \forall S \subset N$. Thus, $v(N) = v(N \backslash S) = k \; \forall S \subset N$. This result in implies that every possible coalition in N must behave like a coalition of dummy agents in a constant-sum trust game and their combinations with other coalitions will yield no value. Hence, the value of the grand coalition is always zero (i.e. $v(N) = k = 0$). This completes the proof.

Our proofs show that any constant-sum trust game is necessarily a zero-sum trust game that represents a special case of an additive trust game. These facts reinforce a notion that a group of agents who do not trust each other will always prefer to work as singleton coalitions. And even if there is some mutual trust between agents, gains from trust synergy are always lost to the trust liability, making it irrational to form any coalition with any other agent. Thus, if one determines that Γ is a constant-sum trust game, then this provides immediate justification for using non-cooperative game theory as the basis for modeling the purely competitive agents.

2.4. Superadditive trust game

In a superadditive game, the value of the union of two disjoint coalitions ($S_1 \cap S_2 = \emptyset$) is never less than the sum of the values of each coalition.

$$v(S_1 \cup S_2) \geq v(S_1) + v(S_2) \; \forall S_1, S_2 \subset N \tag{8}$$

This implies a monotonic increase in the value of any coalition as the coalition gets larger.

$$S \subseteq A \subseteq N \rightarrow v(S) \leq v(A) \leq v(N) \tag{9}$$

This property of superadditivity tells us that the new links that are established between the agents in the two disjoint coalitions are the sources of the monotonic increases. This results in a snowball effect that causes all agents in the game to form the grand coalition (a coalition containing all agents in the game) since the total value of the new trust relationships *between* any two disjoint coalitions must always be positive semi-definite. In other words, the trust synergy between any two disjoint coalitions must always result in a value that is at least as large as their combined individual trust liabilities. Thus, by expanding the definition for trust games and rearranging the terms, we can characterize a superadditive trust game as:

$$s(S_1 \cup S_2) - l(S_1 \cup S_2) \geq s(S_1) - l(S_1) + s(S_2) - l(S_2)$$
$$\{\forall S_1, S_2 \subset N : S_1 \cap S_2 = \emptyset\}$$

$$s(S_1 \cup S_2) - s(S_1) - s(S_2) \geq l(S_1 \cup S_2) - l(S_1) - l(S_2)$$
$$\{\forall S_1, S_2 \subset N: S_1 \cap S_2 = \emptyset\} \quad (10)$$

2.5. Convex trust games

A game is convex if it is supermodular, and this trivially implies superadditivity (when $S_1 \cap S_2 = \emptyset$). Thus, we see that convexity is a stronger condition than superadditivity since the restriction that two coalitions must be disjoint no longer applies.

$$v(S_1 \cup S_2) + v(S_1 \cap S_2) \geq v(S_1) + v(S_2)\ \forall S_1, S_2 \subset N \quad (11)$$

In convex games, the incentives of joining a coalition grow as the coalition gets larger. This means that the marginal contribution of each agent $i \in N$ is non-decreasing.

$$v(S \cup i) - v(S) \leq v(A \cup i) - v(A) \text{ whenever } S \subset A \subset N \backslash i \quad (12)$$

Definition: A *subgame* $v_R: 2^R \rightarrow \mathbb{R}$, where $R \subseteq N$ is not empty, is defined as $v_R(S) = v(S)$ for each $S \subseteq N$. In general, solution concepts that apply to a grand coalition can also apply to smaller coalitions in terms of a subgame.

Definition: Given a game $\Gamma = (N, v)$ and a coalition $R \subseteq N$, the R-marginal game $v_R: 2^{N \backslash R} \rightarrow \mathbb{R}$ is defined by $v_R(S) = v(R \cup S) - v(R)$ for each $S \subseteq N \backslash R$.

Using these definitions, Branzei, Dimitrov, and Tijs proved that a game is convex if and only all of its marginal games are superadditive [6]. We provide their proof here as a means for the reader to readily justify this assertion.

Theorem:

A game $\Gamma = (N, v)$ is convex if and only if for each $R \in 2^N$ the *R-marginal game* $(N \backslash R, v_R)$ is superadditive.

Proof: Suppose (N, v) is convex. Let $R \subseteq N$ and $S_1, S_2 \subseteq N \backslash R$. Then:

$$\begin{aligned} v_R(S_1 \cup S_2) + v_R(S_1 \cap S_2) &= v(R \cup S_1 \cup S_2) + v(R \cup (S_1 \cap S_2)) - 2v(R) \\ &= v\big((R \cup S_1) \cup (R \cup S_2)\big) + v\big((R \cup S_1) \cap (R \cup S_2)\big) - 2v(R) \\ &\geq v(R \cup S_1) + v(R \cup S_2) - 2v(R) \\ &= \big(v(R \cup S_1) - v(R)\big) + \big(v(R \cup S_2) - v(R)\big) \\ &= v_R(S_1) + v_R(S_2) \end{aligned} \quad (13)$$

where the inequality follows from the convexity of v. Hence, v_R is convex (and superadditive as well).

Now, let $S_1, S_2 \subseteq N$ and $R = S_1 \cap S_2$. Suppose that for each $R \in 2^N$, the game $(N \backslash R, v_R)$ is superadditive. If $R = \emptyset$, then the game $(N \backslash \emptyset, v_\emptyset) = (N, v)$ and $v(\emptyset) = 0$; hence, Γ is superadditive. If $R \neq \emptyset$, then because $(N \backslash R, v_R)$ is superadditive:

$$v_R((S_1 \cup S_2) \backslash R) \geq v_R(S_1 \backslash R) + v_R(S_2 \backslash R)$$
$$v(S_1 \cup S_2) - v(R) \geq v(S_1) - v(R) + v(S_2) - v(R)$$

$$v(S_1 \cup S_2) + v(R) \geq v(S_1) + v(S_2)$$
$$v(S_1 \cup S_2) + v(S_1 \cap S_2) \geq v(S_1) + v(S_2) \tag{14}$$

This completes the proof.

By using this characterization in the previous theorem and expanding it to our definition of a trust game, we can state a necessary requirement to produce a convex trust game: that the marginal trust synergy between any two coalitions must always result in a value that is at least as large as their marginal trust liability.

$$s_R((S_1 \cup S_2)\backslash R) - l_R((S_1 \cup S_2)\backslash R) \geq s_R(S_1\backslash R) - l_R(S_1\backslash R) + s_R(S_2\backslash R) - l_R(S_2\backslash R)$$
$$\{\forall S_1, S_2 \subset N: S_1 \cap S_2 = R\}$$

$$s_R((S_1 \cup S_2)\backslash R) - s_R(S_1\backslash R) - s_R(S_2\backslash R) \geq l_R((S_1 \cup S_2)\backslash R) - l_R(S_1\backslash R) - l_R(S_2\backslash R) \tag{15}$$
$$\{\forall S_1, S_2 \subset N: S_1 \cap S_2 = R\}$$

3. Trust game model

In the previous section, we characterized different classes of trust games without explicitly defining a trust game model. In this section, we provide a general model for trust games that conforms to the theoretical constructions in the previous section and can be adapted to a wide variety of applications.

3.1. Modeling trust synergy and trust liability

The attitude of trustworthiness agents have toward other agents in a trust game is managed in an $|N| \times |N|$ matrix T.

$$T = [t_{i,j}]_{|N|\times|N|} = \begin{cases} t_{i,j} = 1, & i = j \\ t_{i,j} \in [0,1], & i \neq j \end{cases} \tag{16}$$

This matrix is populated with values $t_{i,j}$ that represent the probability that agent j is trustworthy from the perspective of agent i. The values $t_{i,j}$ can also be interpreted as the probabilities that agent i will allow agent j to interact with him, since rational agents would prefer to interact with more trustworthy agents.

The manner in which $t_{i,j}$ is evaluated depends on an underlying trust model. We make no assumption about the use of a particular trust model, as the choice of an appropriate model may be application-specific. We also make no assumption about the spatial distribution of the agents in a game – therefore, this matrix should not necessarily imply the structure of a communications graph.

We provide a general model for trust synergy and trust liability that can be adapted for a variety of applications. Our model makes use of a symmetric matrix Σ to manage potential trust synergy and a matrix Λ to manage potential trust liability. Σ is symmetric because we assume that agents mutually agree on the benefits of a synergetic interaction.

$$\Sigma = \left[\sigma_{i,j}\right]_{|N|\times|N|} = \begin{cases} \sigma_{i,j} = 0, & i = j \\ \sigma_{i,j} = \sigma_{j,i} \geq 0, i \neq j \end{cases} \tag{17}$$

$$\Lambda = \left[\lambda_{i,j}\right]_{|N|\times|N|} = \begin{cases} \lambda_{i,j} = 0, i = j \\ \lambda_{i,j} \geq 0, i \neq j \end{cases} \tag{18}$$

As with the T matrix, we make no assumptions about how Σ and Λ are calculated, since the meaning of their values may depend on the application. For example, the calculations for $\sigma_{i,j}$ and $\lambda_{i,j}$ between two agents may not only take into account each agent's individual intrinsic attributes – it may also factor in externalities (i.e. political climate, weather conditions, pre-existing conditions, etc.) that neither agent has direct control over.

Definition: The total value of the trust synergy in a coalition is defined as the following set function:

$$s(S) = \sum_{i,j \in S} \sigma_{i,j}\, t_{i,j} t_{j,i} \; \forall i > j \tag{19}$$

Trust synergy is the value obtained by agents in a coalition as a result of being able to work together due to their attitudes of trust for each other. The set function $s(S)$ assumes that the events "agent i allows agent j to interact" and "agent j allows agent i to interact" are independent. This is reasonable since agents are assumed to behave as independent entities within a trust game (i.e. no agent is controlled by any other agent). Therefore, we treat the product $t_{i,j}t_{j,i}$ as the relative strength of a trust-based synergetic interaction, which justifies the use of the summation. The value for $\sigma_{i,j}$ serves as a weight for a trust-based synergetic interaction.

Definition: The total value of the trust liability in a coalition is defined as the following set function:

$$l(S) = \sum_{i,j \in S} \lambda_{i,j} t_{i,j} \; \forall i \neq j \tag{20}$$

Trust liability can be thought of as the vulnerability that agents in a coalition expose themselves to due to their attitudes of trust for each other. We treat the product $\lambda_{i,j}t_{i,j}$ as a measure for agent i's exposure to unfavorable trust-based interactions from agent j. A high amount of trust can expose agents to high levels of vulnerability. But each agent can regulate its exposure to trust liability by adjusting $t_{i,j}$. Changes to $t_{i,j}$, however, also influence the benefits of trust synergy.

3.2. Modeling the trust game

We define the trust game (also known as the total value of the trust payoff in a coalition) as the difference between its trust synergy and trust liability.

$$v(S) = \sum_{\substack{i,j \in S \\ \forall i > j}} \sigma_{i,j}\, t_{i,j} t_{j,i} - \sum_{\substack{i,j \in S \\ \forall i \neq j}} \lambda_{i,j} t_{i,j} \tag{21}$$

$$v(S) = \sum_{\substack{i,j \in S \\ \forall i > j}} t_{i,j} t_{j,i} \left(\sigma_{i,j} - \frac{\lambda_{i,j}}{t_{j,i}} - \frac{\lambda_{j,i}}{t_{i,j}} \right)$$

The factorization shows us that the first factor $(t_{i,j}t_{j,i})$ will always be greater than or equal to zero while the second factor can be either positive or negative. Hence, by isolating the second factor and recognizing that trust values equal to 1 produce the smallest possible reduction in the second factor, we can state the condition that guarantees the potential for two agents to form a trust-based pair coalition.

Proposition 1: Any two agents $i, j \in N$ will never form a trust-based pair coalition if $\sigma_{i,j} < \lambda_{i,j} + \lambda_{j,i}$. Otherwise, the potential exists for agent i and j to form a trust-based pair coalition.

Proposition 2: If two agents can never form a trust-based pair coalition, then the best strategy for both agents is to never trust each other (i.e. $t_{i,j} = t_{j,i} = 0$).

In general, proposition 1 does not extend to trust-based coalitions larger than two due to the complex coupling of trust dynamics between different agents as coalitions grow larger. For example, two agents who may produce a negative trust payoff value as a pair may actually realize a positive trust payoff with the addition of a third agent. This situation occurs if both agents have positive trust relationships with the third agent that outweighs their own negative trust relationship. Such a situation is common in real world scenarios, and justifies the importance of various trusted third parties, such as escrow companies, website authentication services, and couples therapists.

In light of this, we can mathematically justify a condition similar to proposition 1 that is valid for coalitions of any size – but only for a special type of trust game.

Theorem: A trust-based coalition $S \subseteq N$ will never form if:

$$\sum_{\substack{i,j \in S \\ \forall i>j}} \sigma_{i,j} < \sum_{\substack{i,j \in S \\ \forall i \neq j}} \frac{\lambda_{i,j}}{t_{j,i}} \tag{22}$$

$$\{\forall i,j \in S : t_{i,j}t_{j,i} = k\}$$

Proof: Let $S \subseteq N$ and $t_{i,j}t_{j,i} = k$ for all $i, j \in S$. Then, by substituting k into the trust model:

$$v(S) = \sum_{\substack{i,j \in S \\ \forall i>j}} \sigma_{i,j}\, k - \sum_{\substack{i,j \in S \\ \forall i \neq j}} \frac{\lambda_{i,j}}{t_{j,i}} k \tag{23}$$

$$v(S) = k\left(\sum_{\substack{i,j \in S \\ \forall i>j}} \sigma_{i,j} - \sum_{\substack{i,j \in S \\ \forall i \neq j}} \frac{\lambda_{i,j}}{t_{j,i}}\right)$$

Because k is a constant that is always greater than or equal to zero, we can clearly see that the second factor affects whether or not $v(S)$ is positive or negative. Hence, if the second term in the second factor is larger than the first term in the second factor, then a coalition S will never form. This completes the proof.

3.3. Incorporating context into a trust game

In practice, trust is often defined relative to some context. Context allows individuals to simplify complex decision-making scenarios by focusing on more narrow perspectives of situations or others, avoiding the potential for inconvenient paradoxes.

Coalitional trust games can also be defined relative to different contexts using the multi-issue representation [7], where we use the words "context" and "issue" interchangeably.

Definition: A multi-issue representation is composed of a collection of coalitional games, each known as an issue, $(N_1, v_1), (N_2, v_2), \cdots, (N_k, v_k)$, which together constitute the coalitional game (N, v) where

- $N = N_1 \cup N_2 \cup \cdots \cup N_k$
- For each coalition $S \subseteq N$, $v(S) = \sum_{i=1}^{k} v_i(S \cap N_i)$

This approach allows us to define an arbitrarily complex trust game that can be easily decomposed into simpler trust games relative to a particular context. A set of agents in one context can overlap partially or complete with another set of agents in another context. And one can choose to treat the coalitional game in one big context, or the union of any number of contexts based on some decision criteria.

3.4. Altruistic and competitive contribution decomposition

In the analysis of a trust-based coalition, it may sometimes be useful to understand the manner in which different subsets of a coalition contribute to its payoff value. One way to do this is to use a framework developed by Arney and Peterson where measures of cooperation are defined in terms of altruistic and competitive cooperation [8]. The unifying concept in the framework is a *subset team game*, a situation or scenario in which the value of a given outcome (as perceived by a team subset) can be measured.

Definition: Given a game $\Gamma = (N, u)$ and a non-empty coalition $R \subseteq S \subseteq N$, the *subset team game* $u_R: 2^R \to \mathbb{R}$ associates a valued payoff $u_R(S)$ perceived by the agents in R when the agents in S cooperate.

The authors limit the application of the framework to games where more agents in a coalition lead to more successful outcomes. Thus, adding more agents to a coalition should never reduce the coalition's payoff value. Also, the payoff value perceived by a coalition should not be smaller than the payoff value perceived by a subset of the same coalition. We refer to these two properties as *fully-cooperative* and *cohesive*, respectively.

Definition: A subset team game is fully-cooperative if $u_A(B) \leq u_A(C)$ for all $A \subseteq B \subseteq C \subseteq N$.

Definition: A subset team game is *cohesive* if $u_A(C) \leq u_B(C)$ for all $A \subseteq B \subseteq C \subseteq N$.

The authors show that in a fully-cooperative and cohesive game, the marginal contribution of a subset team is equal to the sum of the competitive and altruistic contributions of the subset team.

Definition: Given a payoff function $u_R(S)$ in a subset team game, the *total marginal contribution* of $R \subseteq S$ to a team S is $m_R(S) = u_S(S) - u_{S\backslash R}(S\backslash R)$. If the game is both cohesive and fully-cooperative, then the *competitive contribution* of R is $c_R(S) = u_S(S) - u_{S\backslash R}(S)$ and the *altruistic contribution* is $a_R(S) = u_{S\backslash R}(S) - u_{S\backslash R}(S\backslash R)$. Note that the total marginal contribution decomposes as $m_R(S) = c_R(S) + a_R(S)$.

In order to use these definitions within a trust game, we must first show they relate to the coalition game classes described in Section 3.

Theorem: A subset team game that is both fully-cooperative and cohesive is a convex game.

Proof:

First, we prove the fully-cooperative case. If $u_A(B) \le u_A(C)$ such that $A \subseteq B \subseteq C \subseteq N$, then following inequalities are also true:

$$u_A(B) \le u_A(B \cup i) \qquad A \subseteq B \subseteq N\backslash i \tag{24}$$

$$u_A(C) \le u_A(C \cup i) \qquad A \subseteq C \subseteq N\backslash i \tag{25}$$

$$u_A(B \cup i) \le u_A(C \cup i) \qquad A \subseteq B \subseteq C \subseteq N\backslash i \tag{26}$$

Since the system of inequalities shows that the contribution of an additional agent in a coalition is always non-decreasing, it is trivially true that:

$$u_A(B \cup i) - u_A(B) \le u_A(C \cup i) - u_A(C) \; A \subseteq B \subseteq C \subseteq N\backslash i \tag{27}$$

Next, we prove the cohesive case. If $u_A(C) \le u_B(C)$ such that $A \subseteq B \subseteq C \subseteq N$, then following inequalities are also true:

$$u_A(C) \le u_{A\cup i}(C) \qquad A \subseteq C \subseteq N\backslash i \tag{28}$$

$$u_B(C) \le u_{B\cup i}(C) \qquad B \subseteq C \subseteq N\backslash i \tag{29}$$

$$u_{A\cup i}(C) \le u_{B\cup i}(C) \qquad A \subseteq B \subseteq C \subseteq N\backslash i` \tag{30}$$

Since the system of inequalities show that the contribution of an additional agent in the accessing coalition subset is always non-decreasing, it is trivially true that:

$$u_{A\cup i}(C) - u_A(C) \le u_{B\cup i}(C) - u_B(C) \; A \subseteq B \subseteq C \subseteq N\backslash i \tag{31}$$

This completes the proof.

It is important to note that the additional agent i for both cases is never already inside either coalition B or C. If it was, then the proof would be invalid, as one could easily demonstrate counter examples under cases where an agent $i \in C\backslash B$.

Now that we have shown that a convex subset team game is fully-cooperative and cohesive, we may decompose the total marginal contribution of a set of agents into both altruistic and competitive contributions whenever a trust game is convex. To do so, we must define a value function $u_R(S)$ that utilizes the trust game payoff value function $v(S)$.

Definition: Given a game $\Gamma = (N, u)$ and a non-empty coalition $R \subseteq S \subseteq N$, the *subset trust game* $u_R: 2^R \to \mathbb{R}$ associates a trust payoff value $u_R(S)$ perceived by the agents in R when the agents in S cooperate:

$$u_R(S) = v(R) + \sum_{i \in R, j \in S \setminus R} v(\{i,j\}) \quad R \subseteq S \subseteq N \tag{32}$$

The rationale behind this payoff function in is that the payoff has to be from the perspective of the agents in R. The agents in R can factor in the values related to relationships between themselves (first term) and relationships between agents in R and agents in S (second term). But they cannot factor in values related to relationships between the agents in $S \setminus R$, since agents in R are assumed to have no direct knowledge of what is happening between the $S \setminus R$ agents.

Using the payoff function $u_R(S)$, we can calculate R's altruistic contribution $a_R(S)$ and competitive contribution $c_R(S)$ in a coalition S.

$$a_R(S) = \sum_{i \in R, j \in S \setminus R} v(\{i,j\}) \quad R \subseteq S \subseteq N \tag{33}$$

$$c_R(S) = v(S) - v(S \setminus R) - \sum_{i \in R, j \in S \setminus R} v(\{i,j\}) \quad R \subseteq S \subseteq N \tag{34}$$

$$m_R(S) = a_R(S) + c_R(S) = v(S) - v(S \setminus R) \quad R \subseteq S \subseteq N \tag{35}$$

4. Convoy trust game

In this section, we present an example of cooperative trust for a specific application: the convoy. Our primary purpose here is to demonstrate how one could use the theory in this chapter to model specific scenarios involving trust. We define the *convoy trust game*, which describes a cooperative game where the agents intend to move forward together in a single file. This type of game can be naturally adapted to the analysis of traffic patterns, leader-follower applications, hierarchical organizations, or applications with sequential dependencies. Our goal in this section is to understand how trust-based coalitions will form under this type of scenario.

4.1. 4-Agent convoy case

We begin with a simple convoy scenario that models a four-agent convoy, $N = \{1,2,3,4\}$, which intends to move together in a single file. The value of each index into N represents the agent's position in the convoy. For this scenario, we interpret the trust synergy in the coalition to represent the agents in the coalition moving forward. Thus, we set the values in the trust synergy matrix Σ equal to the number of agents that will move forward if the two agents are moving forward (inclusive of the two agents). We interpret the trust liability in coalition to represent the vulnerability of agents in the coalition to stop moving. Thus, we set the values in the trust liability matrix Λ equal to the number of agents that can prevent a particular agent from moving forward in a agent coalition pair.

Definition: The values in Σ and Λ for a 4-convoy trust game are:

$$\Sigma = \begin{bmatrix} 0 & 2 & 3 & 4 \\ 2 & 0 & 3 & 4 \\ 3 & 3 & 0 & 4 \\ 4 & 4 & 4 & 0 \end{bmatrix} \quad \Lambda = \begin{bmatrix} 0 & 0 & 0 & 0 \\ 1 & 0 & 1 & 1 \\ 2 & 2 & 0 & 2 \\ 3 & 3 & 3 & 0 \end{bmatrix}$$

4.2. 4-Agent convoy trust game analysis

First, let us analyze this game as an additive trust game. While there are infinitely many solutions for T that conform to the additive game, the most obvious solution is the extreme situation where no agent trusts any other agent – or, when T is the identity matrix ($T = I$). In this case, no agent will ever affect another agent, either positively or negatively. Thus, each agent will ultimately form a singleton coalition and fail to work cooperatively with any other agent.

Next, let us analyze another extreme situation where every agent completely trusts every other agent – or, when $T = [1]_{4\times4}$. As such, we can enumerate the trust payoff values for each possible coalition.

$$v(\{1,2\}) = 1;\ v(\{1,3\}) = 1; v(\{1,4\}) = 1;\ v(\{2,3\}) = 0;\ v(\{2,4\}) = 0;$$

$$v(\{3,4\}) = -1;\ v(\{1,2,3\}) = 2;\ v(\{1,2,4\}) = 2;\ v(\{1,3,4\}) = 1;$$

$$v(\{2,3,4\}) = -1;\ v(\{1,2,3,4\}) = 2;$$

These results provide us an interesting insight, in that all agents behind the lead agent find higher values of trust payoff with the lead agent than with the nearest agent. As such, as long as the lead agent is a member of a trust-based coalition in this game, there will be no incentive for any other agent to abandon the coalition. Thus, the agents ultimately form the grand coalition. Note, however, that the formation of a grand coalition does not imply that the trust game is superadditive or convex. This assertion is justified with the observation that $v(\{3,4\}) \ngeq v(\{3\}) + v(\{4\}) = 0$.

In order to form a convex 4-convoy trust game, we must satisfy the conditions that ensure that all trust payoff values in any coalition are at least as large as any sub-coalition – or that the marginal trust synergy is always greater than or equal to the marginal trust liability. While, again, there are infinitely many solutions for T that conform to convex game, the games with the highest trust payoff actually have either one of the following trust matrices (see next section for proof):

$$T_1 = \begin{bmatrix} 1 & 1 & 1 & 1 \\ 1 & 1 & 0 & 0 \\ 1 & 0 & 1 & 0 \\ 1 & 0 & 0 & 1 \end{bmatrix} \quad T_2 = \begin{bmatrix} 1 & 1 & 1 & 1 \\ 1 & 1 & 0 & 1 \\ 1 & 0 & 1 & 0 \\ 1 & 1 & 0 & 1 \end{bmatrix} \quad T_3 = \begin{bmatrix} 1 & 1 & 1 & 1 \\ 1 & 1 & 1 & 0 \\ 1 & 1 & 1 & 0 \\ 1 & 0 & 0 & 1 \end{bmatrix} \quad T_4 = \begin{bmatrix} 1 & 1 & 1 & 1 \\ 1 & 1 & 1 & 1 \\ 1 & 1 & 1 & 0 \\ 1 & 1 & 0 & 1 \end{bmatrix}$$

T_1, T_2, T_3, and T_4 are modified versions of $[1]_{4\times4}$ and all produce the same results in the trust payoff value function. The main modification ensures that agents 3 and 4 have no trust toward each other since the sum of their individual trust liabilities always outweigh the trust synergy they create. The following is the enumeration of the trust payoff values for the 4-convoy trust game with the highest trust payoff:

$$v(\{1,2\}) = 1;\ v(\{1,3\}) = 1;\ v(\{1,4\}) = 1;\ v(\{2,3\}) = 0;\ v(\{2,4\}) = 0;$$

$$v(\{3,4\}) = 0; v(\{1,2,3\}) = 2;\ v(\{1,2,4\}) = 2;\ v(\{1,3,4\}) = 2;$$

$$v(\{2,3,4\}) = 0;\ v(\{1,2,3,4\}) = 3;$$

The deep insight we gain from analyzing optimal trust matrices and payoff value results is that all agents behind the lead agent need only trust the lead agent in the convoy to move forward, provided the lead agent trusts every other agent to follow it. This echoes the intuition seen in Jean-Jacques Rousseau's classic "stag hunt" game, where there is no incentive for any player to cheat by not cooperating as long as each player can trust others to do the same [9].

We can use anecdotal evidence found in our experiences in automobile traffic jams to verify our understanding of the theoretical result. Drivers in traffic lanes (coalitional convoys) rarely place a significant amount of trust in neighboring drivers to justify the value of the traffic lane (as the model corroborates). In fact, in the event a driver becomes stuck in a traffic jam, he likely will not feel betrayed by the driver directly in front. Instead, he will unconsciously begin gauging the coalitional value of the traffic jam by considering his level of trust in the lead driver in the traffic jam, whether in visible range or not. In most cases, the driver monitors the traffic flow or listens to traffic reports to gauge his trust for the lead driver. He may also unconsciously consider other drivers in the traffic jam and estimate their trust perceptions of the traffic jam to gauge the coalition's value. In the event a driver cannot accurately gauge the value of the traffic jam, he may choose to leave the traffic jam and attempt to join another traffic coalition (lane) with a higher payoff value. These types of driver behaviors are generally not performed when the trust for the lead driver to move forward is high. Yet, these behaviors feel necessary when the trust lessens since they attempt to resolve coalitional and environmental uncertainties.

4.3. N-Agent convoy optimal solution proof

We conclude by generalizing the convoy trust game for any number of agents and prove the solution for the highest payoff trust-based coalition. Our proof shows that all agents behind the lead agent in a convoy need only trust the lead agent, and no other agent, to move forward so long as the lead agent trusts every other agent to follow it.

Definition: The generalized values in Σ and Λ for a convoy trust game with $|N|$ agents are:

$$\Sigma = \left[\sigma_{i,j}\right]_{|N|\times|N|} = \begin{cases} \sigma_{i,j} = 0, & i = j \\ \sigma_{i,j} = \max(\{i,j\}), & i \neq j \end{cases} \tag{36}$$

$$\Lambda = [\lambda_{i,j}]_{|N|\times|N|} = \begin{cases} \lambda_{i,j} = 0, & i = j \\ \lambda_{i,j} = i-1, & i \neq j \end{cases} \quad (37)$$

Theorem: The convoy trust game that produces the grand coalition with highest payoff value has a trust matrix that conforms to the following construction:

$$T = [t_{i,j}]_{|N|\times|N|} = \begin{cases} t_{i,j} = 1, & i = j \\ t_{i,j} = 1, & i \neq j, \min(\{i,j\}) = 1 \\ t_{i,j} = t_{j,i} \in \{0,1\}, & i \neq j, \min(\{i,j\}) = 2 \\ t_{i,j} = 0, & i \neq j, \min(\{i,j\}) > 2 \end{cases} \quad (38)$$

Proof:

Suppose we generalize the values in Σ and Λ. According to proposition 1, two agents $i, j \in N$ will never form a trust-based coalition pair if $\sigma_{i,j} < \lambda_{i,j} + \lambda_{j,i}$. Thus, by substitution:

$$\max(\{i,j\}) < (i-1) + (j-1)$$

$$\max(\{i,j\}) < i + j - 2 \quad (39)$$

We see that if i is the maximum value, then $0 < j - 2$. Similarly, if j is the maximum value, then $0 < i - 2$. Thus, the inequalities tell us that any agent behind the second agent will never form a trust-based coalition with any other agent behind the second agent. Therefore, by proposition 2, the best strategy for these agents is to have no trust for each other; hence $t_{i,j} = 0$ when $\min(\{i,j\}) > 2$ for $i \neq j$.

The equalities above alsto tell us that a trust-based coalition formation is possible with the lead agent and the second agent. Using the definition of the trust game model, the trust payoff values for a coalition in the convoy trust game is:

$$v(S) = \sum_{\substack{i,j \in S \\ \forall i > j}} t_{i,j} t_{j,i} \left(\max(\{i,j\}) - \frac{i-1}{t_{j,i}} - \frac{j-1}{t_{i,j}} \right) \quad (40)$$

We may now define the trust payoff values for any pair of agents as:

$$v(\{i,j\}) = t_{i,j} t_{j,i} \left(\max(\{i,j\}) - \frac{i-1}{t_{j,i}} - \frac{j-1}{t_{i,j}} \right) \quad (41)$$

Let us first analyze coalition formation with the lead agent. If $i = 1$, then $\max(\{i,j\}) = j$. Therefore, the payoff value for a pair coalition between i and j is:

$$v(\{1,j\}) = t_{1,j} t_{j,1} \left(j - \frac{j-1}{t_{1,j}} \right)$$

$$v(\{1,j\}) = j t_{1,j} t_{j,1} - j t_{j,1} + t_{j,1}$$

$$v(\{1,j\}) = t_{j,1} (j t_{1,j} - j + 1) \quad (42)$$

By inspection, we see that the highest trust payoff value is achieved when both the lead agent and any other agent completely trust each other (i.e., when $t_{1,j} = t_{j,1} = 1$). However, to justify this assertion, we must also show this is true when $j = 1$. If $j = 1$, then $\max(\{i,j\}) = i$. Therefore, the payoff value for a pair coalition between i and j is:

$$v(\{i,1\}) = t_{i,1}t_{1,i}\left(i - \frac{i-1}{t_{1,i}}\right)$$

$$v(\{i,1\}) = it_{i,1}t_{1,i} - it_{i,1} + t_{i,1}$$

$$v(\{i,1\}) = t_{i,1}\left(it_{1,i} - i + 1\right) \tag{43}$$

Again, by inspection, we confirm that the highest trust payoff is achieved when both the lead agent and any other agent completely trust each other. Therefore, $t_{i,j} = 1$ when the $\min(\{i,j\}) = 1$ for $i \neq j$.

Now, we analyze coalition formation with the second agent. If $i = 2$, then $\max(\{i,j\}) = j$. Therefore, the payoff value for a pair coalition between i and j is:

$$v(\{2,j\}) = t_{2,j}t_{j,2}\left(j - \frac{1}{t_{j,2}} - \frac{j-1}{t_{2,j}}\right)$$

$$v(\{2,j\}) = t_{2,j}t_{j,2}j - t_{2,j} - jt_{j,2} + t_{j,2}$$

$$v(\{2,j\}) = t_{j,2}\left(jt_{2,j} - j + 1\right) - t_{2,j} \tag{44}$$

The highest trust payoff that can be achieved with the second agent is equal to zero, and this only occurs when both agents either have complete trust in each other (i.e., when $t_{2,j} = t_{j,2} = 1$) or no trust in each other (i.e., when $t_{2,j} = t_{j,2} = 0$). Any other combination of trust values will produce negative trust payoff values. However, to justify this assertion, we must also show this is true when $j = 2$. If $j = 2$, then $\max(\{i,j\}) = i$. Therefore, the payoff value for a pair coalition between i and j is:

$$v(\{i,2\}) = t_{i,2}t_{2,i}\left(i - \frac{i-1}{t_{2,i}} - \frac{1}{t_{i,2}}\right)$$

$$v(\{i,2\}) = it_{i,2}t_{2,i} - it_{i,2} + t_{i,2} - t_{2,i}$$

$$v(\{i,2\}) = t_{i,2}\left(it_{2,i} - i + 1\right) - t_{2,i} \tag{45}$$

By inspection, we confirm that the highest trust payoff that can be achieved with the second agent is equal to zero. Therefore, $t_{i,j} = t_{j,i} \in \{0,1\}$ when $\min(\{i,j\}) = 2$ for $i \neq j$.

To complete the proof, we simply state our assumption that each agent fully trusts itself, since it is impossible for an agent to diverge from a singleton coalition. Therefore, $t_{i,j} = 1$ when $i = j$. This completes the proof.

Author details

Dariusz G. Mikulski
U.S. Army Tank-Automotive Research Development and Engineering Center (TARDEC), Warren, MI, USA

Acknowledgement

The author would like to acknowledge the Ground Vehicle Robotics group at the U.S. Army Tank-Automotive Research Development, and Engineering Center (TARDEC) in Warren, MI for their basic research investment, which resulted in the development of the cooperative trust game theory in this chapter. Furthermore, the author would like to thank his academic advisors, Dr. Edward Gu (Oakland University) and Dr. Frank Lewis (University of Texas in Arlington) for their insight and advice, which tremendously helped to guide the research to a successful outcome.

5. References

[1] N. Luhmann, "Familiarity, Confidence, Trust: Problems and Alternatives," *Trust: Making and Breaking Cooperative Relations*, pp. 94-108, 1988.
[2] B. Adams and R. Webb, "Trust in Small Military Teams," in *7th International Command and Control Technology Symposium*, 2002.
[3] C. McLeod. (2006) Trust. [Online]. http://plato.stanford.edu/entries/trust/
[4] S. D. Ramchurn, D. Huynh, and N. R. Jennings, "Trust in Multi-Agent Systems," *The Knowledge Engineering Review*, vol. 19, no. 1, pp. 1-25, 2004.
[5] Y. Shoham and K. Leyton-Brown, "Teams of Selfish Agents: An Introduction to Coalitional Game Theory," in *Multiagent Systems: Algorithmic, Game-Theoretic, and Logical Foundations*. Cambridge: Cambridge University Press, 2009, pp. 367-391.
[6] R. Branzei, D. Dimitrov, and S. Tijs, "A new characterization of convex games," in *Tiburg University, Center of Economic Research*, 2004.
[7] V. Conitzer and T. Sandholm, "Computing Shapley values, manipulating value division schemes, and checking core membership in multi-issue domains," in *AAAI Conference on Artificial Intelligence*, 2004.
[8] D. C. Arney and E. Peterson, "Cooperation in social networks: communication, trust, and selflessness.," in *26th Army Science Conference*, Orlando, FL, 2008.

[9] A. Dixit and B. Nalebuff, "Prisoners' Dilemmas and How to Resolve Them," in *The Art of Strategy*. New York: W.W. Norton and Company, 2008, pp. 64-101.

Chapter 12

A Graphical Game for Cooperative Neighbourhoods of Selfishly Oriented Entities

Antoniou Josephina, Lesta Papadopoulou Vicky,
Libman Lavy and Pitsillides Andreas

Additional information is available at the end of the chapter

http://dx.doi.org/10.5772/54394

1. Introduction

The chapter presents a graphical game of selfishly oriented players, inspired by considering a dense urban residential area where each home unit has its own IEEE 802.11 based wireless access point (AP), deployed without any coordination with other such units. The motivation for this game is to provide a framework where coordination between the game players is desired even though in the real world environment the APs lack any management regarding the efficient utilization of the communication channels and furthermore, it is quite common for a terminal served by one of the APs to be within the signal range of multiple alternative APs. This may not be desirable since APs can be in competition for the same communication resource (radio channel), and since the current standards dictate that at any given time every terminal must be rigidly associated with one particular AP, this situation results in increased interference and consequently a low utilization efficiency of the radio resource, when same or overlapping channels are selected by neighbouring APs.

The graphical game aims to motivates cooperation of the players, i.e. the APs, to overcome the resulting interference because of the unmanaged dense deployment of the APs. In fact, it would be much better for individual APs that are in physical proximity to each other to form groups, where one member of the group would serve the terminals of all group members in addition to its own terminals, so that the other access points of the group can be silent or even turned off, thereby reducing interference and increasing overall *Quality of Experience* (QoE). In this chapter, these groups include only members whose signal strength is sufficient to serve all group members, so that the access point that would be responsible for serving the terminals of a particular group or *neighbourhood* could change on a rotating basis, to allow all group members to equally serve and be served. Since there is no centralized entity that can control the APs and force them to form cooperative groups, the creation of such groups must be able to arise from a distributed process where each AP makes its own decisions independently and rationally for the benefit of itself and its terminals. *Graphical game theory* [2] is an appropriate tool to model such decentralized, topology-dependent schemes.

In the case of cooperation, i.e. when an AP serves the client of one or more neighbouring APs, in addition to its own clients, it is assumed that there is always enough resource for all clients served, otherwise the particular AP would not be considered as a neighbour. It is reasonable to assume that all APs have an active own set of clients that they serve when active whether they are in cooperation with a neighbour or not. The basic motivation behind the proposed approach, is the inefficiency of wireless communication caused by interference when multiple closely located APs, using the same or overlapping channels operate at the same time. If the clients of all APs are served by only one AP (even if these alternate from time to time), the interference between them is avoided and they receive service much closer to the theoretical limit of the radio resource. Hence, the proposed model esentially aims to capture how the APs may enter cooperation agreements in order to share the serving of each other's clients, in order to eliminate the interference factor caused by each other's transmission from their clients' experience.

In this chapter, we model the idea of cooperative neighbourhhoods as a graphical game and show that there exists motivation towards the cooperation of individual APs, each represented in the model by a dual nature node, a node encapturing both a server (AP) and a client. The strategical decision of such a unit to voluntarily participate in a group where members serve clients of neighbouring nodes in addition to their own, has the property that a unit is more likely to gain more in terms of QoE, than a unit defecting from such cooperation. We will henceforth refer to the proposed graphical game model as the *cooperative-neighbourhood game*.

2. Related work

2.1. Wireless deployments in urban environments

The density of wireless networks in urban residential areas is on the rise with more and more home networks being deployed in quite close proximity, enabled by the low cost and easy deployment of off-the-shelf IEEE 802.11 hardware and other personal wireless technologies. It is not uncommon for a wireless station to be within range of dozens of APs [3], competing for the limited number of channels offered by the IEEE 802.11 wireless standard. In this sense, urban areas are becoming similar to campus-like environments; however, in organizations and campuses experts can carefully control and manage interference of access points by planning the setup of the network in advance [4]. On the other hand, wireless networks in urban residential environments have a number of characteristics that make their deployment more challenging. For instance, the network is unplanned, thus aspects of planning such as coverage and interference cannot be controlled. Furthermore, deployments are mostly spontaneous, resulting in uneven density of deployment, the network lacks aspects such as efficient placement of access points, troubleshooting and adapting to network changes such as traffic load, as well as security issues. The authors of [3] use the term chaotic deployments or chaotic networks to refer to such a collection of wireless networks which are unplanned and unmanaged. However, they do mention advantages of such chaotic networks, for instance easily enabling new techniques to determine location [5] or providing near ubiquitous wireless connectivity [6]. The main disadvantage of these chaotic deployments is that interference can significantly affect end-user performance, while being hard to detect [7]. In this chapter we consider a solution based on "virtualization" among the interfering APs, where APs serve each others' clients. The security implications of allowing association of clients across APs from multiple owners are being addressed, e.g. in [8]. In this chapter we focus on the *incentive* aspect of a particular model of cooperation through the use of graphical game theoretic tools, and propose a framework to ensure that the APs are indeed motivated to provide service to each others' clients. In this chapter we focus on the theoretical framework and ignore protocol-specific implementation details.

2.2. Strategical decision-making in network environmnents

In this chapter, we consider the interactions between the individual units in dense urban deployments of wireless networks, represented by dual nature vertices in a graph. Describing and analysing interactions between independent, *selfish* entities is a situation that makes a good candidate to be modeled using the theoretical framework of Game Theory. Game Theory provides appropriate models and tools to handle multiple, interacting entities attempting to make a decision, and seeking a solution state that maximizes each entity's utility, i.e. each entity's *quantified satisfaction*. Game Theory has been extensively used in networking research as a theoretical decision-making framework, e.g. for cooperative resolution of interactive networking scenarios [1], for routing [9, 10], congestion control [11, 12], resource sharing [13, 14], and heterogeneous networks [15, 16].

In [17] the authors address the issue of cooperative neighbourhoods by concentrating on the *Prisoner's Dilemma/Iterated Prisoner's Dilemma* game model and proposing a group strategy to motivate adjacent neighbours into cooperation. The Prisoner's Dilemma and Iterated Prisoner's Dilemma [18] have been a rich source of research since the 1950s. In particular, the publication of Axelrod's book in 1984 [19] was the main driver that boosted the concept to the attention of other areas outside of game theory, as a model for promoting cooperation. The empirical results of the Iterated Prisoner's Dilemma (IPD) tournaments organized by Axelrod have influenced the game theory, machine learning and evolutionary computation communities, showing how features such as adaptivity and group play can result in gains at indiviual level. In fact, adaptive players, learning from the games in which they are involved, are more likely to survive than non-adaptive players in evolutionary IPD [20] and group strategies performed extremely well and defeated well-known strategies in round-robin competitions in the 2004 and 2005 IPD tournaments [21]. In this chapter, the idea of cooperative neighbourhoods is revisited by using the idea of a graph to set the neighbourhood map and defining a game on that graph where cooperation can result in gains for the individual nodes, shown through the usage of the graphical game model in selected examples.

2.3. Graphical games

Graphical games are one-shot games that model multi-player interaction in situations where restrictions and influences among the player population may exist. In fact, graphical games are a more efficient representation of one-shot multi-player games that cannot efficiently be represented with the normal form representation of a game (where rows and columns in a table represent the available actions and resulting payoffs of the game players). In addition to providing a better representation for multi-player games, graphical games are meant to capture locality and how the game is affected by the player's positioning.

Graphical games adopt a simple graph-theoretic model, where a game is represented by a graph G in which players are identified with vertices. A player of vertex i has payoffs that are entirely specified by the actions of i and those of its neighbour set in G, i.e. the set of vertices that have a direct link to i. Thus the graph of a game defines structural constraints over players' strategic influences towards other players. To fully describe the graphical game, in addition to the graph itself, the numerical payoff functions to each player must be specified.

There are quite a few gains in the use of graphical games. As mentioned before, in large population games, the graph form simplifies the representation of the game through a rich language in which to state and explore the computational benefits of various restrictions, e.g. topological restrictions, to the game payoffs. Furthermore, the simplified modelling of complex interactive situations as games through the use of the graph-theoretic language, is a tool of studying problems in various disciplines and understanding them through a simple but powerful theoretical framework. As such, we use graphical games in this chapter for modelling and analyzing a multi-entity problem characterized by complex interactions, from the communication networking field. [2]

3. Introducing the scenario

Currently, dense residential deployments of home wireless networks consist of uncoordinated APs that serve their terminals individually. The APs do not form groups and share the communication channel, which is an unmanaged common resource, resulting in a low utilization efficiency due to the competition between the APs and the interference it causes. This interference can be reduced if the APs can form groups according to their location, such that any APs belonging to the same group can serve any terminal associated with any of the other group members.

We consider that it is possible for an AP to recognize its *neighbourhood* from the signals it receives, having a knowledge of the required signal strength thresholds that would serve its terminals in a satisfying manner, i.e. with the required perceived QoE. In such a neighbourhood only one of the APs needs to assume the role of a leader, while the others can remain silent, and thereby minimize the interference and improve the overall QoE for all terminals involved. The role of the leader can be assumed on a rotating basis. Of course, in order to take part in a cooperative neighbourhood, the APs need to be motivated to act cooperatively, i.e. have an incentive to be silent or turned off while it is the turn of another AP to serve, and to serve everyone's terminals once their own turn comes. We show how such a distributed logic can be motivated and sustained in the neighbourhood using a graphical game model where each AP can make an independent decision whether or not to cooperate in such manner with its neighbours.

The interactions in a cooperative neighbourhood can be modelled as a game between the participating units, represented as vertices on a graph that captures through its links the signal received at each node from the neighbouring nodes. Given that each member of the neighbourhood has two choices at any given time: (a) to cooperate with one or more neighbours or (b) serve only its own clients, the graphical game model captures the utility of each node according to the decisions made by the whole neighbourhood, including itself, and provides a framework for scheduling the activation and deactivation times for the servers (APs) of cooperating nodes[1]. Which of the two behaviours to select in each round depends on

[1] Note that units may be a part of more than one neighbourhood, i.e. receive a good signal from peers that are in different neighbourhoods, making the scheduling task more challenging. This is part of future work.

the strategy of behaviour that a player has decided to follow during the game. The strategy of each *player*, i.e. of each unit, is selected such that it results in the highest possible payoff for the particular player, regardless of the strategies selected by the other neighbouring nodes. We refer to such interaction as a *cooperative-neighbourhood game*.

The chapter uses graphical game theory tools to define a network of APs represented using graph theory, and the relatiosnhip between any two neighbouring APs is represented in terms of the QoE that an AP's client, based on the signal strength received from its neighbouring AP, perceives, if served by the neighbouring AP. In other words, the signal strength that the client of an AP perceives from its neighbour when all other signals are off (which may not always be the case), is captured by the proposed model as the weight on the link joining the two corresponding vertices. Thus, the model represents each AP as a node on a graph and includes weights on the incoming links of each node, based on the signal(s) received at the node from the particular neighbour. We assume that weight is a measurable quantity, which may be obtained with a successive interchange between a node and its neighbours. However, the practical details of such an interchange are protocol design specific, and are out of the scope of this chapter. In the rest of the chapter the term weight represents the received signal strength at a node from a neighbouring node, when all other neighbouring node signals are off.

4. The graphical game model of a cooperative neighbourhood

4.1. The graph

We consider a set of nodes $V = \{v_1, v_2, v_3, ...v_n\}$ located on a plane. Each node is considered as an entity comprising of one server and a constant number of clients. Without loss of generality we assume that each server serves one client, and consequently each node employs dual functionality, i.e. both the functionality of a server and the functionality of a client. Therefore, a server node serves its client node, using broadcast transmission. At a time t, we say that node v_i is active or *ON*, if its server is broadcasting (to its own client). Otherwise, we say that the node is inactive or *OFF*.

Consider two nodes, v_i, v_j, $i \neq j$, which are close enough to detect each other's signals. In particular, node v_i may receive information from node v_j when it is *OFF* while node v_j is *ON*, i.e. when node v_j broadcasts. If the quality of the signal received by node v_i from node v_j is above some lower bound value assumed by the node v_i, we may consider that there exists a directed edge from node v_j to node v_i, denoted as (v_j, v_i). Moreover, we can quantify the quality of the received signal by having a weight $w(v_j, v_i)$ associated with the directed edge (v_j, v_i). Normalizing this value we consider that $w(v_j, v_i) \in (0,1)$. Finally, this value is analogous to the quality of the received signal, i.e. better received quality corresponds to value of $w(v_j, v_i)$ close to 1. Summing up, this reasoning motivates the following mathematical definition for the two nodes:

Definition 1. Consider two nodes v_i and v_j such that the two nodes can detect each other's signals and the node v_i can receive information from node v_j when it is OFF while node v_j is ON, of quality above some lower bound value. Then, we assume that there exist a directed edge from node v_j to node v_i, denoted as (v_j, v_i) of a positive weight $w(v_j, v_i) \in (0,1)$. The weight is analogous to the quality of the received signal at node v_i.

Thus, more formally, using the set of nodes V, we define a weighted graph $\mathcal{G} = (V, \overrightarrow{E}, \overrightarrow{W})$, where $\overrightarrow{W} : (V \times V) \rightarrow [0,1]$ and $\overrightarrow{E} \ni (v_i, v_j)$, $i \neq j$, if and only if node v_j can receive the signal of node v_i when v_i is *ON* and v_j is *OFF*. The measurement of the quality of the received signal by node v_i's client once it is *OFF* and cooperates with node v_j, which is *ON*, using edge (v_i, v_j), is given by weight $w(v_i, v_j)$. Thus, $w(v_i, v_j)$ is positive if and only if $(v_i, v_j) \in \overrightarrow{E}$. We assume that if $w(v_i, v_j) > 0$, then $w(v_j, v_i) > 0$ for all nodes $v_i, v_j \in V, i \neq j$.

4.2. The time

We consider a basic unit of time period T, e.g. 1 hour, and we split the time period T into x smaller time slots $T_1, T_2, \ldots T_x$ such that for each $T_k \in T$ there exists at least one node that in a group of colocated nodes that may alternate between the *ON* and *OFF* node, remains *ON* for the whole time slot T_k. So, $\bigcup_{T_k} = T$ and $T_k \bigcap T_l = \emptyset, k \neq l$, i.e. the sum of the time slots is the time period T and no two time slots overlap. By $|T_k|$ we denote the time elapsed from the beginning of time slot T_k until the end of the time slot.

Fix a time slot T_k. Then, for any node $v_i \in V$,

$$Mode(T_k, v_i) = \begin{cases} 0, & \text{if node } v_i \text{ is } OFF \text{ in time slot } T_k \\ 1, & \text{if node } v_i \text{ is } ON \text{ in time slot } T_k \,. \end{cases} \tag{1}$$

So, for node v_i the time period T can be partitioned into two sets $ON_T(v_i)$ and $OFF_T(v_i)$, where $ON_T(v_i) = \{T_k \mid Mode(T_k, v_i) = ON\}$ and $OFF_T(v_i) = \{T_k | Mode(T_k, v_i) = OFF\}$.

4.3. Cooperative and non-cooperative neighbours

Within a given time period T, node v_i may be in *agreement* or in *cooperation* with some of its neighbouring nodes. For any node v_j, being in agreement with node $v_i, i \neq j$ means that node v_j broadcasts only when v_i does not broadcast and serves the client of node v_i in addition to its own client, during the time that node v_i is *OFF*. The set of the neighbours of node v_i that are in agreement during time T, is denoted by $Coop_T(v_i)$, while the complete set of neighbours of node v_i is denoted by $Nei_T(v_i)$, where $Coop_T(v_i) \subseteq Nei_T(v_i)$ and $Mode(T_k, v_i) = 1 - Mode(T_k, v_j)$. So, $ON_T(v_i) = OFF_T(v_j)$ and $ON_T(v_j) = OFF_T(v_i)$ for each $v_j \in Coop_T(v_i)$. On the other hand, the set of neighbours with which v_i is not in agreement with is denoted as $NCoop_T(v_i)$, where $NCoop_T(v_i) \subseteq Nei_T(v_i)$ and it may be that $Mode(T_k, v_i) = Mode(T_k, v_j)$ where $v_j \in NCoop_T(v_i)$.

4.4. Interference and quality of experience

4.4.1. Received Signal

We are interested in measuring the QoE received by the client of node v_i through the signal strength received at node v_i, at any time slot T_k during the time period T. The received QoE, considering both cooperative and non-cooperative neighbours during time period T, is approximated, for simplicity, as the summation of the edge weights of all interfering nodes. We denote this quantity as $recS_{T_k}(v_i)$. We also distinguish two kinds of signals received at node v_i:

- the signal received when node v_i is *ON* denoted as $ONrecS_{T_k}(v_i)$, and
- the signal received when node v_i is *OFF* denoted as $OFFrecS_{T_k}(v_i)$.

At any time T_k the node is either *ON* or *OFF*. Thus,

$$recS_{T_k}(v_i) = \begin{cases} OFFrecS_{T_k}(v_i), & \text{if node } v_i \text{ is } OFF \\ ONrecS_{T_k}(v_i), & \text{if node } v_i \text{ is } ON\,. \end{cases} \tag{2}$$

4.4.2. *ON operation*

Fix a time slot $T_k \in T$. When node v_i is *ON* and none of its neighbours broadcast at the same time, the client of node v_i receives the broadcast signal in the best quality (no interference) and hence experiences the best possible QoE. We assume that top QoE is measured by a unit. Therefore, we set the experienced QoE by the node's client functionality to be equal to 1 in this case.

When node v_i is *ON*, none of its neighbours in cooperation, i.e. set $Coop_T(v_i)$ are *ON* at the same time. However, some of its neighbours not in cooperation may be *ON*, i.e. set $NCoop_T(v_i)$. In the second case, interference occurs and the signal received by node v_i is degraded causing a degraded QoE. We consider the degradation to be analogous to the strength of the signal of neighbour v_j received at node v_i and is captured by the weight $w(v_j, v_i)$ of edge (v_j, v_i). This degradation is also increased as more than one non-cooperative neighbours broadcast at the same time as node v_i. For simplicity, we consider the degradation to be analogous to the strengths of their signals received at node v_i.

Thus,

$$ONrecS_{T_k}(v_i) = max\{0, 1 - \textstyle\sum_{v_j \in Nei_{T_k}(v_i)} w(v_j, v_i) \cdot Mode(T_k, v_j)\} \tag{3}$$

4.4.3. *OFF operation*

On the other hand, when node v_i is *OFF*, the quality of the signal received at node v_i, and hence the QoE experienced, depends on the number of neighbours in cooperation with node v_i, that are *ON* at time T_k and serving the clients of the nodes in cooperation that are *OFF*, including the client of node v_i. If there exists one such neighbouring node v_j that is *ON*, the quality of the signal is captured by the weight $w(v_j, v_i)$ of edge (v_j, v_i). However, if there exist more than one neighbouring nodes not in cooperation with node v_i that are *ON* at the same time, this results in interference received at node v_i, which degrades the experienced QoE at the client of node v_i. This chapter considers the QoE degradation to be analogous to the strengths of the received signals.

We assume that node v_i is tuned to the strongest signal of its neighbours in set $Coop_T(v_i)$. Note that the quality of this signal is also degraded by the sum of the received signals from cooperative and non-cooperative neighbours that are *ON* at the same time as node v_i is. The set of neighbouring nodes that are *ON* at the same time as node v_i are considered to be non-cooperative.

Thus,

$$OFFrecS_{T_k}(v_i) = max\{0, (max_{v_j \in Coop_{T_k}(v_i)}\{w(v_j, v_i) \cdot Mode(T_k, v_j)\} - \sum_{v_h \in Nei_{T_k}(v_i),\, v_h \neq m} w(v_h, v_i) \cdot Mode(T_k, v_h))\}, \tag{4}$$

where $m \in V$ and $w(m,i) = max_{j \in Coop_\sigma(i)}\{w(j,i) \cdot Mode(T_k, j)\}$. So, summing up for the time period T:

$$recS_T(v_i) = \sum_{Mode(T_k,v_i)=ON} ONrecS_{T_k}(v_i) \cdot |T_k| + \sum_{Mode(T_k,v_i)=OFF} OFFrecS_{T_k}(v_i) \cdot |T_k|$$

4.5. The cooperative neighbourhood game

We consider an one-shot strategic game resulting from the described scenario in which the players of the game are the nodes (servers). Given the decisions of the nodes whether to operate in *ON* or *OFF* operation during each time slot $T_k \in T$, the utility of player v_i is the received signal of v_i's client during time period T, given by $recS_T(v_i)$.

Thus, more formally we define:

Definition 2. The game $G(V, E, W)$ is defined as follows. The set of players of the game is the set V. For simplicity, we define $V = \{1, \ldots, n\}$. A profile σ of the game is associated with the basic time period T of the scenario described. T is split into time slots $T_1, T_2, \ldots T_x$, such that $\bigcup_{T_k} = T$ and $T_k \bigcap T_l = \emptyset, k \neq l$, T_k corresponding to the smallest time slot where we may have alterations between *ON* and *OFF* operations of the nodes.

The strategies of the players in a profile σ are defined as follows:

The strategy of player (node) i is given by $\sigma_i = (Mode_\sigma(T), Coop_\sigma(i))$, where $Mode_\sigma(T)$, a vector of 0s and 1s, such that $Mode_\sigma(T_k, i) = 0$ *or* 1, based on whether node i operates in *ON* or *OFF* mode during time slot T_k, for each $T_k \in T$. $Coop_\sigma(v_i) \subseteq Nei_\sigma(v_i)$ is the set of neighbouring nodes of node i, with which node i has decided to cooperate with in σ. Cooperation means that for each such cooperative neighbour j of node i, $Mode_\sigma(T_k, i) = 1 - Mode_\sigma(T_k, j)$ for each $T_k \in T$, and the two nodes are in agreement to serve each other's client.

For player (node) i denote,

$$maxCoop_\sigma(T_k, i) = \{m \in V | Mode(T_k, m) = 1, w(m,i) = max_{j \in Coop_\sigma(i)}\{w(j,i) \cdot Mode(T_k, j)\}$$

Then, the utility of player (node) i, representing the signal received as expressed in equations (2), (3), (4), is given by:

$$U_\sigma(i) = \sum_{T_k \in ONU_T(i)} ONU_{T_k}(i) \cdot |T_k| + \sum_{T_k \in OFFU_T(i)} OFFU_{T_k}(i) \cdot |T_k| \tag{5}$$

where,

$$ONU_{T_k}(i) = max\{0, 1 - \sum_{j \in Nei(i)} w(j,i) \cdot Mode(T_k, j)\} \tag{6}$$

and

$$OFFU_{T_k}(i) = max\{0, w(maxCoop_\sigma(T_k, i), i) - \sum_{h \in Nei(i), h \neq maxCoop_\sigma(T_k,i)} w(h,i) \cdot Mode(T_k, h)\} \tag{7}$$

5. A usage example

In this section, we consider a simple scenario to demonstrate the interaction between two interacting neighbours. We juxtapose the situation where the two neighbours do not cooperate versus the case where the two neighbours cooperate, and we look for the time split between *ON* and *OFF* time slots for which cooperation is beneficial. We consider two case studies for the two nodes, where in the first case the time split between *ON* and *OFF* times is such that the *ON* time for one of the nodes is much larger than its *OFF* time. In the second case, we select equal *ON* and *OFF* times. The motivation behind the usage example is to determine the role of a particular time split towards the decision of the nodes to cooperate. We discover that in both cases there exists a motivation to cooperate, which leads to the generalization of these findings for the cooperation of two nodes in the subsequent section.

5.0.1. Case 1

The example first considers the case of the two nodes not in cooperation. Each node is *ON*, with its own client receiving the maximum signal from its own server, i.e. 1, while simultaneously the adjacent node is *ON* causing a continuous interference.

Consider the case where $w(2,1) = 0.7$, and $w(1,2) = 0.6$. Hence according to the utility defined in equation (5), $ONU_T(1) = 1 - w(2,1) = 1 - 0.7 = 0.3$ and $OFFU_T(1) = 0$. Also, $ONU_T(2) = 1 - w(1,2) = 1 - 0.6 = 0.4$ and $OFFU_T(2) = 0$.

In the case that the two nodes are in cooperation, then they alternate between states of *ON* and *OFF* times, while each node serves the client of its neighbour during its *ON* time, while its client is being served by the neighbour's server, during its *OFF* time. The utilities for nodes 1 and 2 are given in equation (5). Given that we want to consider a time split of non-equal times, we arbitrarily select the first time split of the total time period T, to be such that $T = T_1 + T_2$, where $T_1 = 0.2$, $T_2 = 0.8$. Note that in this setting $MaxCoop_\sigma(T_1, 2) = 1$ and $MaxCoop_\sigma(T_2, 1) = 2$. Therefore,

$$\begin{aligned} U_\sigma(1) &= ONU_{T_1}(1) + OFFU_{T_2}(1) \\ &= 1 \times 0.8 + 0.7 \times 0.2 = 0.94 \end{aligned} \tag{8}$$

$$\begin{aligned} U_\sigma(2) &= ONU_{T_1}(2) + OFFU_{T_2}(2) \\ &= 1 \times 0.2 + 0.6 \times 0.8 = 0.68 \end{aligned} \tag{9}$$

Therefore, we observe that even for the given weights, the cooperation option results in improved utility for both nodes 1, 2, compared to the non-cooperative values, therefore, the use of cooperation results in important gains.

5.0.2. Case 2

In this section, we consider a similar case scenario with two interacting nodes 1 and 2 but we modify the proposed time split of the total time period T, to be such that $T = T_1 + T_2$ and $T_1 = T_2$, where $T_1 = 0.5$, $T_2 = 0.5$.

In the case of non-cooperation, as previously, each node is *ON*, with its own client receiving the maximum signal from its own server, i.e. 1, while simultaneously the adjacent node is *ON* causing a continuous interference. Since $w(2,1) = 0.7$, and $w(1,2) = 0.6$, then similarly to case 1, according to equation (5), $ONU_T(1) = 1 - w(2,1) = 1 - 0.7 = 0.3$ and $OFFU_T(1) = 0$. Also, $ONU_T(2) = 1 - w(1,2) = 1 - 0.6 = 0.4$ and $OFFU_T(2) = 0$.

In the case of cooperation, note that as in case 1, $MaxCoop_\sigma(T_1,2) = 1$ and $MaxCoop_\sigma(T_2,1) = 2$. Hence,

$$\begin{aligned} U_\sigma(1) &= ONU_{T_{0.9}}(n_1) + OFFU_{T_{0.1}}(n_2) \\ &= 1 \times 0.5 + 0.7 \times 0.5 = 0.85 \end{aligned} \tag{10}$$

$$\begin{aligned} U_\sigma(2) &= ONU_{T_1}(2) + OFFU_{T_2}(2) \\ &= 1 \times 0.5 + 0.6 \times 0.5 = 0.8 \end{aligned} \tag{11}$$

We observe again that cooperation results in improved utility for both nodes 1, 2, compared to the non-cooperative utility values. Therefore, we have seen that for the both cases, the cooperation is the most profitable option. Next, we discuss and prove that for weights above the value of 0.5, cooperation is the most profitable option regardless of the particular time split between the *ON* and *OFF* times for the situation of two neighbouring nodes.

6. Nash Equilibrium for two neighbouring nodes

For ease of exposition, we consider the simplest case where the *cooperative-neighbourhood* game is played between two adjacent nodes 1 and 2, where the signal broadcast by each node's server is received by the other node's client in addition to the broadcasting node's own client. Let the weights on their incident edges denoted by $w(1,2)$ and $w(2,1)$. Remember that the utility of each node is defined in equations (5), (6) and (7).

Note that if a node decides to cooperate with at least one neighbour, then it is *OFF* for some of the time. On the other hand, if it does not cooperate with any neighbour at any time, then it is *ON* for the whole time T. In the following we compare values of utility in cooperation and non-cooperation of the two nodes and show necessary conditions in order to get a Nash Equilibrium for the two nodes.

6.1. Two interacting neighbouring nodes

Next, we compare the utilities of node 1 and node 2 in both the case where they cooperate and the case they do not cooperate, and we show necessary conditions for a cooperative profile in order to be a Nash Equilibrium.

6.1.1. Case 1: No cooperation of both nodes

In case 1, we consider that both nodes are *ON* for the whole duration of time period T, i.e. they do not cooperate with each other. Let σ be the resulting profile. Then, by equation (5),

$$\begin{aligned} U_\sigma(1) &= ONU_T(1) \\ &= (1 - w(2,1) \cdot Mode(T,2)) \cdot T \\ &= (1 - w(2,1)) \cdot T \end{aligned} \tag{12}$$

Also,

$$\begin{aligned} U_\sigma(2) &= ONU_T(2) \\ &= (1 - w(1,2) \cdot Mode(T,1)) \cdot T \\ &= (1 - w(1,2)) \cdot T \end{aligned} \tag{13}$$

6.1.2. Case 2: Cooperation of the two nodes

We consider here the case where the two nodes agreed to cooperate in T. Assume without loss of generality that the two nodes agreed to split T into two parts, T_1 and T_2, such that $T_1 \bigcap T_2 = \emptyset$ and $T = T_1 \bigcup T_2$ and $Mode(T_1, 1) = 1 = 1 - Mode(T_1, 2)$ and $Mode(T_2, 2) = 1 = 1 - Mode(T_2, 1)$. Let σ be the resulting profile. Then, by equation (5),

$$\begin{aligned} U_\sigma(1) &= ONU_{T_1}(1) \cdot T_1 + OFFU_{T_2}(2) \cdot T_2 \\ &= T_1 + w(2,1) \cdot Mode(T_2, 2) \cdot T_2 \\ &= T_1 + w(2,1) \cdot T_2 \end{aligned} \tag{14}$$

Similarly, we show that,

$$U_\sigma(2) = T_2 + w(1,2) \cdot T_1 \tag{15}$$

6.2. Equilibrium for the two nodes

We prove,

Theorem 6.1. *Assume two interacting nodes 1, 2 with $w(1,2)$ and $w(2,1)$ the weights of the link between them. Then cooperation when the two nodes split T into two sets T_1, T_2 such that $T_1 \bigcap T_2 = \emptyset$ and $T = T_1 \bigcup T_2$ is a Nash Equilibrium for the nodes if both $w(1,2)$ and $w(2,1) \geq 0.5$.*

Proof. We first show that if a profile σ satisfies the requirements of the theorem then it is a Nash Equilibrium. Assuming cooperation, the utilities of the nodes are given by equations (14) and (15) above. We now show that any unilateral alternations of any of the two nodes do not increase their utilities. Each of the two nodes has two possible alternations:

1. To increase its ON period.
2. To decrease its ON period.

We first consider the second option, i.e. for the node to decrease its ON period. Let t be the increased time in which node 1 is OFF. Let σ' be the resulting profile.

Let T_1', T_2' be the new split of T. Consider first node 1. Then, $|T_1|' = |T_1| - |t|, |t| > 0$ and $|T_2|' = |T_2| + |t|$. Note that during time period t, node 1 will receive:

$$\begin{aligned} U_{\sigma'}(1) &= ONU_{T_1'}(1) \cdot T_1' + OFFU_{T_2'}(2) \cdot T_2' \\ &= (|T_1| - |t|) + w(2,1) \cdot Mode(T_2, 2) \cdot T_2 + w(2,1) \cdot Mode(t, 2) \cdot t \\ &= (|T_2| - |t|) + w(2,1) \times 1 \cdot T_2 + w(2,1) \cdot 0 \\ &= |T_1| - |t| + w(2,1) \cdot T_2 \\ &= U_\sigma(1) - |t| < U_\sigma(1), \text{by equation (14) since } |t| > 0. \end{aligned} \tag{16}$$

Thus node 1 has no gain if it changes according to the second option. Similarly, we can show that $U_{\sigma'}(2) < U_{\sigma}(2)$. Therefore, the second option of a node decreasing its ON period does not result in some gain for any of the two nodes 1 or 2.

Next, we consider the first option of a node to increase its ON period. Let T_1', T_2' be the new split of T, $T_1' \bigcup T_2' = T$, $T_1' \bigcap T_2' = \emptyset$. Assume that there exists some $t > 0$, such that without loss of generality, node 1 increases its ON time period to $|T_1|' = |T_1| + |t|$ and $|T_2|' = |T_2| + |t|$. Note that nothing changes for node 2, i.e. $Mode(T_1, 2) = OFF$ and $Mode(T_2, 2) = ON$. Also, $Mode(t, 2) = ON$.

Then the utility of node 1 in σ' by equation (5) is,

$$\begin{aligned} U_{\sigma'}(1) = & \quad ONU_{T_1'}(1) \cdot T_1' + OFFU_{T_2'}(2) \cdot T_2' \\ & = 1 \cdot T_1 + (1 - w(2,1)) \cdot t + w(2,1) \cdot (|T_2| - |t|) \\ = & \quad (|T_1| + w(2,1) \cdot T_2) + |t| - 2|t| \cdot w(2,1) \\ = & \quad U_{\sigma}(1) + |t| - 2|t| \cdot w(2,1), \qquad \text{by equation (15) since } |t|>0. \end{aligned} \tag{17}$$

In order for σ' to be a better response for mode 1, it must be that:

$$U_{\sigma'}(1) > U_{\sigma}(1)$$

$$U_{\sigma'}(1) - U_{\sigma}(1) > 0, by\ equation\ (17)$$

$$U_{\sigma}(1) + |t| - 2|t| \cdot w(2,1) - U_{\sigma}(1) > 0$$

$$2|t| \cdot w(2,1) < |t|$$

Therefore,

$$w(2,1) < \frac{1}{2}$$

A contradiction, since $w(2,1) > \frac{1}{2}$ by assumption.

Similarly, we can show that for node 2 to benefit from increasing its ON time, it should be that $w(1,2) < \frac{1}{2}$, a contradiction by assumption.

Thus, if any of the nodes 1, 2 unilaterally alters its strategy to the first option of increasing its ON time, then profile σ' does not result to a better response than the strategy in σ. Therefore, σ is a Nash Equilibrium.

The theorem is now complete. □

7. An example game profile

Let Figure 1 illustrate the graph of connected nodes representing a neighbourhood of access points, where access points that are neighbours, i.e. that can serve each other's clients, are joined in the given graph through bidirectional links, the weights for each link given.

Let a time split $T = T_1 \bigcup T_2 \bigcup T_3 \bigcup T_4$, where $T_i \bigcap T_j = \emptyset$ and $T_1 = T_2 = T_3 = T_4$. Assuming $|T| = 1$, $|T_1| = |T_2| = |T_3| = |T_4| = \frac{1}{4}$, let the following mode operations as illustrated in Table 1.

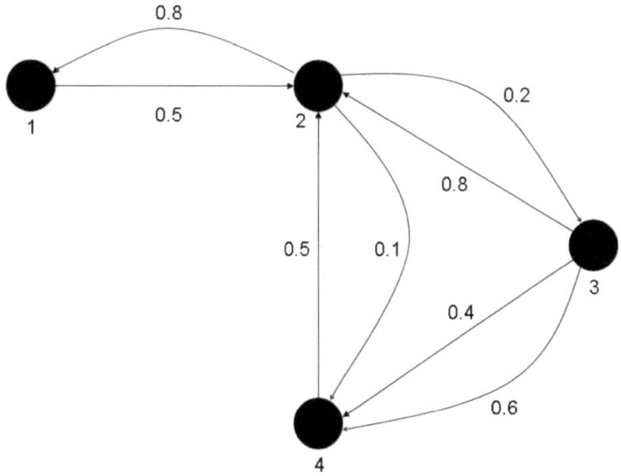

Figure 1. An example graph of the cooperative neighbourhood game

Node ID	Mode Operation
Node 1	Mode(T_1,1) = Mode(T_2,1) = Mode(T_3,1) = 0
	Mode(T_4,1) = 1
Node 2	Mode(T_1,2) = 1
	Mode(T_2,2) = Mode(T_3,2) = Mode(T_4,2) = 0
Node 3	Mode(T_1,3) = Mode(T_2,3) = 0
	Mode(T_3,3) = Mode(T_4,3) = 1
Node 4	Mode(T_1,4) = Mode(T_2,4) = 0
	Mode(T_3,4) = Mode(T_4,4) = 1

Table 1. Mode operations on Nodes 1–4

Next, Table 2 presents the cooperative and non-cooperative sets of neighbours for each node, based on the example and considering that nodes which have a link between them are potential cooperative neighbours.

Node ID	Cooperative Set	Non-Cooperative Set
Node 1	$Coop_\sigma(1) = \{2\}$	$NCoop_\sigma(1) = \{\emptyset\}$
Node 2	$Coop_\sigma(2) = \{1,3,4\}$	$NCoop_\sigma(2) = \{\emptyset\}$
Node 3	$Coop_\sigma(3) = \{2\}$	$NCoop_\sigma(3) = \{4\}$
Node 4	$Coop_\sigma(4) = \{2\}$	$NCoop_\sigma(4) = \{3\}$

Table 2. Cooperative and Non-Cooperative Sets for Nodes 1–4

We now compute the utilities of nodes 1, 2, 3 and 4, starting from node 1. Note that for node 1,

$$\begin{aligned} maxCoop_\sigma(T_1,1) &= \{2\} \\ maxCoop_\sigma(T_2,1) &= maxCoop_\sigma(T_3,1) = \{\varnothing\} \end{aligned}$$

Thus,

$$\begin{aligned} U_\sigma(1) &= ONU_{T_4}(1)\cdot|T_4| + OFFU_{T_1}(1)\cdot|T_1| + OFFU_{T_2}(1)\cdot|T_2| + OFFU_{T_3}(1)\cdot|T_3| \\ &= max\{0,(1-\textstyle\sum_{i\in\varnothing} w(i,1))\cdot Mode(T_4,j)\}\cdot|T_4| \\ &+ max\{0,w(2,1)-\textstyle\sum_{h\in\{2\},h\neq 2} w(h,1)\cdot Mode(T_1,h)\}\cdot|T_1| \\ &+ max\{0,0-\textstyle\sum_{h\in\{2\},h\neq\{\varnothing\}} w(h,1)\cdot Mode(T_2,h)\}\cdot|T_2| \\ &+ max\{0,0-\textstyle\sum_{h\in\{2\},h\neq\{\varnothing\}} w(h,1)\cdot Mode(T_3,h)\}\cdot|T_3| \\ &= 1\cdot|T_4| + w(2,1)\cdot|T_1| \\ &= 0.25 + 0.80\cdot 0.25 = 0.45 \end{aligned}$$

We proceed to compute the utility of node 2. Note first that,

$$\begin{aligned} maxCoop_\sigma(T_2,2) &= \varnothing \\ maxCoop_\sigma(T_3,2) &= \{3\} \\ maxCoop_\sigma(T_4,2) &= \{3\} \end{aligned}$$

Thus,

$$\begin{aligned} U_\sigma(2) &= ONU_{T_1}(2)\cdot|T_1| + OFFU_{T_2}(2)\cdot|T_2| + OFFU_{T_3}(2)\cdot|T_3| + OFFU_{T_4}(2)\cdot|T_4| \\ &= max\{0,1-\textstyle\sum_{j\in\{1,3,4\}} w(j,2)\cdot Mode(T_1,j)\}\cdot|T_1| \\ &+ max\{0,0-\textstyle\sum_{h\in 1,3,4,h\neq\{\varnothing\}} w(h,1)\cdot Mode(T_2,h)\}\cdot|T_2| \\ &+ max\{0,w(3,2)-\textstyle\sum_{h\in 1,3,4,h\neq\{3\}} w(h,2)\cdot Mode(T_3,h)\}\cdot|T_3| \\ &+ max\{0,w(3,2)-\textstyle\sum_{h\in 1,3,4,h\neq\{3\}} w(h,2)\cdot Mode(T_4,h)\}\cdot|T_4| \\ &= |T_1| + max\{0,w(3,2)-(w(4,2)\cdot 1 + w(1,2)\cdot 0)\}\cdot|T_3| \\ &+ max\{0,w(3,2)-(w(4,2)\cdot 1 + w(1,2)\cdot 1)\}\cdot|T_4| \\ &= 0.25 + max\{0,(0.8-0.5)\}\cdot 0.25 + 0 \\ &= 0.25 + 0.30\cdot 0.25 = \tfrac{13}{40} \end{aligned}$$

Next, we proceed to node 3. Note that,

$$maxCoop_\sigma(T_1,3) = \{2\}$$
$$maxCoop_\sigma(T_2,3) = \emptyset$$

Thus,

$$\begin{aligned}
U_\sigma(3) &= ONU_{T_3}(3)\cdot|T_3| + ONU_{T_4}(3)\cdot|T_4| + OFFU_{T_1}(3)\cdot|T_1| + OFFU_{T_2}(3)\cdot|T_2| \\
&= max\{0, 1-\textstyle\sum_{j\in\{2,4\}} w(j,3)\cdot Mode(T_3,j)\}\cdot|T_3| \\
&+ max\{0, 1-\textstyle\sum_{j\in\{2,4\}} w(j,3)\cdot Mode(T_4,j)\}\cdot|T_4| \\
&+ max\{0, w(2,3)-\textstyle\sum_{h\in 2,4,h\neq 2} w(h,3)\cdot Mode(T_1,h)\}\cdot|T_1| \\
&+ max\{0, 0-\textstyle\sum_{h\in 2,4,h\neq\{\emptyset\}} w(h,3)\cdot Mode(T_2,h)\}\cdot|T_2| \\
&= max\{0,(1-w(4,3)\cdot 1)\}\cdot|T_3| + max\{0,(1-w(4,3)\cdot 1)\}\cdot|T_4| \\
&+ max\{0,(w(2,3)-w(4,3)\cdot 0)\}\cdot|T_1| \\
&+ max\{0,(0-\textstyle\sum_{h\in\{2,4\},h\neq\{\emptyset\}}\cdot 0)\}\cdot|T_2| \\
&= w(1-w(4,3))\cdot|T_3| + (1-w(4,3))\cdot|T_4| + w(2,3)\cdot|T_1| + 0 \\
&= (1-0.6)(0.25)+(1-0.6)(0.25)+(0.2)(0.25) = 1\cdot\tfrac{1}{4}
\end{aligned}$$

Finally, we compute the utility of node 4. Note that,

$$maxCoop_\sigma(T_1,4) = \{2\}$$
$$maxCoop_\sigma(T_2,4) = \emptyset$$

Thus,

$$\begin{aligned}
U_\sigma(4) &= ONU_{T_3}(4)\cdot|T_3| + ONU_{T_4}(4)\cdot|T_4| + OFFU_{T_1}(4)\cdot|T_1| + OFFU_{T_2}(4)\cdot|T_2| \\
&= max\{0, 1-\textstyle\sum_{j\in\{2,3\}} w(j,4)\cdot Mode(T_3,j)\}\cdot|T_3| \\
&+ max\{0, 1-\textstyle\sum_{j\in\{2,3\}} w(j,4)\cdot Mode(T_4,j)\}\cdot|T_4| \\
&+ max\{0, w(2,3)-\textstyle\sum_{h\in\{2,3\},h\neq 2} w(h,4)\cdot Mode(T_1,h)\}\cdot|T_1| \\
&+ max\{0, 0-\textstyle\sum_{h\in\{2,3\},h\neq\{\emptyset\}} w(h,4)\cdot 0)\}\cdot|T_2| \\
&= w(1-w(3,4))\cdot|T_3| + (1-w(3,4))\cdot|T_4| + w(2,4)\cdot|T_1| + 0 \\
&= (1-0.4)(0.25)+(1-0.4)(0.25)+(0.1)(0.25) = 1.3\cdot\tfrac{1}{4}
\end{aligned}$$

Now, let's investigate whether profile σ is a Nash Equilibrium. In order for σ to be a Nash Equilibrium for player i, i's action in σ must be a best response action (to the current actions of the best of the players).

Consider player 2. Assume that it increases its *ON* operation and that it remains *ON* during time slot T_4. Then let σ' be the modified profile. Note that the rest of the players act as in σ.

$$\begin{aligned}
U_{\sigma'}(2) &= ONU_{T_1}(2)\cdot|T_1| + OFFU_{T_2}(2)\cdot|T_2| \\
&+ OFFU_{T_3}(2)\cdot|T_3| + ONU_{T_4}(2)\cdot|T_4| \\
&= max\{0, 1-\textstyle\sum_{j\in\{1,3,4\}} w(j,2)\cdot Mode(T_1,j)\}\cdot|T_1| \\
&+ max\{0, 0-\textstyle\sum_{h\in 1,3,4,h\notin\varnothing} w(h,1)\cdot Mode(T_2,h)\}\cdot|T_2| \\
&+ max\{0, w(3,2)-\textstyle\sum_{h\in 1,3,4,h\neq 3} w(h,2)\cdot Mode(T_3,h)\}\cdot|T_3| \\
&+ max\{0, 1-\textstyle\sum_{j\in 1,3,4} w(j,2)\cdot Mode(T_2,j)\}\cdot|T_4| \\
&= |T_1| + max\{0, w(3,2)-(w(4,2)\cdot 1 + w(1,2)\cdot 0)\}\cdot|T_3| \\
&+ max\{0, 1-w(1,2)\cdot 1 - w(3,2)\cdot 1 - w(4,2)\cdot 1\}\cdot|T_4| \\
&= 0.25 + (0.8-0.5)\cdot 0.25 + 0 \\
&= 0.25 + 0.30\cdot 0.25 = \tfrac{13}{40} = U_{\sigma}(2)
\end{aligned}$$

It follows that player 2 in σ' gains the same as σ. However, this does not imply that σ is a Nash equilibrium, since there must be another alternation of the player (or some other player) which results in a higher utility.

Similarly, the reader may attempt to evaluate alternate profiles of the game, by switching the operations mode of one node and re-evaluating the utilities of the players in a similar fashion as demonstrated above. We do not attempt to show the reader how to find the Nash Equilibrium for the particular topology at this time. However, such generalization remains within our future work goals.

8. Conclusion

The chapter has presented the use of graphical game theory towards the investigation and resolution of the following communication networking problem: Interference may arise using interactions between wireless access points that operate in the same geographical region without any coordination. Using a graphical game theoretic model, it has been shown that the players are motivated to create alliances with their neighbours so as to serve their terminals jointly and in a coordinated manner, leading to the decrease or even elimination of interference for all cooperating neighbours, an important achievement given the potential gains in Quality of Experience that the proposed framework can provide. The theoretical analysis shows the value of the cooperation for the interacting neighbours to form such a cooperative neighbourhood.

The cooperative neighbourhood model is represented as a game on a graph where the outgoing links of each node represent the signal strength (or interference depending on whether the node receiving the signal is *ON* or *OFF*) received from a neighbour, and consequently quantifies the *satisfaction* that may result from cooperation. The theoretical analysis and the equilibrium between two players interacting in a cooperative neighbourhood, is just the first step towards the resolution of the game in a more generalized way. This first step towards the more generalized cooperative neighbourhood solution, has shown us that graphical games are a promising approach of modelling such interference situations.

In addition to the generalization of the theoretical work, the authors recognize that there exist more practical protocol design issues such as synchronization and dynamic node update issues, and are to be looked at as a part of future work issues. Furthermore, the practical application of the weight setting in a real world scenario also remains in the scope of future work. On the theoretical side, we plan to specifically seek equilibriums in more complex graphical representations of cooperative neighbourhoods represented as multi-player games. The considered example profile of a cooperative neighbourhood game, which simplifies the proposed cooperation in a four-node topology, illustrates the management of the *ON* and *OFF* times of each server (AP) and how a profile can be modified. However, future work plans to further investigate the generation of equilibrium profiles in larger and more complex topologies, e.g. with partial neighbourhood relationships among the APs (i.e. where ranges of the individual APs overlap only partially), giving more general cooperative solutions.

Author details

Antoniou Josephina
School of Computing & Mathematics, UCLan (University of Central Lancashire) Cyprus, Pyla, Cyprus

Lesta Papadopoulou Vicky
Department of Computer Science and Engineering, European University Cyprus, Nicosia, Cyprus

Libman Lavy
School of Computer Science and Engineering, University of New South Wales, Sydney, Australia

Pitsillides Andreas
Department of Computer Science, University of Cyprus, Nicosia, Cyprus

9. References

[1] J. Antoniou, and A. Pitsillides, *Game Theory in Communication Networks: Cooperative Resolution of Interactive Networking Scenarios*, ISBN: 987-1-4398-4808-1, CRC Press, Taylor & Francis Group, Boca Raton, 2013.

[2] N. Nisan, T. Roughgarden, E. Tardos and V. V. Vazirani, *Algorithmic Game Theory*, ISBN: 978-0-521-87282-9, Cambridge University Press, New York, USA, 2007.

[3] A. Akella, G. Jedd, S. Seshan and P. Steenkiste, *Self-Management in Chaotic Wireless Deployments*, In ACM MobiCom, pp. 185-199, 2005.

[4] A. Hills, *Large-Scale Wireless LAN Design*, IEEE Communications vol. 39, no. 11, pp. 98-104, November 2001.

[5] Intel Research Seattle, Place Lab *A Privacy-Observant Location System*, http://placelab.org/, 2004.

[6] O. A. Dragoi and J. P. Black, *Enabling Chaotic Ubiquitous Computing*, Technical Report CS-2004-35, University of Waterloo, Canada, 2004.

[7] P. A. Frangoudis, D. I. Zografos and G. C. Polyzos, *Secure Interference Reporting for Dense Wi-Fi Deployments*, Fifth International Student Workshop on Emerging Networking Experiments and Technologies, pp. 37-38, 2009.

[8] J. Hassan, H. Sirisena and B. Landfeldt, *Trust-Based Fast Authentication for Multiowner Wireless Networks*, IEEE Transactions on Mobile Computing, vol. 7, no. 2, pp. 247-261, 2008.
[9] A. van de Nouweland, P. Borm, W. van Golstein Brouwers, R. Groot Briunderink and S. Tijs, *A Game Theoretic Approach to Problems in Telecommunication*, Management Science, vol. 42, no. 2, pp. 294-303, February 1996.
[10] A. Orda, R. Rom and N. Shimkin, *Competitive Routing in Multiuser Communication Networks*, IEEE/ACM Transactions on Networking, vol. 1, no. 5, pp. 510-521, 1993.
[11] A. de Palma, *A Game Theoretic Approach to the Analysis of Simple Congested Networks*, The American Economic Review, vol. 82, no. 2, pp. 185-199, 2005.
[12] L. Lopez, A. Fernandez and V. Cholvi, *A Game Theoretic Comparison of TCP and Digital Fountain based protocols*, Computer Networks, vol. 51, pp. 3413–3426, 2007.
[13] S. Rakshit and R. K. Guha, *Fair Bandwidth Sharing in Distributed Systems: A Game Theoretic Approach*, IEEE Transactions on Computers, vol. 54, no. 11, pp. 1384–1393, November 2005.
[14] H. Yaiche, R. R. Mazumdar and C. Rosenberg, *A game theoretic framework for bandwidth allocation and pricing in broadband networks*, IEEE/ACM Transactions on Networking, vol. 8, no. 5, pp. 667–678, 2000.
[15] J. Antoniou, I. Koukoutsidis, E. Jaho, A. Pitsillides, and I. Stavrakakis, *Access Network Synthesis in Next Generation Networks*, Elsevier Computer Networks Journal Elsevier Computer Networks Journal, vol. 53, no. 15, pp. 2716-2726, October 2009.
[16] J. Antoniou, V. Papadopoulou, V. Vassiliou, and A. Pitsillides, *Cooperative User-Network Interactions in next generation communication networks*, Computer Networks, vol. 54, no. 13, pp. 2239-2255, September 2010.
[17] J. Antoniou, L. Libman, and A. Pitsillides, *A Game-Theory Based Approach To Reducing Interference In Dense Deployments of Home Wireless Networks*. In proceedings of 16th IEEE Symposium on Computers and Communications (ISCC 2011), June 2011.
[18] G. Kendall, X. Yao and S. Y. Chong, *The Iterated Prisoner's Dilemma: 20 Years On*, Advances In Natural Computation Book Series, vol. 4, World Scientific Publishing Co., 2009.
[19] R. M. Axelrod, *The Evolution of Cooperation*, BASIC Books, New York, USA, 1984.
[20] M. Nowak, A. Sasaki, C. Taylor and D. Fudenberg, *Emergence of cooperation and evolutionary stability in finite populations*, Letters to Nature, vol. 428, April 2004.
[21] W. M. Grossman, *New Tack Wins Prisoner's Dilemma*, http://www.wired.com/culture/lifestyle/news/2004/10/65317, October 2004.
[22] A. Rogers, R. K. Dash, S. D. Ramchurn, P. Vytelingum and N. R. Jennings, *Coordinating team players within a noisy Iterated Prisoner's Dilemma tournament*, Elsevier Theoretical Computer Science, vol. 377, pp. 243-259, 2007.
[23] G. Taylor, *Iterated Prisoner's Dilemma in MATLAB* , Archive for the "Game Theory" Category "http://maths.straylight.co.uk/archives/category/game-theory", March 2007.

Section 4

Game Theory in Transdisciplinary Contexts

Chapter 13

Geometrical Exploration of Quantum Games

David Schneider

Additional information is available at the end of the chapter

http://dx.doi.org/10.5772/53931

1. Introduction

With nearly one decade of life, the theory of quantum games has nowadays become a very reach and prolific field of research. Multiplayer and multistrategy setups, a quantum approach to Evolutionary Game Theory, quantum game-like simulations of market phenomena, quantum duels, the effect of decoherence and noise during the implementation of a quantum game, are just some of the many scenarios where the quantum aspects of games were analyzed [2–4, 6, 7, 11–13, 16–18, 23]. Here we do not attempt to review the vast universe of quantum game theory, but just to give a comprehensive and didactic introduction to the pioneering work of Eisert [5]. This introduction will allow us to expose a recent approach developed to geometrically understand quantum games [20, 21], which represents the main focus this chapter. For a complete review of quantum games, we refer the reader to [9].

The theory of quantum games started in 1999 with the seminal papers by Meyer [15] and Eisert [5]. As usual, the first question that should be addressed is: why quantizing games? In their paper of 1999 , Eisert et at. outline the main reasons which make game theory a suitable framework for quantization. The first motivation relies on the probabilistic nature of the theory of games. Having a probabilistic background, it turns out to be natural an extension of game theory into the quantum probability domain. Moreover, as a game can be expressed as a setup where players exchange information with a 'referee', it becomes a perfect model to study quantum information. Finally, in [24] it was shown that the problem of the Optimal Cloning can be expressed in terms of a strategic game.

2. Eisert's quantum games

As Eisert's work relies on the quantization of the Prisoner's dilemma, we will devote a few lines to describe the features of this game (for a further description, see [10]). The Prisoner's dilemma is a standard example of a non-zero sum game, and it involves two parties, Alice (A) and Bob (B), who have to choose among the options 'cooperate' (C) and 'defect' (D). According to the decisions that the players make, they receive a payoff (see table 1). The dilemma arises from the fact that, although rational reasoning forces both players to defect, mutual cooperation represents a much better option for them. In game theory, these two outcomes of the game, 'defect-defect' and 'cooperate-cooperate', are referred as Nash Equilibrium (NE) and Pareto Optimal (PO), respectively, and the fact that they can not be reached simultaneously, should be regarded as the theoretical origin of the conflict.

	Bob: C	Bob: D
Alice: C	$(3,3)$	$(0,5)$
Alice: D	$(5,0)$	$(1,1)$

Table 1. Bi-matrix representation of the Prisoner's dilemma. The first entry corresponds to the payoff of Alice, and the second entry to the payoff of Bob.

To understand the main ideas behind Eisert's theory, it is necessary to introduce mixed strategies. In a mixed strategy game, players do not decide which strategy they are going to play, but have the freedom to chose a probability distribution for the available strategies. For instance, in a 2×2 game like the Prisoner's dilemma, players can choose a value for a parameter p to be the probability for the strategy D. Accordingly, the set of strategies for Alice becomes a random variable with a Bernoulli distribution with parameter $p_A \in [0,1]$. Namely,

$$S_A \sim B(p_A), \text{ with } P^A(S_A) = \begin{cases} 1 - p_A & S_A = C \\ p_A & S_A = D \end{cases} \tag{1}$$

and the same for Bob. As the strategies are now probabilistic, the goal of the players is to maximize the mean value of the payoff function calculated over all possibles outcomes of the game, which for Alice reads

$$\begin{aligned} \bar{\$}_A(p_A, p_B) &= \sum_{S_A, S_B} \$_A(S_A, S_B) P(S_A, S_B) \\ &= \sum_{S_A, S_B} \$_A(S_A, S_B) P^A(S_A) P^B(S_B) \end{aligned} \tag{2}$$

Nash equilibria are defined in an equivalent manner as in the pure strategy game, namely, as the vector $(p_A^*, p_B^*) \in [0,1]^{\otimes 2}$ such that

$$\begin{aligned} \bar{\$}_A(p_A^*, p_B^*) \geq \bar{\$}_A(p_A, p_B^*) \quad \forall p_A \in [0,1] \\ \bar{\$}_B(p_A^*, p_B^*) \geq \bar{\$}_B(p_A^*, p_B) \quad \forall p_B \in [0,1] \end{aligned} \tag{3}$$

For instance, for the Prisoner's dilemma the mean value of the payoff function is given by $\bar{\$}_A(p_A, p_B) = 3 + 2p_A - 3p_B - p_A p_B$, and the game accounts for the single NE $(p_A^*, p_B^*) = (1,1)$, which corresponds to the joined strategy (D,D) of the pure strategy game. A more interesting example concerns the Chicken game, which can be represented by the following bi-matrix

The mean value of the payoff function reads now $\bar{\$}_A(p_A, p_B) = 3 + 2p_A - p_B - 3p_A p_B$, and the NE are represented by the pure strategies $(p_A^*, p_B^*) = (1,0)$ and $(p_A^{**}, p_B^{**}) = (0,1)$, corresponding to (D,C) and (C,D) respectively, and the mixed strategy NE $(p_A^{***}, p_B^{***}) = (2/3, 2/3)$.

In order to introduce Eisert's scheme for quantizing games, it is worth to give a formal framework to the mixed strategy games. This can be done by considering a vector

	Bob: C	Bob: D
Alice: C	(3,3)	(2,5)
Alice: D	(5,2)	(1,1)

Table 2. Bi-matrix representation of the Chicken game.

representing the probabilities for the players to chose the available strategies. For instance, for Alice this vector reads

$$x_A = \begin{bmatrix} p^A(C) \\ p^A(D) \end{bmatrix} = \begin{bmatrix} 1-p_A \\ p_A \end{bmatrix} \tag{4}$$

and it is defined in a vector space spanned by the canonical vectors

$$p_A = 0 \sim \begin{bmatrix} 1 \\ 0 \end{bmatrix} \quad p_A = 1 \sim \begin{bmatrix} 0 \\ 1 \end{bmatrix} \tag{5}$$

The strategies are represented by stochastic matrices acting on the initial 1-player state $p_A = 0$,

$$M_A = \begin{bmatrix} 1-p_A & p_A \\ p_A & 1-p_A \end{bmatrix} \tag{6}$$

so that

$$x_A = M_A.\begin{bmatrix} 1 \\ 0 \end{bmatrix} \tag{7}$$

With this construction, the action of the players can be regarded as a stochastic process in which the players receive a biased coin (with all the probability in the side "C"), and change the probabilities by applying the stochastic matrix to the initial vector. For example, the pure strategies are represented by the matrices

$$\tilde{C} = M_A(0) = \begin{bmatrix} 1 & 0 \\ 0 & 1 \end{bmatrix} \text{ and } \tilde{D} = M_A(1) = \begin{bmatrix} 0 & 1 \\ 1 & 0 \end{bmatrix} \tag{8}$$

Finally, the payoff assignment can be defined through a matrix in the 2-players space

$$\mathbb{S}_A = \begin{bmatrix} \$_A(C,C) & 0 & 0 & 0 \\ 0 & \$_A(C,D) & 0 & 0 \\ 0 & 0 & \$_A(D,C) & 0 \\ 0 & 0 & 0 & \$_A(D,D) \end{bmatrix} \tag{9}$$

so that the average payoff is computed as the trace of $\mathbb{S}_A$ times the state of probabilities of the players $\rho = x_A \otimes x_B$.

$$\text{Tr}[\mathbb{S}_A.x_A \otimes x_B] = \bar{\$}_A(p_A, p_B) \tag{10}$$

The procedure of quantization is now straightforward, as the probabilities need to be simply replaced by amplitude probabilities and the stochastic matrices by unitary operators,

- $x \to |x\rangle = a\,|0\rangle + b\,|1\rangle$ ($|a|^2 + |b|^2 = 1$), initial state: $|0\rangle$
- $M \to \hat{U}(\theta, \phi, \eta) \sim \begin{bmatrix} e^{i\phi}\cos(\theta/2) & e^{i\eta}\sin(\theta/2) \\ -e^{-i\eta}\sin(\theta/2) & e^{-i\phi}\cos(\theta/2) \end{bmatrix}$

After quantization, the pure strategies C and D of the classical game become the identity and the spin flip operators, respectively,

$$\hat{C} = \hat{U}(0,0,0) \sim \begin{bmatrix} 1 & 0 \\ 0 & 1 \end{bmatrix} \text{ and } \hat{D} = \hat{U}(\pi,0,0) \sim \begin{bmatrix} 0 & 1 \\ -1 & 0 \end{bmatrix} \tag{11}$$

The corresponding payoff assignment is the operator whose matrix elements are those of equation 9

$$\begin{aligned} \hat{\$}_A = & \$_A(C,C) \left|00\right\rangle \left\langle 00\right| + \$_A(C,D) \left|01\right\rangle \left\langle 01\right| \\ & + \$_A(D,C) \left|10\right\rangle \left\langle 10\right| + \$_A(D,D) \left|11\right\rangle \left\langle 11\right| \end{aligned} \tag{12}$$

so that

$$\mathrm{Tr}[\hat{\$}_A \left|\psi_f\right\rangle \left\langle \psi_f\right|] = \left\langle \psi_f\right| \hat{\$}_A \left|\psi_f\right\rangle \tag{13}$$

with $\left|\psi_f\right\rangle = \hat{U}_A \otimes \hat{U}_B \left|00\right\rangle$.

There are two important things to be pointed out. First, the procedure relies on an usual formalism for quantization, so it represents an very acceptable framework to study quantum phenomena. And second, the new game entitles the classical game, so the latter can be reobtained as its classical limit. As a matter of fact, by the identification $p = \sin^2(\theta/2)$ it turns out that both Eisert's quantum game and the mixed strategy classical game are the same. This fact makes necessary to include any new ingredient in the theory. This new ingredient, with no classical counterpart, is the entanglement between the states of the two players represented by the unitary operator $\hat{J}(\gamma) = \exp(\mathrm{i}\gamma D \otimes D/2)$.

Figure 1 outlines the circuital representation of Eisert's protocol. The final state $\left|\psi_f\right\rangle = \hat{J}^\dagger(U_A \otimes U_B)\hat{J}\left|00\right\rangle$ is the state by means of which the expectation value of the operator $\hat{\$}_A$ is calculated.

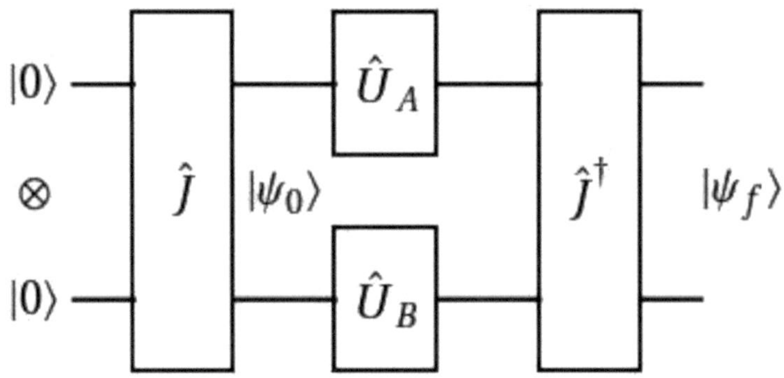

Figure 1. Eisert's quantization protocol for 2-player games.

Eisert's results relies on the so called two-parameter operator

$$\hat{U}(\theta,\phi,\eta) \sim \begin{bmatrix} e^{\mathrm{i}\phi}\cos(\theta/2) & \sin(\theta/2) \\ -\sin(\theta/2) & e^{-\mathrm{i}\phi}\cos(\theta/2) \end{bmatrix} \tag{14}$$

The authors argue that it proves to be sufficient to restrict the space of strategies to this set of unitary operators, with $\theta \in [0, \pi]$ and $\phi \in [0, \pi/2]$. For $\phi = 0$ the game again reduces to the mixed strategy Prisoner's dilemma, meaning a unique NE for the pure joined strategy 'defect-defect'. However, the extra degree of freedom given by the parameter ϕ makes the game behave differently. Eisert et al. choose $\hat{J} = \exp(i\pi\hat{D} \otimes \hat{D}/4)$, which makes the initial state maximal entangled, and show that 'defect-defect' ceases now to be a NE. For instance, the players can improve by taking the strategy

$$\hat{Q} \equiv \hat{U}(0, \pi/2) = \begin{bmatrix} i & 0 \\ 0 & -i \end{bmatrix}. \tag{15}$$

By computing the payoff expectation value for Alice [1], it turns out that $\$_A(\hat{D}, \hat{Q}) = 5$, meaning that

$$\$_A(\hat{D}, \hat{D}) < \$_A(\hat{Q}, \hat{D}). \tag{16}$$

The main result of [5], however, concerns the emergence of a new NE given by the outcome '$\hat{Q}-\hat{Q}$', for which $\$_A = \$_B = 3$. This solution fulfills not only the NE condition, provided that

$$\$_A(\hat{U}(\theta, \phi), \hat{Q}) = \$_B(\hat{Q}, \hat{U}(\theta, \phi)) = \cos^2(\theta/2)(3\sin^2\phi + \cos^2\phi) \leq 3, \tag{17}$$

but it is also an outcome of the game which rewards Alice and Bob as good as the mutual cooperation (Pareto Optimal condition). In that sense, this is a version of the game that on one hand encompasses the classical Prisoner's dilemma, and on the other hand has the intriguing feature of making the players able to perform an optimal decision.

3. A periodic point-based method to analyze Nash equilibria in Eisert's quantum games

In this section, we introduce a periodic point-based method designed to explore the strategy space in order to identify those strategies which fulfill the NE condition. The general problem concerns two functions $f_A(x, y)$ and $f_B(x, y)$ such that $f_B(x, y) = f_A(y, x)$, and every point (x^*, y^*) satisfying the generalized NE definition,

$$f_A(x^*, y^*) \geq f_A(x, y^*) \quad \forall x \tag{18}$$

$$f_B(x^*, y^*) \geq f_B(x^*, y) \quad \forall y. \tag{19}$$

The map associated to the game is defined as the (eventually multivalued) function that, for a given value of the second argument of $f_A(x, y)$, picks the (eventually multiple) value of the first argument which makes $f_A(x, y)$ maximal[2]. Specifically,

$$M(y) = x, \tag{20}$$

for every x satisfying

$$f_A(x, y) \geq f_A(x', y) \quad \forall x'. \tag{21}$$

Following the previous definition, a pair (x^*, y^*) satisfying equation 18 must be obviously a pair for which

$$M(y^*) = x^*. \tag{22}$$

[1] From now on, we will use the symbol '\$' to refer to the expectation value $\langle\$\rangle$.

[2] In game theory language, this map is called 'best response correspondence'.

Equation 19, instead, can be rearranged by means of the symmetry relationship of the two functions. This procedure leads to an equivalent equation as the previous one, but with the roles of x^* and y^* inverted,

$$M(x^*) = y^*. \tag{23}$$

By simply applying the map to both sides of 22 and combining with equation 23, one obtains

$$M^2(y^*) = y^*. \tag{24}$$

According to equation 24, all NE of the game defined by the functions $f_A(x,y)$ and $f_B(x,y)$ can be extracted from the 2-periodic points of M, and are of the form $(y^*, M(y^*))$.

For $M(y^*) = y^*$, y^* turns out to be a fixed point of M. We will refer to this solution of the form (y^*, y^*) as a fixed point NE, and to those for which $M(y^*) \neq y^*$ as 2-cycle NE.

3.1. A simple example

As the classical Prisoner's dilemma (even in its mixed-strategy version) accounts for a fixed point NE only, it is not a suitable example to illustrate the method implementation. The simplest example which shows how the method picks fixed point and 2-cycle NE concerns the mixed-strategy version of the Chicken game, derived from the bi-matrix of table 2.

Renaming $p \equiv p_A$ and $q \equiv p_B$, the mean value of the payoff function reads $\bar{\$}_A(p,q) = 3 + 2p - q - 3pq$. The map M is therefore an eventually multivalued function of Bob's mixed strategy q[3].

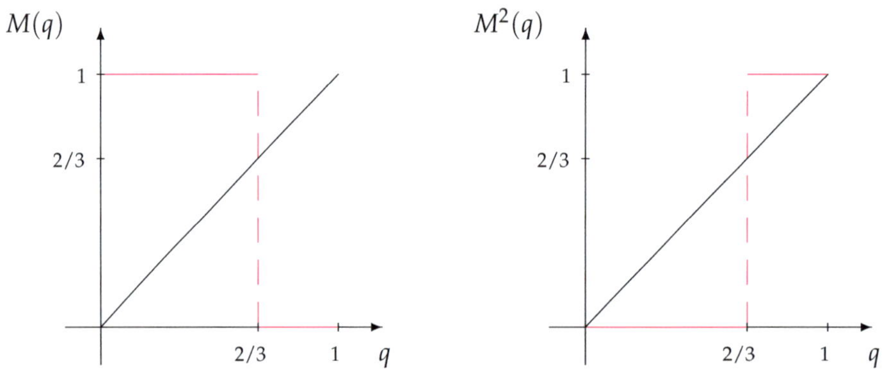

Figure 2. First (left) and second (right) iterations of the maximization map for the mixed-strategy Chicken game. Attached to each plot, a line at 45° which intersects each function at its fixed points. The point $q = 2/3$ is mapped both under M and M^2 into the whole interval $[0,1]$ (dashed lines).

Figure 2 illustrates the shape of the first and the second iterations of the map M. A straight line at 45° is attached to each plot to graphically identify the fixed points and the 2-period orbits of M, which in general are obtained from the plot of M^2. The latter map has three fixed points, namely, $q_1^* = 0$, $q_2^* = 2/3$ and $q_3^* = 1$. The corresponding NE are deduced by taking each of these fixed points and computing the corresponding value $M(q^*)$. Proceeding

[3] The maximization map associated to the mixed-strategy classical Prisoner's dilemma is the constant and idempotent function $M(q) = M^2(q) = 1\ \forall q \in [0,1]$, as can be deduced from Alice's payoff function $\bar{\$}_A(p,q) = 3 + 2p - 3q - pq$. Hence, the game has not periodic orbits except the fixed point $q = 1$. This fact implies a sole NE for $p = q = 1$, namely, for the pure joined strategy 'defect-defect'.

like this, one immediately recognizes the two NE related to the pure strategy game, given by $(p_1^*, q_1^*) = (0,1)$ and $(p_3^*, q_3^*) = (1,0)$. According to our previous nomenclature, these are 2-cycle NE. The remaining fixed point of M^2 is also a fixed point of M ($p_2^* = q_2^* = 2/3$), and it corresponds therefore to a fixed point NE. Of course, this NE is not present in the pure strategy game.

3.2. Topological aspects of Eisert's game's maps

Now the mechanism for identifying NE is described, our next goal is to extend it for the case of the 2-parameter quantum Prisoner's dilemma. For that purpose, we consider the maximization map for that problem, obtained from equation 20 by taking $x = \hat{U}(\theta', \phi')$, $y = \hat{U}(\theta'', \phi'')$, and $f_A = \$_A$. Accordingly, we write now

$$M(\hat{U}(\theta'', \phi'')) = \hat{U}(\theta', \phi'), \tag{25}$$

for every $\hat{U}(\theta', \phi')$ satisfying

$$\$_A(\hat{U}(\theta', \phi'), \hat{U}(\theta'', \phi'')) \geq \$_A(\hat{U}(\theta''', \phi'''), \hat{U}(\theta'', \phi'')) \quad \forall \hat{U}(\theta''', \phi''').$$

Following the steps of Eisert's protocol, one gets the following expression for Alice's payoff function,

$$\begin{aligned} \$_A(\hat{U}(\theta', \phi'), \hat{U}(\theta'', \phi'')) &= 3[\cos(\phi' + \phi'') \cos(\theta'/2) \cos(\theta''/2)]^2 \\ +5\left[\sin(\phi') \cos(\theta'/2) \sin(\theta''/2) - \cos(\phi'') \cos(\theta''/2) \sin(\theta'/2)\right]^2 \\ + \left[\sin(\phi' + \phi'') \cos(\theta'/2) \cos(\theta''/2) + \sin(\theta'/2) \sin(\theta''/2)\right]^2. \end{aligned} \tag{26}$$

By inserting this function in the definition above, however, it is not difficult to check that for M to be a well defined map, we are forced to take in to account the following extended strategy set,

$$S = \{U(\theta, \phi) | -\pi \leq \theta \leq \pi \text{ and } -\pi/2 \leq \phi \leq \pi/2\}. \tag{27}$$

The latter definition ensures that every point in the M domain is mapped into the same region, namely, that the map M is actually an endomorphism[4]. As it is usual for bidimensional maps, we will make a planar representation of the M domain. This representation deserves however a detailed explanation, because it will be crucial for the analysis of the periodic points of the map. We stress that the map given by equations 25, 26 and 27 is defined on a compact surface which can not be continuously embedded in a tridimensional manifold. This special surface is called projective plane, and it is central in 2-dimensional algebraic topology [14]. The construction of a projective plane follows a square's edges identification scheme similar to that employed to construct a torus or a sphere (see figure 3). As in the case of a torus, the scheme follows an identification of opposite edges, however the edges are inverted before the identification. The important fact is that this inversion makes it possible to repeat the construction starting from a 2-edges-polygon, as it is also the case in the construction of a sphere. Nevertheless, as the edges have to be also inverted in the 2-edges diagram of the projective plane, the 'North' and 'South' poles merge ultimately into the same point[5].

[4] Moreover, the region in the parameter space defined by means the set of strategies S turns out to be the smallest region which guaranties the endomorphic character of M. Of course, this set encompass the set of strategies defined in [5], but it is not however a redundant set.

[5] Equivalently, a projective plane can be constructed starting from a square and generating a Möbius strip. If in a second step we map the (single) edge of the Möbius strip into one point (as we would map the two edges of a cylinder into the two points that would result in the poles of a sphere), the compact surface we obtain is actually a projective plane.

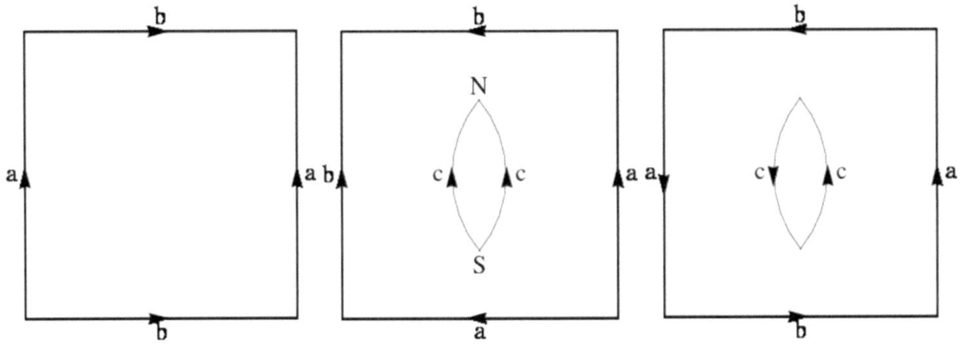

Figure 3. The square's edges identification scheme for the construction of a torus (left), a sphere (middle), and a projective plane (right). Attached to the latter two pictures, an equivalent construction performed with 2-edges-polygons to emphasizes the fact that, when the procedure is accomplished for a projective plane, the poles of the sphere (denoted by N and S) merge into the same point.

Following an analogy with the planar representation of the earth surface, and regarding θ and ϕ as the latitudes and longitudes, respectively, we may say that the map M is defined on a domain whose extreme longitudes ($\phi = \pm\pi/2$) have to be identified after a reflexion by means the equator ($\theta = 0$), so that the extreme latitudes (the 'lines' representing the poles, $\theta = \pm\pi$) result in two representations of the same point.

We are now in the position to explain the numerical exploration of the map of equation 25. However the topological aspects of the map domain are more complicated, there are many features in the analysis analogous to those outlined for the very simple example of the the previous section. When possible, we will refer to those similarities to make the picture clearer. Figure 4 depicts the first and second iterations under M of the projective plane first quadrant (except for the line segment $\phi = \tilde{\phi} = \arccos(1/5)/2$). Five line segments for five different values of the parameter ϕ (with $\theta \in [0, \pi]$) are mapped under M into five different curves at the second quadrant, respectively. The latter curves are in turn mapped into five different curves at the fourth quadrant. The following remarks summarize the overall behavior of the map,

- For $\phi < \tilde{\phi} = \arccos(1/5)/2$, the map image of every line segment $\phi = \phi_0$ starts at $\theta = \pi$ (a piece of the projective plane pole), whereas for $\phi > \tilde{\phi}$, the map images start at the points with coordinates $\phi = -\phi_0, \theta = 0$.
- All map images of the line segments $\phi = \phi_0$ finish at the point with coordinates $\phi = -\pi/2, \theta = 0$ (which is the same as that with coordinates $\phi = \pi/2, \theta = 0$).
- For $\phi < \tilde{\phi}$, the second iterations of the line segments $\phi = \phi_0$ start and finish at the point with coordinates $\phi = \pi/2$, $\theta = 0$; whereas for $\phi > \tilde{\phi}$ the corresponding iterations start at the points with coordinates $\phi = \phi_0$, $\theta = 0$, and finish at the points with coordinates $\phi = \pi/2, \theta = 0$.
- As ϕ goes to $\pi/2$, the map images of the line segments $\phi = \phi_0$ converge to the point with coordinates $\phi = -\pi/2$, $\theta = 0$, whereas the second iterations converge to the point with coordinates $\phi = \pi/2, \theta = 0$ (which is the same as the previous one).

In figure 5, we sketch the behavior of the map for points on the line segment $\phi = \tilde{\phi}$. In the open interval $(0, \pi]$, the segment in the first quadrant maps into the solid curve at the second

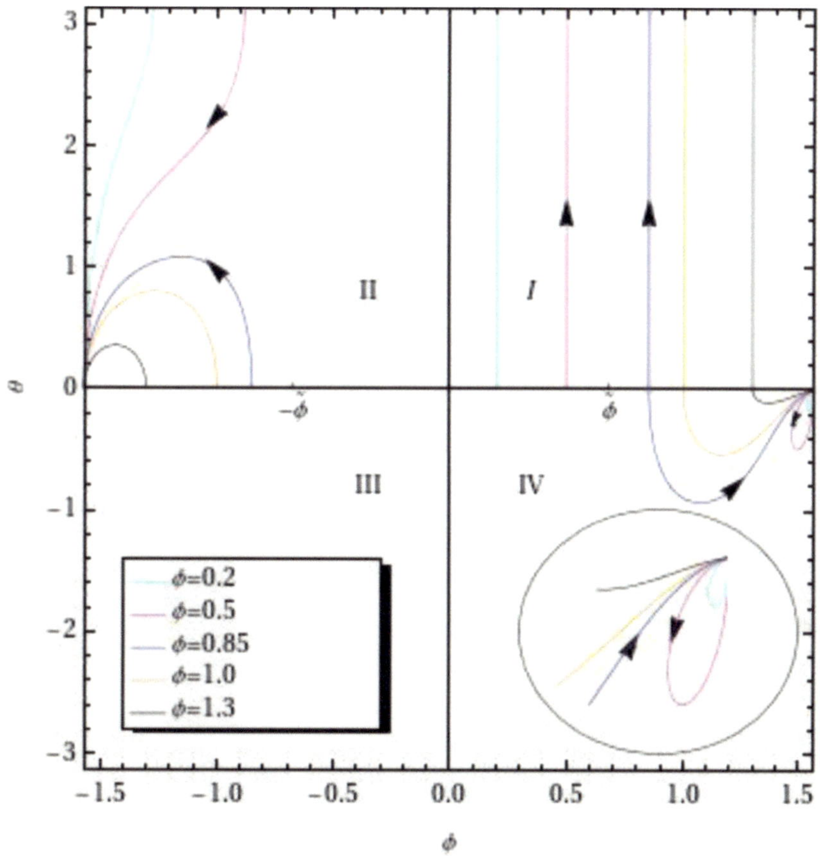

Figure 4. For the first quadrant of the extended strategy set, the first and second iterations of the maximization map of the 2-parameter quantum Prisoner's dilemma. The line segments corresponding to different values of the parameter ϕ are mapped into several curves at the second quadrant, which are in turn mapped into the corresponding curves at the fourth quadrant. $\tilde{\phi} = \arccos(1/5)/2 \simeq 0.685$.

quadrant in a similar fashion as for the remaining values of ϕ. However, for $\theta = 0$ the map becomes a multivalued function having its image at all points on the dotted line $\phi = -\tilde{\phi}$, with $\theta \in [-\pi, \pi]$. This behavior is exactly the same as for the strategy $q = 2/3$ in the mixed strategy Chicken game.

Finally, the second iteration of the map for the line segment $\phi = \tilde{\phi}$ at the first quadrant is depicted in figure 6. The black-solid curve at the second quadrant corresponds to the first iteration for $\theta \in (0, \pi]$ (compare with figure 5), which is mapped into the black-solid curve at the fourth quadrant. We split the first iteration of the point with coordinates $\phi = \tilde{\phi}, \theta = 0$ (namely, the dotted line $\phi = -\tilde{\phi}$ of figure 5) in three parts: the green segment $\theta > 0$, which is mapped into the green curve at the first quadrant, the red segment $\theta < 0$, which is mapped into the red curve at the fourth quadrant, and the point $\theta = 0$, which again makes the map a multivalued function, having its image in all points on the dotted line $\phi = \tilde{\phi}$, with $\theta \in [-\pi, \pi]$. This is again comparable with the behavior of the mixed strategy $q = 2/3$ in the previous section, when iterated for a second time.

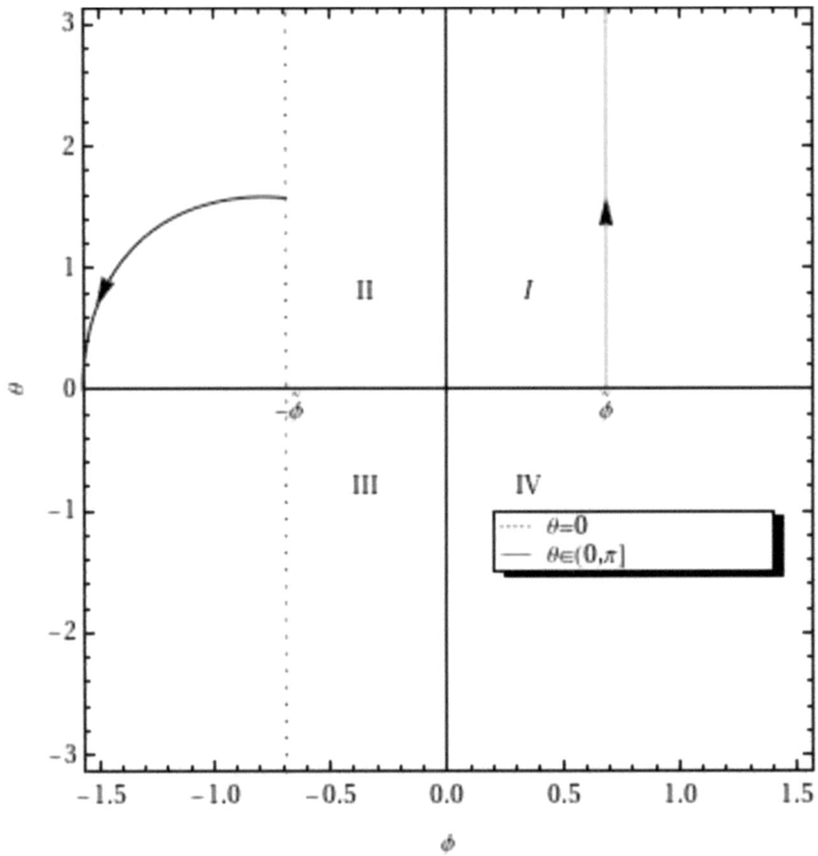

Figure 5. First iteration of the line segment $\phi = \tilde{\phi}$ (with $\theta \in [0, \pi]$) for the 2-parameter quantum Prisoner's dilemma. The point with coordinates $\theta = 0$, $\phi = \tilde{\phi}$ is mapped into the entire (dotted) line $\phi = -\tilde{\phi}$.

According to the previous statements, only points with coordinates $\theta = 0$, $\tilde{\phi} \leq \phi \leq \pi/2$ represent 2-periodic points of the map. By repeating the analysis for the fourth quadrant one obtains exactly the same outcome, whereas for the second and third quadrant the 2-periodic points turn out to be represented by the points with coordinates $\theta = 0$, $-\pi/2 \leq \phi \leq -\tilde{\phi}$.

We can summarize the results of this section by arguing that the 2-periodic orbits of the map related to the 2-parameter quantum Prisoner's dilemma, in its extended strategy version, lay on the projective plane equator $\theta = 0$, excluding those points with coordinates $|\phi| < \tilde{\phi}$. Moreover, these periodic points are such that $M(\theta, \phi) = (\theta, -\phi)$. Finally, as coordinates $\theta = 0$, $\phi = -\pi/2$, and $\theta = 0$, $\phi = \pi/2$ represent the same point in the projective plane, this point turns out to be actually a fixed point of the map.

3.3. Nash equilibria in the extended strategy 2-parameter quantum Prisoner's dilemma

In what follows, we will check the connection between the periodic points of the map discussed above and the NE of the quantum game. As it was already mentioned, a joined

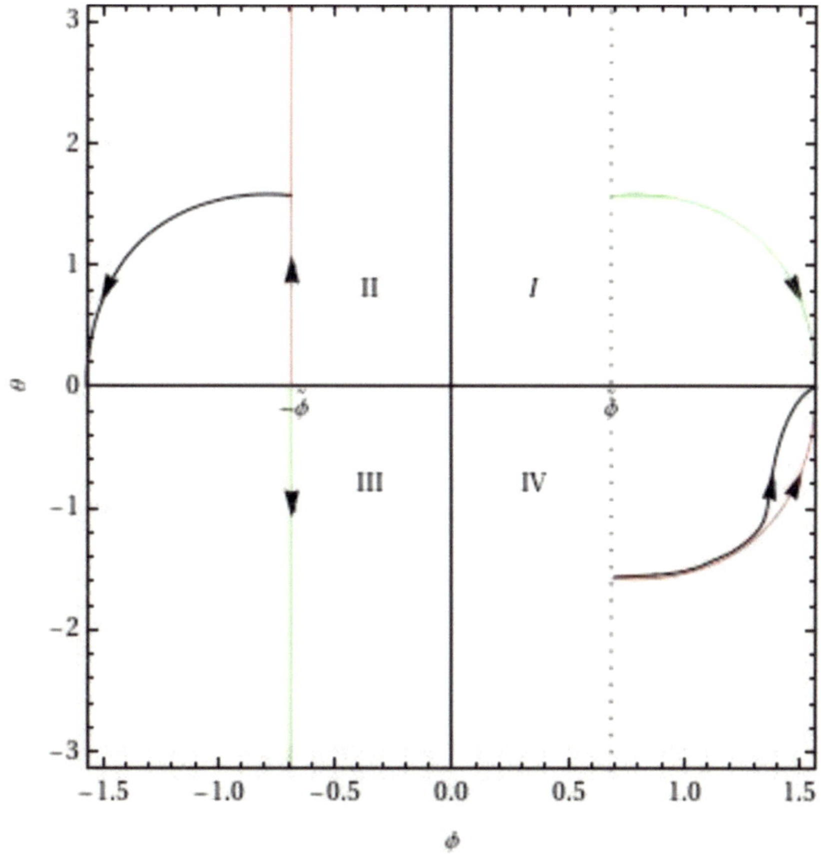

Figure 6. Second iteration of the line segment $\phi = \tilde{\phi}$ (with $\theta \in [0, \pi]$). The solid-black curve at the second quadrant (map image of the line segment $\phi = \tilde{\phi}$, with $\theta \in (0, \pi]$) is mapped into the solid-black curve at the fourth quadrant. The first iteration of the point with coordinates $\phi = \tilde{\phi}, \theta = 0$ (represented by the red and green line segments at the second and third quadrants) is mapped into the red and green curves at the fourth and first quadrants. Finally, the point with coordinates $\phi = -\tilde{\phi}, \theta = 0$, belongs also to the image under M of the point with coordinates $\phi = \tilde{\phi}, \theta = 0$, and it is mapped into the entire line $\phi = \tilde{\phi}$.

strategy fulfilling NE conditions is related to both periodic points in a 2-periodic orbit (namely, to one particular periodic point, and to its iteration). So let us define the quantum strategies

$$\hat{R}_{\pm} \equiv \hat{U}(0, \pm\alpha)$$

with $\alpha \in [0, \pi/2]$, and consider the functions defined as follows:

$$f_A^{\pm}(\theta, \phi) \equiv \$_A(\hat{R}_{\mp}, \hat{R}_{\pm}) - \$_A(\hat{U}(\theta, \phi), \hat{R}_{\pm}).$$

By replacing 26 in the previous definition, and after some manipulations, one obtains

$$f_A^{\pm}(\theta, \phi) = 3 - \cos(\theta/2)^2[2 + \cos(2(\alpha \pm \phi))] - 5\cos(\alpha)^2 \sin(\theta/2)^2. \tag{28}$$

Figure 7 shows the contours of $f_A^{+}(\theta, \phi)$ for different values of the parameter α. It turns out that the this function is positive-semidefinite whenever $\tilde{\phi} \leq \alpha \leq \pi/2$, and it has positive and negative regions for $0 \leq \alpha < \tilde{\phi}$. In addition, equation 28 implies that $f_A^{+}(\cdot, \phi) = f_A^{-}(\cdot, -\phi)$,

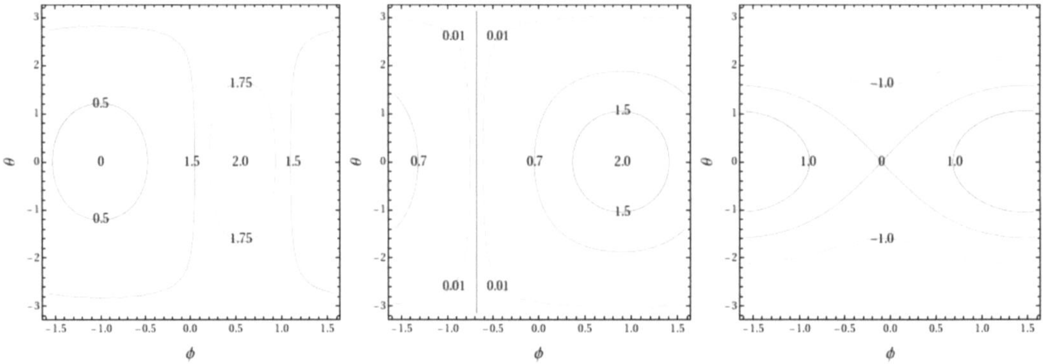

Figure 7. Contours of the function $f_A^+(\theta,\phi)$ for different values of the parameter α. Left: $\alpha = 1$, middle: $\alpha = \tilde{\phi}$, right: $\alpha = 0.1$.

meaning that the previous statements applies in the same way for the function f_A^-. Now, according to the symmetry relationship between $\$_A$ and $\$_B$, we can rewrite equation 28 as follows,

$$f_A^+(\theta',\phi') = \$_A(\hat{R}_-,\hat{R}_+) - \$_A(\hat{U}(\theta',\phi'),\hat{R}_+)$$
$$f_A^-(\theta'',\phi'') = \$_B(\hat{R}_-,\hat{R}_+) - \$_B(\hat{R}_-,\hat{U}(\theta'',\phi'')).$$

The latter observation, and the fact that f_A^+ and f_A^- behave with the parameter α as discussed above, imply in conclusion that

$$\$_A(\hat{R}_-,\hat{R}_+) \geq \$_A(\hat{U}(\theta',\phi'),\hat{R}_+) \quad \forall(\theta',\phi')$$
$$\$_B(\hat{R}_-,\hat{R}_+) \geq \$_B(\hat{R}_-,\hat{U}(\theta'',\phi'')) \quad \forall(\theta'',\phi'')$$

if and only if $\tilde{\phi} \leq \alpha \leq \pi/2$. Being the argument in the previous paragraphs symmetric under the interchange of the strategies $\hat{R}_+$ and $\hat{R}_-$, we assert that the outcomes of the game given by the joined strategies '$\hat{R}_- - \hat{R}_+$' and '$\hat{R}_+ - \hat{R}_-$' represent Nash Equilibria provided that $\tilde{\phi} \leq \alpha \leq \pi/2$. These Nash Equilibria are of diagonal type only in the case in which $\hat{R}_+$ and $\hat{R}_-$ coincide (up to a global phase), what happens for the special value $\alpha = \pi/2$. The latter strategy corresponds to the Eisert's operator $\hat{Q}$.

To give a possible interpretation for the NE observed in the extended strategy quantum Prisoner's dilemma, let consider the 2-player game given by the following bi-matrix

	Bob : $\hat{C}$	Bob : $\hat{D}$	Bob : $\hat{Q}$
Alice : $\hat{C}$	$(3,3)$	$(0,5)$	$(1,1)$
Alice : $\hat{D}$	$(5,0)$	$(1,1)$	$(0,5)$
Alice : $\hat{Q}$	$(1,1)$	$(5,0)$	$(3,3)$

where $\hat{C}$, $\hat{D}$ and $\hat{Q}$ are the quantum strategies of the Eisert's game. This reduced game, being different from the 2-parameter quantum Prisoner's dilemma, summarizes in a simple manner the ideas behind Eisert's result. Namely, if the strategy $\hat{Q}$ were not present, a possible agreement (to "C-ooperate") of the parties before they make their decisions would make no sense. As departing from the strategy $\hat{C}$ would keep improving their payoffs, the dilemma would persists as before the agreement. However, the table clearly shows that the situation is different if after the agreement they decide to "Q-ooperate" (namely, to simultaneously choose the strategy $\hat{Q}$). Now, they can not be further better off by departing from the agreement solution (on the contrary, it would be a self destructive behavior). Of course, we have not said much up to this point, but just showed in a concise fashion what Eisert's new NE exactly means.

Now, we can follow a similar idea and construct a reduced game which extracts the features present in the extended strategy quantum game. This is accomplished by including two more strategies in the normal form, which mean the $\hat{R}_-$ and $\hat{R}_+$ strategies for some specific, but arbitrary, value $\tilde{\phi} \leq \alpha < \pi/2$. We choose for example the value $\alpha = \pi/4$.

	Bob: $\hat{C}$	Bob: $\hat{D}$	Bob: $\hat{Q}$	Bob: $\hat{R}_+$	Bob: $\hat{R}_-$
Alice: $\hat{C}$	(3,3)	(0,5)	(1,1)	(2,2)	(2,2)
Alice: $\hat{D}$	(5,0)	(1,1)	(0,5)	(2.5,2.5)	(2.5,2.5)
Alice: $\hat{Q}$	(1,1)	(5,0)	(3,3)	(2,2)	(2,2)
Alice: $\hat{R}_+$	(2,2)	(2.5,2.5)	(2,2)	(1,1)	(3,3)
Alice: $\hat{R}_-$	(2,2)	(2.5,2.5)	(2,2)	(3,3)	(1,1)

It is straightforward to check from the table above that the three solutions '$\hat{Q} - \hat{Q}$', '$\hat{R}_+ - \hat{R}_-$' and '$\hat{R}_- - \hat{R}_+$' behave in this 5-strategy game as optimal NE, exactly as '$\hat{Q} - \hat{Q}$' does in the 3-strategy game associated to the Eisert's quantum game. However, we stress here the importance of the previously mentioned agreement. That agreement is not usually included in the story of the Prisoner's dilemma (see for example [10]), although it does not change the nature of the game (at both the classical and quantum levels). However, in our extended strategy version of the quantum game that agreement is crucial because both players could otherwise destroy themselves not by defection, but just by ignorance (for example, they could play the strategy '$\hat{R}_+ - \hat{R}_+$' receiving a payoff of 1 each. Or any other combination of $\hat{Q}$, $\hat{R}_+$ and $\hat{R}_-$ different from the NE outcomes).

The following table is constructed just to show that for $|\alpha| < \tilde{\phi}$ (in this case we choose $\alpha = \pi/6$) '$\hat{R}_+ - \hat{R}_-$' and '$\hat{R}_- - \hat{R}_+$' do not fulfill the NE conditions, and hence the corresponding reduced game is not suitable to depict the features of the extended strategy quantum game.

	Bob: $\hat{C}$	Bob: $\hat{D}$	Bob: $\hat{Q}$	Bob: $\hat{R}_+$	Bob: $\hat{R}_-$
Alice: $\hat{C}$	(3,3)	(0,5)	(1,1)	(2.5,2.5)	(2.5,2.5)
Alice: $\hat{D}$	(5,0)	(1,1)	(0,5)	(3.75,1.25)	(3.75,1.25)
Alice: $\hat{Q}$	(1,1)	(5,0)	(3,3)	(1.5,1.5)	(1.5,1.5)
Alice: $\hat{R}_+$	(2.5,2.5)	(1.25,3.75)	(1.5,1.5)	(1,1)	(3,3)
Alice: $\hat{R}_-$	(2.5,2.5)	(1.25,3.75)	(1.5,1.5)	(3,3)	(1.5,1.5)

4. The periodic point method in different scenarios

The aim of this section is to explore how the periodic point method gives rise to different outcomes when applied to other 2×2 quantum games. For that purpose we extend the previous analysis to the Chicken game and to a symmetrized version of the Battle of the Sexes.

4.1. The extended strategy quantum Chicken game

A possible normal form of the Chicken game was given in section 3. As already mentioned, the mixed-strategy game has three NE, namely, the two NE of the pure strategy game and a third NE represented by the outcome $p_A = p_B = 2/3$. Curiously, when applied to the quantum version of this game, the periodic point procedure gives rise to a similar picture as that of Figure 4. Line segments at constant values of ϕ in the first quadrant are mapped into curves at the second quadrant, which are in turn mapped into curves at the forth quadrant. Hence, the set of strategies of equation 27 should be considered again as the smallest space which guaranties the endomorphic behavior of M. Moreover, the first and second iterations of the first quadrant can be obtained after smooth deformations of the curves displayed in Figure 4, as well as the iterations of the remaining quadrants (not shown). The latter remarks imply that the maps associated to the quantum Chicken game and to the quantum Prisoner's dilemma are topological equivalent, so that the results outlined in section 3 are immediately applicable to the quantum Chicken game. The set of NE is again given by the joined strategies '$\hat{R}_+ - \hat{R}_-$' and '$\hat{R}_- - \hat{R}_+$' with $\tilde{\phi} \leq \alpha \leq \pi/2$ (for the new value $\tilde{\phi} = \arccos(-1/3)/2 \simeq 0.955$), and all these strategies are optimal in the sense of Eisert's solution $\hat{Q} - \hat{Q}$. It is interesting to observe that, even when both games (the Prisoner's dilemma and the Chicken) are totally different at the classical level, they show the same quantum behavior.

4.2. The extended strategy quantum symmetrized Battle of the Sexes

We finally present a symmetrized version of the Battle of the Sexes game. As the well known Battle of the Sexes, this game has two NE in its pure classical presentation and it faces both players to a problematic decision. However, the two NE are now located at the off-diagonal positions. The normal form of the game is given in the following bi-matrix[6] However the

	Bob: C	Bob: D
Alice: C	$(0,0)$	$(1,2)$
Alice: D	$(2,1)$	$(0,0)$

Table 3. Bi-matrix representation of the simmetrized Battle of the Sexes game.

payoff function of the mixed-strategy game is given now by $\bar{\$}_A = 2\,p_A + p_B - 3\,p_A\,p_B$, the corresponding 1-dimensional map is exactly the same as that of the mixed-strategy Chicken game, and it hence accounts for the same periodic orbits. Thus, this version of the Battle of the Sexes has the same NE distribution as the Chicken game.

Now, when we apply Eisert's quantization protocol and construct the 2-dimensional map similar to that discussed for the quantum Prisoner's dilemma, we unexpectedly find that we do not need any longer to consider the space of strategies S. Specifically, the space of strategies given by

$$S' = \{U(\theta,\phi)| \ -\pi \leq \theta \leq \pi \text{ and } 0 \leq \phi \leq \pi/2\} \tag{29}$$

proves now to be the smallest set which guaranties the endomorphic property of the map. This fact is demonstrated in Figure 8. In the left panel we iterate ϕ-lines in the upper semiplane (solid lines). We observe that the whole upper semiplane is mapped into the lower semiplane (dashed curves), which in turn is mapped again into the upper semiplane (dotted curves). That means that the mapping procedure perfectly works when restricted to this space. In addition, we conclude from Figure 8 that the map we are exploring does not account for fixed points. Moreover, restricting our analysis to the upper semiplane, we identify one possible

[6] To make the comparison easier to follow, we keep 'C' and 'D' to label the strategies, instead of the usual 'O' and 'T' (for 'Opera' and 'TV', respectively).

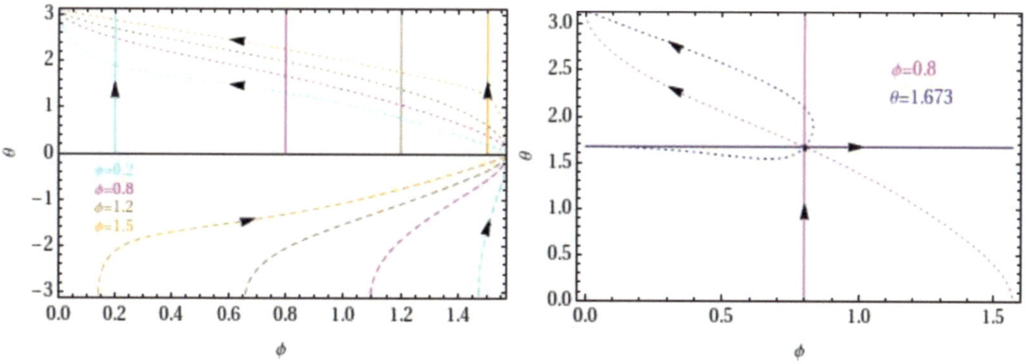

Figure 8. Maximization map associated to the quantum symmetrized Battle of the Sexes game. Left: for the upper semiplane, the first and second iterations (dashed and dotted curves, respectively) of the constant ϕ lines (solid lines). Right: Second iterations of the lines $\phi = 0.8$ (for $\theta > 0$) and $\theta = 1.673$. See the text for the explanation.

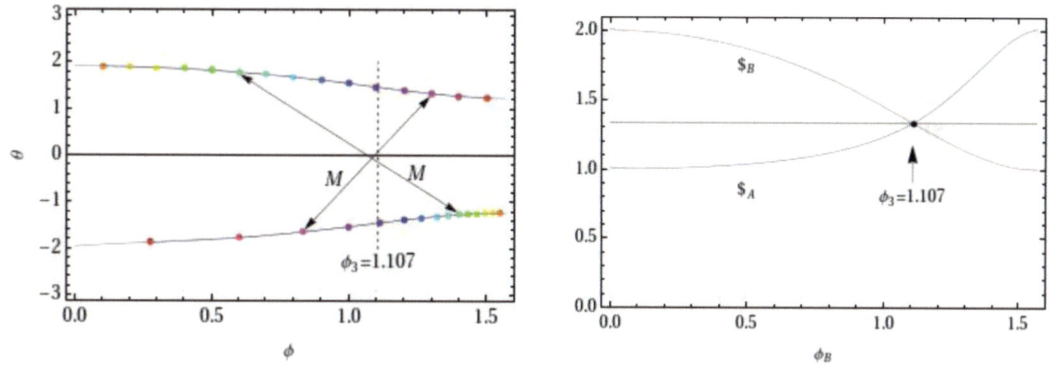

Figure 9. Left: Maximization map associated to the quantum symmetrized Battle of the Sexes game. Right: As a function of Alice's ϕ parameter, payoffs of Alice and Bob in the set of NE (see the text).

2-periodic point for each value of the parameter ϕ, namely, the point where each dotted curve intersects the corresponding solid line. To test whether these candidates actually represent 2-periodic points, we peak the values of the parameter θ where the intersections occur and iterate the squared map for these values of θ in the whole range of ϕ. In the right panel of Figure 8 we show this procedure for $\phi = 0.8$. The double intersection shows that our candidates iterate (under M^2) in points with the same coordinate θ, being hence fixed points of the squared map. As a conclusion, we find a continuous set of primitive 2-periodic orbits. Each orbit in this set is composed by one point in the upper semiplane and a second point in the lower semiplane. Left panel of Figure 9 depicts the organization of the orbits in the set, within the analysis discussed so far[7]. The two members of each cycle are located in two branches symmetric under a reflection around $\theta = 0$, however the correspondence between both members is not simply given by an inversion of the parameter θ, but that related to the colored points in the graph (such colored points were selected to be equidistant in the upper branch). Yet, for the specific value $\phi_3 = 1.107$, the two members are actually symmetrically located, having coordinates $\phi = \phi_3, \theta = \theta_3$ and $\phi = \phi_3, \theta = -\theta_3$, with $\theta_3 = 1.4596$.

[7] As the map is symmetric under a reflection around $\theta = 0$, the iteration of the lower semiplane gives rise to the same outcome to that depicted in Figure 9.

Now, as each periodic orbit can be parametrized by its ϕ coordinate associated to the member in the upper branch, we can come to the quantum game problem and study the NE as this parameter is varied. Right panel of Figure 9 shows the payoffs of Alice and Bob for each NE in the set, as a function of Alice's ϕ parameter (in the upper branch). It is interesting to observe that the outcomes $\phi_1 = 0, \theta_1 = 1.9$ and $\phi_2 = \pi/2, \theta_2 = -1.24$ play a role similar to that of the strategies C and D in the classical game. Namely, if for example Alice plays $\hat{S}_1 = \hat{U}(\theta_1, \phi_1)$ and Bob plays $\hat{S}_2 = \hat{U}(\theta_2, \phi_2)$, they receive the payoffs $\$_A = 1$ and $\$_B = 2$. Nevertheless, in the present scenario the joined strategy '$\hat{S}_1 - \hat{S}_2$' does fulfill the NE conditions, whereas '$\hat{C} - \hat{D}$' does not. Of course, if we plotted the payoff functions as a function of Bob's ϕ parameter in the upper branch, the conclusions would be symmetrical under the interchange of $\hat{S}_1$ and $\hat{S}_2$ (the same way as an interchange of C and D accounts for a second NE in the classical game, with the payoffs inverted). However, an interchange of the strategies is not the only way to invert the payoffs of the parties. From Figure 9 it turns out that the strategies $\hat{S}_{-1} = \hat{U}(-\theta_1, \phi_1)$ and $\hat{S}_{-2} = \hat{U}(-\theta_2, \phi_2)$ give $\$_A(\hat{S}_{-2}, \hat{S}_{-1}) = 2$ and $\$_B(\hat{S}_{-2}, \hat{S}_{-1}) = 1$, so '$\hat{S}_{-1} - \hat{S}_{-2}$' and '$\hat{S}_{-2} - \hat{S}_{-1}$' represent a second pair of 2-cycle NE with exactly the same characteristics of '$\hat{S}_1 - \hat{S}_2$' and '$\hat{S}_2 - \hat{S}_1$'. Nevertheless, the most interesting feature of the game concerns the emergence of a very special NE given by the strategies $\hat{S}_3 = \hat{U}(\theta_3, \phi_3)$ and $\hat{S}_{-3} = \hat{U}(-\theta_3, \phi_3)$. The joined strategies '$\hat{S}_3 - \hat{S}_{-3}$' and '$\hat{S}_{-3} - \hat{S}_3$' give both players the same payoff $\$_A = \$_B = 4/3$, and are not comparable to some outcome of the classical game. In the classical game we saw that the mixed joined strategy $(p, q) = (2/3, 2/3)$ fulfills the NE conditions, however this NE accounts for a payoff $\bar{\$}_A = \bar{\$}_B = 2/3$, even worse than the lower payoff given by the pure-strategy NE. '$\hat{S}_3 - \hat{S}_{-3}$' and '$\hat{S}_{-3} - \hat{S}_3$', instead, are joined strategies from which players can not increase their payoffs without lessening that of the other party, being therefore PO outcomes of the game. Accordingly, this version of the symmetrized Battle of the Sex game accounts for 2 possible optimal solutions which prevent players from facing with the classical conflict.

Finally, we observe that when restricted to the set of strategies given by Eisert et al. there are no periodic points surviving in the map domain. That means that in the original space of strategies there are no NE. Namely, for any possible decision of Alice, Bob can find a convenient counter-strategy which places Alice in a situation from which she would prefer to deviate.

From the previous discussion, we observe again an interesting fact from the comparison of two games. When we compare the Chicken and the symmetrized Battle of the Sexes, we see that both games display the same NE distribution at the classical level, however they are totally different when the periodic points method is applied to identify NE in the extended space of quantum strategies. From this comparison and that established between the Prisoner's dilemma and the Chicken game we conclude that from the classical game it is not possible to predict a priori what is going to happen at the quantum level, and that from the quantum behavior it is not possible to unambiguously infer the classical substrate of the game. Of course, the NE distribution does not completely defines the category of a game (see [19]).

4.3. Connections between Eisert's games

As we established above, the maps corresponding to the quantum Prisoner's dilemma and the quantum Chicken game are topological equivalent. For $a = \$_A(C, C)$, $b = \$_A(C, D)$,

$c = \$_A(D,C)$ and $d = \$_A(D,D)$, this fact can be explicitly stated by considering the function

$$f(a,b,c) = \frac{1}{2}\arccos\left[\frac{2a-b-c}{c-b}\right], \tag{30}$$

which gives the parameter $\tilde{\phi}$ limiting the region where the $\tilde{\phi}$-lines are iterated (see Figure 4). By changing the value of b we only stretch or shrink that region, but whenever $\tilde{\phi}$ ranges in the interval $(0, \pi/2)$, we do not modify the topological structure of the map. For $a = 3$ and $c = 5$, $\tilde{\phi}$ can escape this interval only if $b \geq 3$, but this would contradict the conditions of the Prisoner's dilemma ($b < d < a < c$) and the Chicken game ($d < b < a < c$). The transition between the two games happens at $b = 1$ ($\tilde{\phi} = \pi/4$), for which the map still preserves the same characteristics[8].

On the other hand, the comparison of the Chicken game and the simmetrized battle of the Sexes game reveals an other interesting feature. This time, we have two games with exactly the same classical ingredients (even when mixed strategies are allowed), but with completely different behaviors at the quantum level. The most surprising thing concerns the symmetrized Battle of the Sexes game, which when studied within Eisert's protocol displays not only a very rich distribution of NE (two of these off-diagonal NE fulfilling the PO condition as well), but also a different domain for the associated map. This intriguing result forces us to establish a connection in terms of geometrical arguments, as we did when compared the Prisoner's dilemma and the Chicken game. To pursue this task, let start by considering the following game

	Bob: C	Bob: D
Alice: C	(3,3)	(2,5)
Alice: D	(5,2)	(3,3)

This game, having not much interest from the classical point of view, when quantized has exactly the same behavior as the Prisoner's dilemma and the Chicken game. Moreover, as it differs from the Chicken in the entry d only, it has the same value for the parameter $\tilde{\phi}$ and thus an associated map which is a copy of that of the quantum Chicken game.

Now, when b starts increasing, the regions where the iterations (dashed and doted curves in Figure 4) are located shrink, and eventually they collapse (as the threshold given by $b = 3$ is reached) to the edges $\phi = \pm\pi/2$ (which are connected in a Möbius strip fashion, according to the projective plane topology). For $b > 3$, $\tilde{\phi}$ ceases to be real, which means that the map can not behave anymore in the same fashion. If we rewrite the table for any value of $b > 3$, we obtain the conditions of the symmetrized Battle of the Sex game ($a = d < b < c$), and an associated map which is topological equivalent to that of Figure 8 (curves in both maps are related by means of smooth deformations). In terms of nonlinear dynamic language, we are here in presence of a bifurcation of codimension 1, the entry b in the payoff matrix being the bifurcation parameter. In our example, when $b < 3$ we have a class of quantum games whose associated map displays (even when they can be different classically) a topological equivalence. However, when $b > 3$ the structure of the maps associated to the quantum game changes abruptly, giving rise to a more complex class of maps. Moreover, within the latter class, each side of the projective plane accounts for an endomorphism on its own (which is a

[8] We chose standard values to present the games. However, the conclusions outlined here are independent of the specific choice.

symmetric repetition of the other endomorphism), a fact that allows us to define the map in just a half piece of the initial domain.

5. A geometrical classification of Eisert's games

The discussion of the previous section allows us to formulate new questions. If we compute the number of different bi-matrices which can be constructed with at least 3 different entries, we find that there are 54 2×2 possible games which can be analyzed according to our procedure. However, we saw that some quantum games can be connected by means of geometrical considerations, and grouped in classes according to the topological structure of the associated maps. It is therefore natural to ask how many classes can be obtained under this classification scheme[9]. Here we only summarizes the main results of the classification scheme and establish, for the most famous examples, to which class each game belongs. For a full description of the the subject see [22].

Class 1 comprises two subclasses. The games in these subclasses are enumerated in table 4. Class 1a includes the Prisoner's dilemma (game 2) and the Chicken game (game 3). The maps within this class are defined on the projective plane (space of strategies S). In both subclasses points of period 2 are located at $\theta = 0$. Specifically, the maps belonging to class 1a have its periodic points at $\theta = 0$ and $\tilde{\phi} \leq |\phi| \leq \pi/2$ whereas those of class 1b have its periodic points at $\theta = 0$ and $0 \leq |\phi| \leq \tilde{\phi}$.

	Class 1a		*Class 1b*
1	$b = d < a < c$	5	$c = d < a < b$
2	$b < d < a < c$	6	$c < d < a < b$
3	$d < b < a < c$	7	$d < c < a < b$
4	$b < a = d < c$	8	$c < a = d < b$

Table 4. Games of Class 1.

Games belonging to class 2 are enumerated in table 2. They include the well known Deadlock game (game 10). The associated map is defined in the space of strategies

$$S''' = \{U(\theta,\phi)|\ 0 \leq \theta \leq \pi \text{ and } 0 \leq \phi \leq \pi/2\}$$

an is simply given by the expression

$$M(\theta,\phi) = (\theta, \pi/2 - \phi).$$

Accordingly, every second iteration is located on the ϕ-line which is iterated. This makes every point in the space of strategies a member of an orbit of period 2 whose second member is obtained by reflecting on the line $\phi = \pi/4$. Points with coordinate $\phi = \pi/4$ are thus the fixed points of the map.

Class 3 comprises two subclasses each composed by a single game. The games are enumerated in table 3. Being the maps defined in the projective plane, periodic points in both classes are located in the region $\tilde{\phi} \leq \phi \leq \pi/2 - \tilde{\phi}$, for

$$\tilde{\phi} \equiv \frac{1}{2} \arccos \left[\frac{b + c - \max\{a,d\}}{b - c} \right]. \tag{31}$$

[9] We stress that this classification relies on the geometrical properties of the map associated to the game, so it needs to be differentiate from other existing classification schemes for games (see [19] for a discussion of this subject).

9	$b < a < c = d$	15	$a < c < b = d$	21	$b = a < c < d$
10	$b < a < c < d$	16	$a < c < b < d$	22	$c = a < b < d$
11	$c < a < b = d$	17	$b < c < a = d$	23	$b < a = c < d$
12	$c < a < b < d$	18	$b < c < a < d$	24	$c < a = b < d$
13	$a < b < c = d$	19	$c < b < a = d$	25	$a = b < d < c$
14	$a < b < c < d$	20	$c < b < a < d$	26	$a = c < d < b$

Table 5. Games of Class 2.

Any of these points is a member of an orbit of period 2 whose second member is given by a reflection on the line $\phi = \pi/4$. In this class, the periodic orbits are symmetric under a reflection on the θ axis.

	Class 3a		*Class 3b*
27	$b < a < d < c$	28	$c < a < d < b$

Table 6. Games of Class 3.

The two subclasses comprised by class 4 are enumerated in table 7. Class 4*a* includes the symmetrzed Battle of the Sexes (game 29). The behavior of the map and the location of the periodic orbits was explained above for the symmetrzed battle of the Sexes. For class 4*b* the picture is the same as for class 4*a* after a reflexion around the axis $\phi = \pi/4$, so the same symmetry holds for the periodic orbits.

	Class 4a		*Class 4b*
29	$a = d < b < c$	33	$a = d < c < b$
30	$a < d < b < c$	34	$a < d < c < b$
31	$d < a < b < c$	35	$d < a < c < b$
32	$a < b = d < c$	36	$a < c = d < b$

Table 7. Games of Class 4.

Games included in class 5 are enumerated in table 8. They are divided in 2 subclasses with a single member each, and have an associated map defined in the space of strategies of the symmetrized Battle of the Sexes game (space of strategies S'). NE are of two different types in each subclass. For class 5*a*, NE are of the same type as in class 4*a* for $\phi < \tilde{\phi}$, and as those of class 2 for $\phi > \tilde{\phi}$, whereas for games in class 5*b* NE are as those of class 4*b* for $\phi > \tilde{\phi}$ and as those of class 2 for $\phi < \tilde{\phi}$. $\tilde{\phi}$ is given in equation 31.

	Class 5a		*Class 5b*
37	$a < b < d < c$	38	$a < c < d < b$

Table 8. Games of Class 5.

Games belonging to class 6 are enumerated in table 9. Again, the games are separated in two subclasses, corresponding to two different maps. Class 6*a* includes the so called Stag Hunt game (games 41 and 42). The associated map, for both subclasses, is defined in the projective plane, and the periodic orbits are located in the axis $\theta = 0$, with no restrictions in the coordinate ϕ.

	Class 6a		Class 6b
39	$d = b < c < a$	46	$d = c < b < a$
40	$d < b < c < a$	47	$d < c < b < a$
41	$b < d < c < a$	48	$c < d < b < a$
42	$b < c = d < a$	49	$c < b = d < a$
43	$b < c < d < a$	50	$c < b < d < a$
44	$b < d < a = c$	51	$c < d < a = b$
45	$d < b < a = c$	52	$d < c < a = b$

Table 9. Games of Class 6.

Games belonging to class 7 are enumerated in table 10. Again, the games are separated in two subclasses, corresponding to two different maps. These maps have a curious behavior. Being defined in the projective plane, the map corresponding to class 7*a* maps all the domain into the line $\phi = \pi/2$, which is in turn mapped into the point $\theta = \pi/2, \phi = 2$. Therefore, Eisert's joined strategy '$\hat{Q} - \hat{Q}$' represents the sole NE of this game. The map corresponding to the class 7*b*, in turn, maps all the projective plane in the axis $\phi = 0$, which in turn is mapped into the origin. The sole NE of the latter game is thus represented by the joined strategy '$C - C$' of the classical game.

	Class 7a		Class 7b
53	$d < a = b < c$	54	$d < a = c < b$

Table 10. Games of Class 7.

6. Conclusions

After introducing Eisert's theory for quantizing games, we analyzed a periodic point-based procedure designed to identify NE in 2×2 quantum games defined on an extended set of strategies. According to our analysis, NE of Eisert's 2-parameter quantum Prisoner's dilemma are located on a segment of a projective plane equator (points with $\theta = 0$, excluding those for which $|\phi| < \tilde{\phi}$). All strategies fulfilling the NE condition are 2-cycle NE except that corresponding to $\phi = \pi/2$. The latter strategy corresponds to Eisert's operator $\hat{Q}$, and it is actually the only strategy which survives if we restrict ourselves to the original set defined in [5]. Now, as $\$_A(0,\phi',0,\phi'') = 3\cos(\phi' + \phi'') + \sin((\phi' + \phi'')$, it turns out that

$$\$_A(\hat{R}_{\pm}, \hat{R}_{\mp}) = 3 \quad \forall \alpha \in [0, \pi/2].$$

Hence, we conclude that all joined strategies '$\hat{R}_+ - \hat{R}_-$' and '$\hat{R}_- - \hat{R}_+$' with $\tilde{\phi} \leq \alpha < \pi/2$ are NE as good as the '$\hat{Q} - \hat{Q}$' one (in the sense of the payoff given to the players). However, in the scenario of the extended strategy set the players have not an obvious choice as they have in the restricted case. This is the reason why Eisert et al. designed a protocol with such a special (somehow artificial) set. Nevertheless, as it is discussed in [1] and [8], the set of strategies adopted in [5] and its corresponding outcome are far from general. Motivated by this fact, we adopted here an alternative but still valid set which was the only one which proved to be suitable for developing the periodic point procedure. Yet, at the end of section 3 we discuss a possible modification of the story behind the game which has not effects on the original quantum game (not even at the classical level), but that makes the new NE be as suitable as the result obtained by Eisert et al.

In section 4 we considered the Chicken game and a symmetrized version of the Battle of the Sexes. As a conclusion of the comparison of this two examples and the Prisoner's dilemma, we observed that the outcome of the quantization procedure has no reminiscences with the classical nature of the game. Specifically, at the classical level the Prisoner's dilemma and the Chicken game display a different NE distribution. In the first case there is just one NE represented by the joined strategy 'defect-defect', whereas in the second case the NE correspond to 'cooperate-defect' and 'defect-cooperate' . Nevertheless, when analyzed within the framework of Eisert's protocol, both games share exactly the same behavior, namely, the continuous set of NE given by the joined strategies '$\hat{R}_+ - \hat{R}_-$' (for $\tilde{\phi} \leq \alpha < \pi/2$) which in addition fulfill the PO condition (being therefore rational solutions of the game). On the contrary, after comparing the Chicken game and the simmetrized Battle of the Sexes, we conclude that games having the same NE distribution at the classical level, can behave in a completely different fashion when Eisert's scheme is implemented in the extended space of strategies.

In section 4.3 we showed that games having different behaviors at the quantum level, can actually be related through bifurcations for maps. Moreover, we demonstrated that the bifurcation parameter is directly obtained from the bi-matrix defining the classical game. In section 5, we grouped every 2×2 games in classes according to the map which is associated to the quantum game, and showed that the number of classes is certainly small. It is therefore natural to ask how all games (or the classes representing them) connect by means of a specific bifurcation. Namely, which are the entries in the payoff matrices which play the role of the bifurcation parameter for any two, arbitrary, games. Or whether it is necessary, to connect two classes through a bifurcation, to include a third one as an intermediate class in the path. This would make it necessary to consider two different entries to account for the bifurcation, changing therefore the bifurcation codimension and making the picture a little more complicated but more interesting. These issues will be addressed in a future work.

Author details

David Schneider
Universidade Estadual de Campinas, Brazil

7. References

[1] Benjamin, S. C. & Heyden, P. M. [2001a]. Comments on "quantum games and quantum strategies, *Phys. Rev. Lett.* 87: 069801–1.

[2] Benjamin, S. C. & Heyden, P. M. [2001b]. Multiplayer quantum games, *Physical Review A* 64: 030301.

[3] Chen, L., Ang, H., Kiang, D., Kwek, L. C. & Lo, C. [2003]. Quantum prisoner dilemma under decoherence, *Physics Letters A* 316: 317–323.

[4] Du, J., Li, H., Xu, X., Shi, M., Wu, J., Zhow, X. & Han, R. [2001]. Entanglement playing a dominating role in quantum games, *Physics Letters A* 289: 9–15.

[5] Eisert, J., Wilkens, M. & M.Lowenstein [1999]. Quantum games and quantum strategies, *Phys. Rev. Lett.* 83: 3077–3080.

[6] Flitney, A. P. & Abbot, D. [2004]. Quantum two and three person duels, *J. Opt. B* 6: S860–6.

[7] Flitney, A. P. & Abbot, D. [2005]. Multiplayer quantum minority game with decoherence, *J. Phys. A: Math. Gen.* 38: 449.

[8] Flitney, A. P. & Hollenberg, L. C. L. [2007]. Nash equilibria in quantum games with generalized two-parameters strategies, *Physics Letters A* 363: 381–388.

[9] H. Guo, J. Z. & Koehler, G. [2008]. A survey of quantum games, *Decis. Support Syst.* 46: 318–332.

[10] Hidalgo, E. G. [2008]. Quantum games and the relationships between quantum mechanics and game theory, *arXiv:0803.0292v1 [quant-ph]* .

[11] Iqbal, A. & Toor, A. H. [2001a]. Entanglement and dynamic stability of nash equilibria in a symmetric quantum game, *Physics Letters A* 286: 245–250.

[12] Iqbal, A. & Toor, A. H. [2001b]. Evolutionarily stable strategies in quantum games, *Physics Letters A* 280: 249–256.

[13] Iqbal, A. & Toor, A. H. [2002]. Quantum mechanics gives stability to a nash equilibrium, *Physical Review A* 65: 022306.

[14] Massey, W. S. [1977]. *Algebraic Topology: An Introduction*, Springer-Verlag, New York.

[15] Meyer, D. [1999]. Quantum strategies, *Phys. Rev. Lett.* 82: 1052–1055.

[16] Nawaz, A. & Toor, A. H. [2006]. Quantum games with correlated noise, *J. Phys. A: Math. Gen.* 39: 9321.

[17] Piotrowski, E. W. & Sladkowski, J. [2002]. Quantum market games, *Phyisica A* 312: 208–216.

[18] Ramzan, M. & Khan, M. K. [2009]. Noise effects in a three-player prisoner's dilemma quantum game, *J. Phys. A: Math. Gen.* 41: 435302 (11pp).

[19] Rosero, A. F. H. [n.d.]. *Classification of Quantum symmetric Nonzero-sum* 2×2 *games in the Eisert's Scheme*, arXiv:quant-ph/0402117v2.

[20] Schneider, D. M. [2011]. A periodic point-based method for the analysis of nash equilibria in 2×2 symmetric quantum games, *J. Phys. A: Math. Theor.* 44: 905301.

[21] Schneider, D. M. [2012]. A new geometrical approach to nash equilibria organization in eisert's quantum games, *J. Phys. A: Math. Theor.* 45: 085303.

[22] Schneider, D. M. [n.d.]. Geometrical-besed classification of eisert's quantum games, In preparation.

[23] Stohler, M. & E.Fischbach [2005]. Non-transitive quantum games, *Fizika B* 14: 235–244.

[24] Werner, R. F. [1998]. Optimal cloning of pure states, *Phys. Rev. A* 58: 1827.

Chapter 14

Physical Realization of a Quantum Game

A.M. Kowalski, A. Plastino and M. Casas

Additional information is available at the end of the chapter

http://dx.doi.org/10.5772/54216

1. Introduction

Game theory is a mathematical methodology for analyzing calculated circumstances, such as it happens in games, where a person's success is based upon the choices of others. More formally, it is the study of mathematical models of conflict and cooperation between intelligent rational decision-makers [1]. Another way of describing this is interactive decision theory [2]. Game theory is mainly used in economics, political science, psychology, logic, and biology. The subject first addressed is called zero-sum games, such that one person's gains exactly equal net losses of the other participant(s). In our days game theory applies to a wide range of class relations, and has developed into an umbrella term for the logical side of science, including both human and non-humans (computers). Classic uses include a sense of balance in numerous games, where each participant develops a tactic that cannot successfully better his/her results.

Mathematical game theory began with E. Borel in his book *Applications aux Jeux de Hasard*. His results were somewhat limited, and the theory regarding the non-existence of blended-strategy equilibrium in two-player games was incorrect. Modern game theory began with an idea regarding the existence of mixed-strategy equilibria in two-person zero-sum games, proved by J. von Neumann, that used Brouwer's fixed-point theorem on continuous mappings into compact convex sets [3]. Game theory was later explicitly applied to biology in the 1970s. Game theory has been widely recognized as an important tool in many fields. Eight game-theorists have won the Nobel Prize in Economic Sciences, and John Maynard Smith was awarded the Crafoord Prize for his application of game theory to biology [4, 5]. As a method of applied mathematics, game theory has also been used to study a wide variety of human and animal behaviors. The use of game theory in the social sciences has expanded, and game theory has been applied to political, sociological, and psychological behaviors as well. Game-theoretic analysis was initially used to study animal behavior by R. Fisher in the 1930s (note that Charles Darwin made a few informal game-theoretic statements). Fisher's work predates the name "game theory", although it shares important features with this field. The developments in economics were later applied to biology largely by J. M. Smith in his book *Evolution and the Theory of Games*. In addition to being used to describe, predict, and explain behavior, game theory has also been employed to develop theories of ethical or normative behavior and to prescribe such behavior. In economics and philosophy, scholars have applied game theory to help in the understanding of good or proper behavior. Game-theoretic arguments of this type actually date back to Plato [4, 5].

The first known use of game theory attempted to describe and model how human populations behave. Some researcher believe that by finding the equilibria of games they can predict how actual human populations will behave when confronted with situations analogous to the game being studied. This particular view of game theory has come under recent criticism. First, it is criticized because the assumptions made by game theorists are often violated. Game theorists may assume players always act in a way to directly maximize their wins (the Homo *economicus* model), but in practice, human behavior often deviates from this model. Explanations of this phenomenon are many: irrationality, new models of deliberation, or even different motives (like that of altruism). Game theorists respond by comparing their assumptions to those used in physics. Thus while their assumptions do not always hold, they can treat game theory as a reasonable scientific ideal akin to the models used by physicists.

2. Introductory notions: representation of games

The processes studied in game theory are well-defined mathematical objects. A game consists of

- a set of players,
- a set of moves (or strategies) available to those players, and
- a specification of payoffs for each combination of strategies, where
- payoffs are numbers which represent the motivations of players. Payoffs may represent profit, quantity, "utility," or other continuous measures (cardinal payoffs), or may simply rank the desirability of outcomes (ordinal payoffs). In all cases, the payoffs must reflect the motivations of the particular player.

Most cooperative games are presented in a characteristic functional form, while so-called extensive and the normal forms are used to define noncooperative games.

The extensive game-form can be used to formalize games with a time sequencing of moves. Games here are played on trees. Each vertex (or node) of the tree represents a point of choice for a player. The player is specified by a number listed by the vertex. The lines out of the vertex represent a possible action for that player. The payoffs are specified at the bottom of the tree. The extensive form can be viewed as a multi-player generalization of a decision tree. If there are two players, player 1 moves first and chooses a specific move. Player 2 watches player 1's move and then chooses a response, etc. The extensive form can also capture simultaneous-move games and games with imperfect information.

The normal (or strategic form) game is usually represented by a matrix which shows the players, strategies, and payoffs. More generally it can be represented by any function that associates a payoff for each player with every possible combination of actions. For two players, one chooses the matrix' rows and the other the columns. Each player has a number of strategies, which are specified by the number of rows and the number of columns. The payoffs are provided in the matrix' interior.

A player's strategy in a game is a complete plan of action for whatever situation might arise; this fully determines the player's behavior. A player's strategy will determine the action the player will take at any stage of the game, for every possible history of play up to that stage. A strategy profile (sometimes called a strategy combination) is a set of strategies for each player which fully specifies all actions in a game. A strategy profile must include one and only one strategy for every player. The strategy concept is sometimes (wrongly) confused with that of a move. A move is an action taken by a player at some point during the play of a game (e.g., in chess, moving white's Knight). A strategy on the other hand is a complete algorithm for playing the game, telling a player what to do for every possible situation throughout the game.

A *pure* strategy provides a complete definition of how a player will play a game. In particular, it determines the move a player will make for any situation he/she could face. A player's strategy set is the set of pure strategies available to that player. A *mixed* strategy is an assignment of a probability to each pure strategy. This allows for a player to randomly select a pure strategy. Since probabilities are continuous, there are infinitely many mixed strategies available to a player, even if their strategy set is finite. Of course, one can regard a pure strategy as a degenerate case of a mixed strategy, in which that particular pure strategy is selected with unit probability and every other strategy with null probability. A totally mixed strategy is a mixed strategy in which the player assigns a strictly positive probability to every pure strategy.

The above elementary paragraphs on game theory miss an extremely important point, namely that players in a game maintain internal models, which in turn contain the models which they expect other players are using, and so on. This basic ingredient (infinite regress in mental model building) might be regarded as posing difficulties whenever comparisons with physical processes are made. In this respect we bypass the issue by following ideas arising already in the 19th Century with Maxwell. One considers the description of natural processes as a game between Nature and the Observer, which need not be a human being. A lucid and detailed example is provided in a celebrated book by Frieden, for instance [6]. We will here effect the associations

- players $\rightarrow$ sets of strategies/payoffs,
- classical probability distributions $\rightarrow$ mixed strategies,
- quantum states $\rightarrow$ quantal strategies,
- players using mixed strategies $\rightarrow$ classical players,
- players using quantum states' strategies $\rightarrow$ quantum players.

3. Elementary quantum notions

Quantum mechanics, also known as quantum physics or quantum theory (QT), is a physics' discipline that deals with physical phenomena where the action is of the order of Planck's constant. QT provides a mathematical description of much of the dual particle-like and wave-like behavior and interactions of energy and matter. It departs from classical mechanics primarily at the atomic and subatomic scales, the so-called quantum realm. In advanced topics of quantum mechanics, some of these behaviors are macroscopic and only emerge at very low or very high energies or temperatures. The name "quantum mechanics" was coined by Planck, and derives from the observation that some physical quantities can change only by discrete amounts, or quanta. For example, the angular momentum of an electron bound to an atom or molecule is quantized. In a quantal context, the wave particle duality of energy and matter and the uncertainty principle provide a unified view of the behavior of photons, electrons and other atomic-scale objects. The mathematical formulations of quantum mechanics are abstract. A mathematical function called the wave-function, or a matrix called the density matrix provide information about the probability amplitude of position, momentum, and other physical properties of a system. Mathematical manipulations of the wave-function (density matrix) involves the mathematics of Hilbert's space. Many of the results of quantum mechanics are not easily visualized in classical terms: For instance, the ground state in the quantum mechanical model is a non-zero energy state that is the lowest permitted energy state of a system, rather than a more traditional system that is thought of as simply being at rest with zero kinetic energy. Aa wave-function changes (density matrix) when a mathematical entity called an Operator is applied to it. In this vein, time-evolution of a systems is conceived as the temporal change of the wave-function (density matrix) caused by the action of a spacial kind of operators called Unitary Operators. These operators are functions of the Hamiltonian operator, that represents the system's energy. The trace of an operator $\hat{O}$, denoted by $Tr\hat{O}$ is an important quantity. Each operator is represented by a square matrix and the trace is the sum of the diagonal elements of that matrix.

Particles in Nature are either bosons or fermions. Bosons are subatomic particles that obey statistical rules called the Bose Einstein ones. Several bosons can occupy the same quantum state. The word boson derives from the name of the Indian physicist Satyendra Nath Bose. Bosons contrast with fermions, which obey a different set of statistical rules, called the Fermi Dirac ones. Two or more fermions cannot occupy the same quantum state. Since bosons with the same energy can occupy the same place in space, bosons are often force carrier particles. In contrast, fermions are usually associated with matter (although in quantum physics the distinction between the two concepts is not clear cut). Bosons may be either elementary, like photons, or composite, like mesons. All observed bosons have integer spin, as opposed to fermions, which have half-integer spin. This is in accordance with the spin-statistics theorem, which states that in any reasonable relativistic quantum field theory, particles with integer spin are bosons, while particles with half-integer spin are fermions.

4. The semi-classical approach to quantum mechanics

This is an approach in which one part of a system is described quantum-mechanically whereas the other is treated classically.

The semiclassical approach has had a long and distinguished history and is a very important weapon in the physics' armory. Indeed, semiclassical approximations to quantum mechanics remain an indispensable tool in many areas of physics and chemistry. Despite the extraordinary evolution of computer technology in the last years, exact numerical solution of the Schrödinger equation is still quite difficult for problems with more than a few degrees of freedom. Another great advantage of the semiclassical approximation lies in that it facilitates an intuitive understanding of the underlying physics, which is usually hidden in blind numerical solutions of the Schrödinger equation. Although semiclassical mechanics is as old as the quantum theory itself, the field is continuously evolving. There still exist many open problems in the mathematical aspects of the approximation as well as in the quest for new effective ways to apply the approximation to various physical systems (see, for instance, [7, 8] and references therein).

5. Attractors and fixed points

The branch of mathematics that studies fixed points and attractors for different systems is called the theory of dynamical systems. An attractor is a set towards which a variable moving according to the dictates of a dynamical system evolves over time. That is, points that get close enough to the attractor remain close even if slightly disturbed. The evolving variable may be represented algebraically as an n-dimensional vector. The attractor is a region in an n-dimensional space. In physical systems, the n dimensions may be, for example, two or three positional coordinates for each of one or more physical entities; in economic systems, they may be separate variables such as the inflation rate and the unemployment rate. If the evolving variable is two- or three-dimensional, the attractor of the dynamic process can be represented geometrically in two or three dimensions. An attractor can be a point, a finite set of points, a curve, a manifold, or even a complicated set with a fractal structure known as a strange attractor. If the variable is a scalar, the attractor is a subset of the real number line. Describing the attractors of chaotic dynamical systems has been one of the achievements of chaos theory. A trajectory of the dynamical system in the attractor does not have to satisfy any special constraints except for remaining on the attractor. The trajectory may be periodic or chaotic. If a set of points is periodic or chaotic, but the flow in the neighborhood is away from the set, the set is not an attractor, but instead is called a repeller.

A fixed point (also known as an invariant point) of a function is a point that is mapped to itself by the function. A set of fixed points is sometimes called a fixed set. Thus, c is a fixed point of the function $f(x)$ if and only if $f(c) = c$. Not all functions have fixed points: for example, if f is a function defined on the real numbers as $f(x) = x + 1$, then it has no fixed points, since x is never equal to $x + 1$ for any real number. In graphical terms, a fixed point means the point $(x, f(x))$ is on the line $y = x$. The example $f(x) = x + 1$ is a case where the graph and the line are a pair of parallel lines. Points which come back to the same value after a finite number of iterations of the function are known as periodic points. Thus, a fixed point is a periodic point with period equal to one. The stability of a fixed point is addressed in mathematics by the so-called stability theory. It investigates the stability of solutions of differential equations and of trajectories of dynamical systems under small perturbations of initial conditions.

6. Our goals

In recent times much attention has been paid to the task of extending game theory concepts [3] to quantum mechanics [9–19]. *Quantum games* may be of interest given that its classical counterpart (CGT) has had such a phenomenal success. Thus, much is to be expected from "quantizing" classical game theory. It is well known that various problems in physics can be usefully thought of as games. Quantum cryptography, for example, is easily reworded as a game between i) individuals who wish to communicate and ii) those who wish to eavesdrop [9]. Quantum cloning has been cast in the guise of a physicist playing a game against nature [10]. Even the cornerstone of physics, measurement processes themselves, may be approached in these terms.

Meyer [12] has pointed out that algorithms conceived for quantum computers may also be regarded as games between classical and quantum agents, and thus viewed in scenarios in which players are a quantum computer and its operator. It has been stated in Ref. [14] that *against this background, it is natural to seek a unified theory of games and quantum mechanics.* We wish here to add, within such philosophy, material to the notion of regarding quantum processes from a game theory viewpoint, *with emphasis on the semiclassical realm.*

Our ideas revolve around the picture developed in [18], un which an open quantum system corresponding to a biophysical Hamiltonian is regarded as a quantum game. Although our subject is quite different the underlying concepts are is the same. We shall regard the temporal evolution of a semiclassical system as a game. The concomitant process is then to be transcribed in term os strategies involving players hoping to optimize their chances.

7. The physical model and its associated game

As stated above, much quantum insight is gained from semiclassical viewpoints. Several methodologies are available (WKB, Born-Oppenheimer approach, etc.) Here we consider two interacting systems: one is classical and the other quantal. This can be done whenever the quantum effects of one of the two systems are negligible in comparison to those of the other one. Examples can be readily found. We just mention Bloch equations [20], two-level systems interacting with an electromagnetic field within a cavity and Jaynes-Cummings semiclassical model [21–26], collective nuclear motion [27], etc.

More recently [28–32], a special bipartite model has been employed with reference to problems in such fields as chaos, wave-function collapse, measurement processes, and cosmology [33]. In a related vein we consider the interaction between a quantum system and a classical one described by a Hamiltonian of the form [34–37]

$$H = H_q + H_{cl} + H_{cl}^{q}, \tag{1}$$

where H_q and H_{cl} stand for quantal and classical Hamiltonians, respectively, and H_{cl}^{q} is an interaction potential. The dynamical equations for the associated quantal variables are the canonical ones, i.e., any operator O evolves as

$$\frac{dO}{dt} = \frac{i}{\hbar}[H, O], \tag{2}$$

and the concomitant mean value as (Ehrenfest's theorem))

$$\frac{d\langle O\rangle}{dt} = \frac{i}{\hbar}\langle [H, O]\rangle. \tag{3}$$

The evolution of the system is dissipative. A dissipative system is a thermodynamically open system which is operating out of, and often far from, thermodynamic equilibrium in an environment with which it exchanges energy and/or matter. A dissipative structure, in turn, is a dissipative system that has a dynamical regime (here provided by Eqs. (3)) that is in some sense in a reproducible steady state. This reproducible steady state may be reached by natural evolution of the system, by artifice, or by a combination of these two.

We will see below that our classical variables obey dissipative equations because of the presence of an $\eta-$term. Without it, no dissipation occurs, because the resulting equations would conserve energy. Accordingly, if we take the classical variables to be a position X and a momentum P_X, we set [34, 35]

$$\frac{dX}{dt} = \frac{\partial \langle H \rangle}{\partial P_X}, \tag{4a}$$

$$\frac{dP_X}{dt} = -(\frac{\partial \langle H \rangle}{\partial X} + \eta P_X). \tag{4b}$$

The energy is taken here to coincide with the quantum expectation value of the Hamiltonian (1). Consequently, the classical equations of motion to be used here are well-defined ones [35].

As anticipated, the parameter $\eta > 0$ is a dissipative one. Through this parameter η, the classical variable is coupled to an appropriate impulse P_X-reservoir (that provides for P_X-growth) and energy is dissipated into this reservoir. Such indirect route allows for a dynamical description in which no quantum rules are violated [34–37]. The commutation-relations are trivially conserved for all time (the quantal evolution is the canonical one), so that one is able to avoid any quantum pitfall [34, 35]. The set of equations derived from (i) Eqs. (3) for variables belonging to the quantal system, and from (ii) (4) for the classical variables, give raise to an autonomous set of first-order coupled differential equations of the form [34–37]

$$\frac{d\vec{u}}{dt} = \vec{F}(\vec{u}), \tag{5}$$

where $\vec{u}$ a "vector" with both classical and quantum components. The reader may go back to Section V and realize that $\vec{F}$ is the function f of that section. If ones considers an arbitrary volume element V_S enclosed by a surface S in the space where evolves this vector, the dissipative η term induces a contraction of V_S [34, 35]

$$\frac{dV_S(t)}{dt} = -\eta V_S(t). \tag{6}$$

If the classical Hamiltonian adopts the general appearance $H_{cl} = \frac{1}{2M} P_X^2 + V(X)$, one easily ascertains that the temporal evolution for the total energy $\langle H \rangle$ is given by [34–37]

$$\frac{d\langle H \rangle}{dt} = -\frac{\eta}{M} P_X^2, \tag{7}$$

whose significance is to be appreciated in the light of Eq. (6).

7.1. Reformulation in game-theory language

According to the theory of dynamical systems, Eqs. (6) and (7) **guarantee the existence of attractors** [34, 35]. In translating the physical problem at hand into a game, our essential ingredients are these attractors.

We invite the reader to imagine a game whose results are in correspondence with the end-points of the possible trajectories that eventually reach one of these attractors:

possible results	$\rightarrow$	different attractors
possible player's payoffs	$\rightarrow$	appropiate quantum operators P

Bets are placed on which attractor will "prevail" [38]. One assumes that players are aware of the details of the underlying dynamical process, i.e., they are cognizant of (1). Thus, the only freedom of choice refers to the initial conditions(IC) for the system of differential equations (5). Remember such mathematical field an initial value problem (also called the Cauchy problem) reduces to an ordinary differential equation together plus the IC for an unknown function (here $\vec{F}$) at a given point in the domain of the solution (here the initial time). In physics or other sciences, modelling a system frequently amounts to solving an initial value problem. In such context, the differential equation is an evolution equation specifying how, given initial conditions, the system will evolve with time. A pivotal role is then played by the initial density matrix, which is the mathematical quantum object ρ containing the available physical information, that weights possible quantum states $|Q_j >$ of the system with the amounts p_j:

$$\rho = \sum_j |Q_j > p_j < Q_j|, \tag{8}$$

where $\sum_j p_j = 1$. We face here a complete-information game. Each player knows all possible strategies and payoffs. We will call "classical" those players who choose probabilities p_j as (mixed) strategies. Quantum players, instead, select quantum states $|Q_j >$ as their strategies [38]. These choices can be made following specified rules, each distinct set of rules leading to a different game, all of them for the same Hamiltonian.

A bit (a contraction of binary digit) is the basic unit of information in computing and telecommunications, being the amount of information stored by a digital device or other physical system that exists in one of two possible distinct states. These may be the two stable states of a flip-flop, two positions of an electrical switch, two distinct voltage or current levels allowed by a circuit, two distinct levels of light intensity, two directions of magnetization or polarization, the orientation of reversible double stranded DNA, etc. Instead, a qubit or quantum bit is a unit of quantum information: the quantum analogue of the classical bit, with additional dimensions associated to the quantum properties of a physical atom. A quantum computation is performed by initializing a system of qubits with a quantum algorithm. The qubit is described by a quantum state in a two-state quantum-mechanical system, which is formally equivalent to a two-dimensional vector space over the complex numbers (see next Section below). One example of a two-state quantum system is the polarization of a single photon: here the two states are vertical polarization and horizontal polarization. In a classical system, a bit would have to be in one state or the other, but quantum mechanics allows the qubit to be in a superposition of both states at the same time, a property which is fundamental to quantum computing.

Quantum players make use of qubits and *classical ones* of bits ([11], [12], [14]). In particular, Meyer [12] considers density matrices of the type (8). Moves that reflect quantum strategies are represented by unitary operators and classical strategies are of a mixed character. Expected payoffs are calculated via (8) using [38]

$$Tr(\rho\ P), \tag{9}$$

where P standing for a convenient ("payoff") operator associated to each player. The above calculation is performed at the attractors' locations for the pertinent initial values, as detailed below in Sect. 9.

8. Two level systems

In quantum mechanics, a two-state system (also known as a TLS or two-level system) is a system which has two possible states. More formally, the Hilbert space of a two-state system has two degrees of freedom, so a complete basis spanning the space must consist of two independent states. An example of a two-state system is the spin of a spin-1/2 particle such as an electron or proton, whose spin can have values $1/2$, $-1/2$ in units of the Planck constant. The physics of a quantum mechanical two-state system is trivial if both states are degenerate, that is, if the states have the same energy. However, if there is an energy difference between the two states, nontrivial dynamics can ensue that often allow for deep insight into physical problems. Here, we consider the following two-level boson Hamiltonian [34, 35]

$$H = E_1 N_1 + E_2 N_2 + \frac{\omega}{2}(P_X^2 + X^2) + \gamma X \left(a_1^\dagger a_2 + a_2^\dagger a_1\right), \tag{10}$$

that represents matter interacting with a single-mode of a electromagnetic field within a cavity. One has $N_1 = a_1^\dagger a_1$ and $N_2 = a_2^\dagger a_2$, the population operators corresponding to the levels one and two, respectively, and we assume $E_2 > E_1$. Here $a_1^\dagger$, a_1 and $a_2^\dagger$, a_2, are the creation and annihilation operators of a boson in the levels "one" and "two", respectively. The electromagnetic field, regarded as classical, is represented by the variables (classical) X and P_X (X's conjugate momentum) [24, 25].

Taking the set $\{\Delta N = N_2 - N_1, O_- = i(a_1^\dagger a_2 - a_2^\dagger a_1), O_+ = (a_1^\dagger a_2 + a_2^\dagger a_1)\}$, where we have introduced the population difference operator ΔN, and applying (3) we obtain ($\hbar = 1$)

$$\frac{d\langle \Delta N\rangle}{dt} = 2\gamma X \langle O_-\rangle, \tag{11a}$$

$$\frac{d\langle O_-\rangle}{dt} = -2\gamma X \Delta N + \omega_0 \langle O_+\rangle, \tag{11b}$$

$$\frac{d\langle O_+\rangle}{dt} = -\omega_0 \langle O_-\rangle, \tag{11c}$$

with $\omega_0 = (E_2 - E_1)$. The mean value $\langle O_-\rangle$ represents a "current" vector and $\langle O_+\rangle$ is the expectation value of the quantal factor of the interaction potential. For the classical variables we obtain

$$\frac{dX}{dt} = \omega P_X, \tag{12a}$$

$$\frac{dP_X}{dt} = -(\omega X + \gamma \langle O_+\rangle + \eta P_X). \tag{12b}$$

Each level's population, $\langle N_1\rangle$ and $\langle N_2\rangle$, can be obtained in the fashion:

$$\langle N_2\rangle(t) = \frac{1}{2}(n + \langle \Delta N\rangle(t)), \tag{13a}$$

$$\langle N_1\rangle(t) = \frac{1}{2}(n - \langle \Delta N\rangle(t)), \tag{13b}$$

where $\langle N\rangle(t) = n$, with n the total number of particles, as $N = N_1 + N_2$ is an motion-invariant of the system. We can also define the Bloch-like quantity I_B as

$$I_B = \left(\Delta N^2 + \langle O_-\rangle^2 + \langle O_+\rangle^2\right)^{1/2}, \tag{14}$$

which is also an invariant of the motion. We consider the five-dimensional space determined by $u = (\langle \Delta N \rangle, \langle O_- \rangle, \langle O_+ \rangle, X, P_X)$. The fixed points or equilibrium points (labelled by the subindex f) of our system of non linear equations can be classified as being of type A or B, respectively, according to whether its X value vanishes or not. Using the invariant I_B we obtain [34, 35]

Type A:

$$\langle \Delta N \rangle_f = -\frac{\omega\,\omega_0}{2\gamma^2}, \tag{15a}$$

$$\langle O_- \rangle_f = 0, \tag{15b}$$

$$\langle O_+ \rangle_f = \pm({I_B}^2 - 4\frac{\omega^2\,\omega_0^2}{\gamma^4})^{1/2}, \tag{15c}$$

$$X_f = -\frac{\gamma}{\omega}\langle O_+ \rangle_f, \tag{15d}$$

$$P_{Xf} = 0, \tag{15e}$$

if $(\omega\,\omega_0)/2\,\gamma^2 < I_B$.

Type (B)

$$\langle \Delta N \rangle_f = \pm I_B, \tag{16a}$$

$$\langle O_- \rangle_f = 0, \tag{16b}$$

$$\langle O_+ \rangle_f = 0, \tag{16c}$$

$$X_f = 0, \tag{16d}$$

$$P_{Xf} = 0. \tag{16e}$$

Studying the stability of these fixed points we can ascertain that those of Type A are stable [34, 35], while those of Type B are stable only when $(\omega\,\omega_0)/2\,\gamma^2 \geq I_B$ together with $\langle \Delta N \rangle_f = -I_B$. The stable fixed points are the only attractors of the system (see the detailed investigation of [34]). For this case, the final population distribution is originated by a flux from the upper to the lower level, independently of the initial conditions, and of the values of the H-parameters. Instead, for the unstable solution, the flux runs towards the upper level, but for this to happen we need that at the initial time the system has to be already found at the fixed point, where of course it remains for ever.

Type B points minimize the quantum energy as well as the total energy. Instead, for Type A only the total energy is minimized, allowing for the quantum energy part to be either increased or not, depending on the initial conditions and on the parameter-values. This fact allows for the final boson-number of the upper level to be greater than the initial ones, i.e.,

$$\langle N_2 \rangle_f - \langle N_2 \rangle(0) = -\frac{1}{2}\left(\Delta N(0) + \frac{\omega\,\omega_0}{2\,\gamma^2}\right) \geq 0, \tag{17}$$

which can happen for

$$\frac{\omega\,\omega_0}{2\,\gamma^2} < -\Delta N(0), \tag{18}$$

with $\Delta N(0) < 0$.

9. Expected payoffs for two level games

On the basis of (17) we define a game with two options and two players: *the populations of each level either increase or decrease*, with the following expected payoffs [38]:

$$P_2 = \langle N_2 \rangle_f - \langle N_2 \rangle(0), \tag{19a}$$
$$P_1 = \langle N_1 \rangle_f - \langle N_1 \rangle(0), \tag{19b}$$

that can be recast in the form (9) as $P_i = Tr[\rho\ (N_i(t \to \infty) - N_i(0))]$ [35]. Using (13) we have

$$P_2 = -P_1 = \langle \Delta N \rangle_f - \langle \Delta N \rangle(0), \tag{20}$$

so that we face a **zero-sum game**, whose physical counterpart is boson-number conservation. Henceforth we need to fix attention only on P_2. According to the stable point character (A or B) we get

Type A

$$P_2 = -\frac{1}{2}\left(\Delta N(0) + \frac{\omega\,\omega_0}{2\,\gamma^2}\right), \tag{21}$$

if $(\omega\,\omega_0)/2\,\gamma^2 < I_B$. Of course, $P_2 \geq 0$, if (18) is verified.

Type B

$$P_2 = -\frac{1}{2}\left(\Delta N(0) + I_B\right), \tag{22}$$

if $(\omega\,\omega_0)/2\,\gamma^2 \geq I_B$. In the last case we always have $P_2 \leq 0$. We remark that, of course, if initially the system is at any fixed point, including those unstable of the type B, it will remain there and $P_2 = P_1 = 0$. Also, the validity-ranges and the payoffs do not depend on the values of the classical variables X and P_X. We proceed next to determine under which initial conditions the system ends-up in one or the other of the two attractors.

10. Game's strategies and initial conditions

The density matrix (8) may represent a game played i) by classical players if we keep fixed the $|Q_j> -$states), ii) between quantum players if $p_1 = 1$ and the remaining p_j vanish, and iii) between both classical and quantum players. If there are several players, the probabilities p_j are expressed as products (of probabilities) y the states $|Q_j>$ as tensor products of quantum states.

We now specialize (8) to the case of the two levels-system (Cf. Eq. 10)). We consider the illustrative instance in which a classical C-player and a quantum Q-one play with two different strategies: a mixed one for player C and a quantum strategy for player Q. This is as follows: *mixed*: select probability p_j, *quantum*: choice of $|Q_j>$ [38]. Matrix (8) expresses a situation in which *C-players* support with probability p_j (o reject with probability $1 - p_j$) the $|Q_j>$-strategy. This scenario motivates one to follow Meyer's approach [12] regarding quantum coins. In his game a coin is hidden within a box, initially heads-up. It can be in alternate fashion manipulated three times by two players C (just once) and Q (twice). The *Q-player* wins if the penny is head up when finally the open the box is open for all to see. C supports Q's strategy with probability p ("leave the coin untouched") and $1 - p$ ("set the coin in the "tails" state") (see [12]). We slightly generalize things in our game. Let the general initial state be

$$|Q>=\sum_{i=0}^{n} \alpha_i \, |n-i, i>, \tag{23}$$

with $\sum_{i=0}^{n} \alpha_i \, \alpha_i^* = 1$. Vectors $|n-i, i>$, represent states with $n-i$ bosons downstairs and i particles in the upper level. They are a basis in the pertinent Fock-space. If Q chooses the strategy $|Q_1>\equiv |Q>$ of (23), the alternative strategies are given by the set of vectors

$$|Q_j>= \pi_j \, |Q>, \tag{24}$$

with π_j operators that acting on $|Q>$ produce all possible permutations among the α_i, generating $(n+1)!$ quantal strategies. These operators can be written as $\pi_j = \prod_{lm} e_{lm}$, with the e_{lm} being "elemental" operators that exchange α_l with α_m. We are lead to

$$\rho = \sum_{j=1}^{(n+1)!} |Q_j> p_j < Q_j|, \tag{25}$$

with $\sum_{j=1}^{(n+1)!} p_j = 1$. Here $\pi_1 = I$, the identity permutation ($|Q_1>=|Q>$).

Eqs. (23) and (24) state that quantum strategies are represented by qubits for $n=1$ (as in [12]), qutrits for $n=2$, and, in general, by qu$n$-its if we deal with n bosons.

Neither Q nor C know what her rival plays. Although the game may be either of sequential or simultaneous nature, it is here more natural to regard it as sequential, with the *Q-player* making the first move.

The matrix version of (25), in terms of the matrix versions of the operators π_j read

$$\bar{\rho} = \sum_{j=1}^{(n+1)!} p_j \, \bar{\pi}_j \, \bar{\rho}_Q \, \bar{\pi}_j^{\dagger}, \tag{26}$$

where $\bar{\rho}_Q$ corresponds to the pure ρ_Q given by

$$\rho_Q = |Q>< Q|. \tag{27}$$

Matrices $\bar{\pi}_j$ can in turn be cast in the fashion $\bar{\pi}_j = \prod_{lm} \bar{e}_{lm}$, i.e., in terms of the elemental matrices $\bar{e}_{lm}$ that arise out of interchanging rows l and m in the identity matrix. In (26) classical strategies are represented by the choice of the p_j, together with the operations implied by the matrices $\bar{\pi}_j$.

11. A bosonic game

Let us discuss in more detail the $n=1-$instance of one Q-player and one C-one, since this is already enlightening enough, as will be seen. The density operator is written

$$\rho = p_1 \, |Q_1>< Q_1| + p_2 \, |Q_2>< Q_2|, \tag{28}$$

with $p_1 + p_2 = 1$ and

$$|Q_1> \; = \; |Q>= \alpha_0 \, |1, 0> + \alpha_1 \, |0, 1>, \tag{29a}$$

$$|Q_2> \; = \; \alpha_1 \, |1, 0> + \alpha_0 \, |0, 1>, \tag{29b}$$

where $|\alpha_0|^2 + |\alpha_1|^2 = 1$. $|1,0>$ stands for our particle being downstairs while it is upwards in $|0,1>$. Matrices $\bar{\pi}_j$ ($j = 1, 2$) read

$$\bar{\pi}_1 = I = \begin{pmatrix} 1 & 0 \\ 0 & 1 \end{pmatrix}, \qquad \bar{\pi}_2 = \begin{pmatrix} 0 & 1 \\ 1 & 0 \end{pmatrix}. \tag{30}$$

Thus, (26) adopts here the appearance

$$\bar{\rho} = \begin{pmatrix} p_1|\alpha_0|^2 + p_2|\alpha_1|^2 & p_1\alpha_0\alpha_1^* + p_2\alpha_1\alpha_0^* \\ p_1\alpha_1\alpha_0^* + p_2\alpha_0\alpha_1^* & p_1|\alpha_1|^2 + p_2|\alpha_0|^2 \end{pmatrix}. \tag{31}$$

We assume now that the *Q-player* places his bet on that the upper level will increase its population for $t \to \infty$. This is a priori the un-likeliest choice. The ensuing payoff will be P_2. Using (31) we find, associated to the type of fixed point (A or B) the payoffs (see (21) - (22)).

Type A

$$P_2 = \frac{1}{2}\left((2\alpha_0^2 - 1)\,(2p_1 - 1) - \frac{\omega\,\omega_0}{2\,\gamma^2}\right), \tag{32}$$

if $(\omega\,\omega_0)/2\,\gamma^2 < I_B$.

Type B

$$P_2 = \frac{1}{2}\left((2\alpha_0^2 - 1)\,(2p_1 - 1) - I_B\right), \tag{33}$$

if $(\omega\,\omega_0)/2\,\gamma^2 \geq I_B$. In this case I_B writes

$$I_B = \left((2\alpha_0^2 - 1)^2\,(2p_1 - 1)^2 + 4\alpha_0^2(1 - \alpha_0^2)\right)^{1/2}. \tag{34}$$

We have above taken α_0 y α_1 to be real, without loss of generality.

In order to gain intuitive understanding we would need to consider the game's version that uses only pure strategies [3]. The first player bets on one of the two levels and places a particle there. A third party (referee) asks the second player (who ignores what choice has been made before) whether she wishes to change or support her partner's bet. Afterwards, the system evolves and ends up in one of the two levels. If the first player bet on level 2, his payoffs would be 1, -1, or 0 according to whether the boson has climbed, descended, or remained in the original place. Since strategies (mixed ones) can be followed using betting-probabilities , these lead to an appropriate expected payoff [3]. Now, if one player follows a Q-strategy, in the $n = 1$ instance this strategy is represented by a qubit and we are led to the expected payoff computed according to (9) which leads to either (32) or (33).

11.1. Game's results

Some results concerning the just discussed issues are illustrated in Fig. 1 (the reader is also directed to Fig. 1 of Ref. [38]). Therein we display the regions corresponding to each type of fixed point (Type A and Type B), which are separated by the curve $I_B = \omega\,\omega_0/(2\gamma^2)$ (solid curve), with I_B given by (34). Zones in which the Q-player either wins or loses are also delimited. The dotted curve is in this case the "separator", being given by

$$(2\alpha_0^2 - 1)\,(2p_1 - 1) - \frac{\omega\,\omega_0}{2\,\gamma^2} = 0, \tag{35}$$

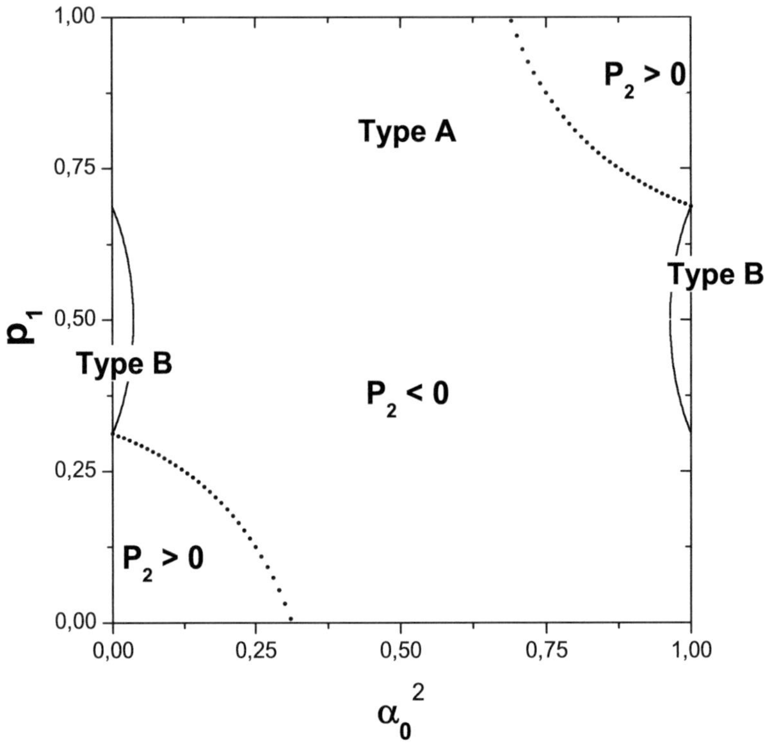

Figure 1. Classical and quantum probabilities p_1 and α_0^2. Regions corresponding to the fixed points of types A and B, separated by the curve $I_B = \omega\,\omega_0/(2\gamma^2)$ (solid line) are delineated (I_B is given by (34)). One sets $\omega\,\omega_0/\gamma^2 = 3/4$. We also depict the zones in which the C-player either wins or loses, represented by positive or negative payoffs, respectively, as given by the P_2 of Eq. (32) or (33). The dotted line separates the two zones and can be calculated using Eq. (35).

that one gets by setting $P_2 = 0$ in (32). We chose as independent parameters α_0^2 and p_1, setting $\omega\,\omega_0/\gamma^2 = 3/4$. Since we wish for the existence of the two types of fixed points for $n = 1$, one needs that $\omega\,\omega_0/\gamma^2 < 2$.

We indeed verify that the strategy corresponding to selecting p_1 in such manner that

$$\frac{1}{2} - \frac{\omega\,\omega_0}{4\,\gamma^2} \leq p_1 \leq \frac{1}{2} + \frac{\omega\,\omega_0}{4\,\gamma^2}\,, \tag{36}$$

guarantees that the C-player will win, independently of the quantal strategy. Moreover, if $p_1 = 1/2$, this holds for any choice of parameters' values. Such "happy" circumstances do not exist for the C-player, no matter its strategic choice.

We detect the existence of a Nash equilibrium-point of type A ("large" coupling) at $p_1 = 1/2$ and $\alpha_0 = 1/\sqrt{2}$. This result mimics the one found for a classical Meyer game of "Penny Flipover" [12], after two rounds. The C-player bets with equal probability on both the Q-strategy and its opposite one. At the same time, the *Q-player*'s strategy (unknown to the C-one) bets evenly on the two alternative options of placing the particle up- or downstairs. Here one has $P_2 = -\omega\,\omega_0/(2\,\gamma^2) < 0$, since $p_1 = 1/2$.

12. Conclusions

We have cast the physics of a semi-classical Hamiltonian (10) in terms of Game Theory. As stated above, this physics is associated to the interaction between matter and a single-mode of a electromagnetic field within a cavity. The concomitant Hamiltonian is a specialization of the Hamiltonian (1).

The interaction-agency is represented as a game between classical and quantum players who bet on initial conditions. The associated dynamics, via the physical system's attractors, is expressed in terms of the game's payoffs (positive o negative). This is clearly appreciated in Fig. 1 for a single-boson example.

In the present context the boson-number conservation is cast in the guise of a zero-sum game. This result can be generalized for any problem with invariants.

Here our game is not a mere abstraction. Given that it represents a physical interaction, we can speak of a "real" game (as far as a physical model can be considered real). In this context, any experimenter can be viewed as a is a classical player. In practical terms, if the experimenter's interest, lies, for instance, in ascertaining that the final ground state population be larger than that of the excited state, she must use a specific strategy to such an end. The best one is choosing to initially prepare the system (in the $n = 1$-case) in such a way that the concomitant level-probabilities are, respectively, p_1 and $1 - p_1$, with p_1 given by (36), depending, of course, on the values of the system's parameters, letting afterwards the system to evolve. Selecting $p_1 = 1/2$ he/she can even neglect possible errors in these values and ensure "victory", thus achieving her goal.

Acknowledgements

This work was partially supported by the project PIP1177 of CONICET (Argentina), and the projects FIS2008-00781/FIS (MICINN) - FEDER (EU) (Spain, EU)

Author details

A.M. Kowalski
La Plata Physics Institute, National University La Plata and Buenos Aires Scientific Research Commission (CIC)

A. Plastino
National University La Plata & CONICET IFLP-CCT, C.C. 727 - 1900 La Plata, Argentina
Universitat de les Illes Balears and IFISC-CSIC, 07122 Palma de Mallorca, Spain

M. Casas
Universitat de les Illes Balears and IFISC-CSIC, 07122 Palma de Mallorca, Spain

13. References

[1] R. B. Myerson (1991) *Game Theory: Analysis of Conflict* (Harvard University Press, 1991).
[2] R. J. Aumann ([1987] 2008) *Game Theory Introduction, The New Palgrave Dictionary of Economics, 2nd Edition,* Edited by S. N. Durlauf and L. E. Blume, Macmillan, NY, 2008).

[3] J. von Neumann, O. Morgenstern, Theory of Games and Economic Behavior, Wiley, New York, 1944.
[4] M. A. Nowak, K. Sigmund, Nature (London) 398 (1999) 367.
[5] R. Dawkins, The Selfish Gene, Oxford University Press, Oxford, 1976.
[6] B. R. Frieden, Physics from Fisher information, Cambridge University Press, Cambridge, UK, 1998.
[7] M. Dimassi, J. Sjoestrand, Spectral Asymptotics in the Semi-Classical Limit, Cambridge University Press, Cambridge, UK, 1999.
[8] M. Brack, R. K. Bhaduri, Semiclassical Physics, Addison-Wesley, Reading, MA, 1997.
[9] A. K. Ekert, Phys. Rev. Lett. 67 (1991) 661.
[10] R. F. Werner, Phys. Rev. A 58 (1998) 1827.
[11] J. Eisert, M. Wilkens, M. Lewenstein, Phys. Rev. Lett. 83 (1999) 3077.
[12] D.A. Meyer, Phys. Rev. Lett. 82 (1999) 1052.
[13] J. Du, H. Li, X. Xu, M. Shi, J. Wu, X. Zhou, R. Han, Phys. Rev. Lett. 88 (2002) 137902.
[14] S.C. Benjamin, P.M. Hayden, Phys. Rev. A 64 (2001) 030301(R).
[15] R. Kay, N.F. Johnson, S.C. Benjamin, J. Phys. A 34 (2001) L547.
[16] Taksu Cheon, Izumi Tsutsui, Physics Letters A 348 (2006) 147.
[17] Badredine Arfi, Physica A 374 (2007) 794.
[18] Jean Faber, Renato Portugal, Luiz Pinguelli Rosa, Physics Letters A 357 (2006) 433.
[19] Edward Jimenez, Douglas Moya, Physica A 348 (2005) 505.
[20] F. Bloch, Phys. Rev. 70 (1946) 460.
[21] P. Meystre, M. Sargent III, Elements of Quantum Optics, Springer-Verlag, New York/Berlin, 1991.
[22] G.F. Bertsch, Phys. Rev. Lett. 95B (1980) 157.
[23] A. Bulgac, Phys. Rev. C 40 (1989) 1073.
[24] P.W. Milonni, J. R. Ackerhalt, H.W. Galbraith, Rev. Lett. 50 (1980) 966.
[25] P.W. Milonni, M.L. Shih, J. R. Ackerhalt, Chaos in Laser-Matter Interactions, World Scientific Publishing Co., Singapore, 1987.
[26] G. Kociuba N. R. Heckenberg, Phys. Rev. E 66 (2002) 026205.
[27] P. Ring, P. Schuck, The Nuclear Many-Body Problem, Springer-Verlag, New York/Berlin, 1980.
[28] L. L. Bonilla, F. Guinea, Phys. Lett. B 271 (1991) 196; Phys. Rev. A 45 (1992) 7718.
[29] R. Blummel, B. Esser, Phys. Rev. Lett. 72 (1994) 3658.
[30] F. Cooper, S. Habib, Y. Kluger, E. Mottola, Phys. Rev. D 55 (1997) 6471.
[31] F. Cooper, J. Dawson, S. Habib, R. D. Ryne, Phys. Rev. E 57 (1998) 1489.
[32] A. M. Kowalski, M. T. Martin, J. Nuñez, A. Plastino, A. N. Proto, Phys. Rev. A 58 (1998) 2596.
[33] D. J. H. Chung, Phys. Rev. D 67 (2003) 083514.
[34] A. M. Kowalski, A. Plastino, A. N. Proto, Phys. Rev. E 52 (1995) 165.
[35] A.M. Kowalski, A. Plastino, A.N. Proto, Physica A 236 (1997) 429.
[36] A.M. Kowalski, M.T. Martin, A. Plastino, O. A. Rosso, International Journal of Modern Physics B 19 (2005) 2273.
[37] A.M. Kowalski, M.T. Martin, A. Plastino, O. A. Rosso, Physica D 233 (2007) 21.
[38] A.M. Kowalski, A. Plastino, Physica A 387 (2008) 5065.

Chapter 15

Nash Equilibrium Strategies in Fuzzy Games

Alireza Chakeri, Nasser Sadati and Guy A. Dumont

Additional information is available at the end of the chapter

http://dx.doi.org/10.5772/54677

1. Introduction

Modern description of game theory is generally considered to have been started with the book "*Theory of Games and Economic Behavior*" [1]. Modern game theory has grown extremely well, in particular after the influential results of Nash [2-4]. It has been widely applied to many problems in economics, engineering, politics, etc.

A game is a model of a situation where two or more groups are in dispute over some issues [5]. The participants in a game are called the players. The possible actions available to players are referred to as strategies. When each player selects a strategy, it will determine an outcome to the game and the payoffs to all players, while tries to maximize his own payoff.

Classical game theory uses the extensive form and the strategic form to explain a game. The extensive form is represented by a game tree, in which the players make sequential actions. However, the strategic form is usually used to describe games with two decision makers, in which players' choices are made simultaneously. Unlike one-player decision making, where optimality has a clear sense, in multi person decision making the optimality is in the form of NE. An NE strategy is a strategy wherein, if a player knows his opponent's strategy, he is totally satisfied and is unwilling to change his strategy.

In classical game theory it is assumed that all data of a game are known exactly by players. However, in real games, the players are often not able to evaluate exactly the game due to lack of information and precision in the available information of the situation. Harsanyi [6] treated imprecision in games with a probabilistic method and developed the theory of Bayesian games. This theory could not entirely solve the problem of imprecision in games, because it was limited to only one possible kind of imprecision. However, in reality, imprecision is of different types and can be modeled by fuzzy sets. The notion of fuzzy sets first appeared in the paper written by Zadeh [7]. This notion tries to show that to what degree an element belongs to a set. The degree, to which an element belongs to a set, is an element of the continuous interval [0, 1] rather than the Boolean values. Using the notion of fuzzy sets, each component in a game (set of players, set of strategies, set of payoffs, etc) can be fuzzified. Initially, fuzzy sets were used by Butnariu [8] in non-cooperative game theory. He used fuzzy sets to represent the belief of each player for strategies of other players. Since then, fuzzy set theory has been used in many non-cooperative [9-15] and cooperative games [16-17].

In this chapter, we will extend the NE set to fuzzy set in games with fuzzy numbers as payoffs. In this regard, using ranking fuzzy numbers, a fuzzy preference relation is constructed over payoffs and then the resultant priorities of payoffs are considered as the grades of being NE. Hence, if a player knows the opponent's strategy, he is satisfied with his own strategy by the degree that this strategy has priority for him. The more priority the players feel for each strategy, the more possible the strategy becomes the game's equilibrium. This generalization shows the distribution of Nash grades in the matrix form of the game. In other words, we can consider strategies with high grades of equilibrium which are not necessarily the equilibrium points. In the proposed approach, the effect of different viewpoints (optimism and pessimism) on the result of the game is also studied.

The remainder of this chapter is organized as follows. Section 2 reviews the backgrounds on fuzzy set theory. This section discusses the basic definition of fuzzy numbers, fuzzy extension principle and ranking fuzzy numbers. Section 3 briefly presents the fuzzy preference relation and a way to obtain priorities from fuzzy preference matrix. In this section, preference ordering of the alternatives and its transformation into fuzzy preference relation is also discussed. Section 4 introduces the proposed algorithm to find the Nash grades for pure and mixed strategies in the matrix form of the game. Two examples and their detailed results are presented in Section 5. Finally, Section 6 contains some concluding remarks.

2. Backgrounds on fuzzy set theory

There are two well-known frameworks for quantifying the lack of knowledge and precision, namely, probabilistic and possibilistic uncertainty type. The probabilistic framework deals with the uncertain events with a probability distribution function (PDF). However, there are some situations in which there is not much information about the PDF of uncertain parameters or they are inherently not repeatable. In possibilistic framework, for each uncertain event $\tilde{M}$, a membership function $\mu_{\tilde{M}}(x)$ is defined which describes how much each element x, of universe of discourse U (the set of all values that x can take) belongs to $\tilde{M}$. Different types of membership functions can be used to describe uncertain values. Fuzzy numbers have also been used in many decision making problems. The following definition of fuzzy numbers is most commonly used [18];

Definition 1: The fuzzy number $\tilde{M}$ is a convex normalized fuzzy set of the real line R such that

1. There is exactly one x_0 with $\mu_{\tilde{M}}(x_0) = 1$.

2. $\mu_{\tilde{M}}(x)$ is piecewise continuous.

$M_\alpha = \{x \in R | M(x) \geq \alpha\}, \forall \alpha \in [0,1]$ is also called α -cut of $\tilde{M}$. From definition 1, α -cuts of a fuzzy number $\tilde{M}$ are closed real intervals, that is

$$M_\alpha = [a_\alpha, b_\alpha] \quad \forall \alpha \in [0,1] \tag{1}$$

2.1. Fuzzy extension principle

Fuzzy extension principle, which was introduced by Zadeh in [19], is an essential principle in fuzzy set theory to generalize the concepts and structures of classical mathematics to fuzzy mathematics as follows [20]

Definition 2: Let $U_1,\ldots,U_n$ are UoDs and $U = U_1 \times \ldots \times U_n$ be their Cartesian product. Also assume that $\tilde{M}_1,\ldots,\tilde{M}_n$ are fuzzy subsets of $U_1,\ldots,U_n$, respectively. Moreover, let $y = f(x_1,\ldots,x_n)$ be a mapping from U to Y. Now if $\tilde{B} = f(\tilde{M}_1,\ldots,\tilde{M}_n)$, then the membership function of $\tilde{B}$ is defined as follows

$$\mu_{\tilde{B}}(y) = \mu_{f(\tilde{M}_1,\ldots,\tilde{M}_n)}(y)$$
$$= \begin{cases} \sup\limits_{\substack{x_1,\ldots,x_n \\ y=f(x_1,\ldots,x_n)}} \min\{\mu_{\tilde{M}_1}(x_1),\ldots,\mu_{\tilde{M}_n}(x_n)\} & f^{-1}(y) \neq \phi \\ 0 & f^{-1}(y) = \phi \end{cases} \tag{2}$$

Based on the above definition, one-dimensional operators and two-dimensional operators on fuzzy numbers can be represented by the following definitions, respectively.

Definition 3: Assume that $\tilde{M}$ is a fuzzy number and $f: R \to R$ is a one-dimensional operator. According to fuzzy extension principle, $f(\tilde{M})$ is a fuzzy set with the following membership function

$$\mu_{f(\tilde{M})}(y) = \begin{cases} \sup\limits_{x, y=f(x)} \mu_{\tilde{M}}(x) & f^{-1}(y) \neq \phi \\ 0 & f^{-1}(y) = \phi \end{cases} \tag{3}$$

In definition 3, if $f(x) = \lambda x$, then the scalar product of the real value λ into fuzzy number $\tilde{M}$ is a fuzzy number with the following membership function

$$\mu_{\lambda \tilde{M}}(x) = \mu_{\tilde{M}}\left(\frac{x}{\lambda}\right) \tag{4}$$

Definition 4: Assume that $\tilde{M}$ and $\tilde{N}$ are fuzzy numbers and $f: R \times R \to R$ is a two-dimensional operator. According to fuzzy extension principle, $\tilde{M} \otimes \tilde{N}$ is a fuzzy set with the following membership function

$$\mu_{\tilde{M}\otimes\tilde{N}}(z) = \sup_{z=x\otimes y} \min\left(\mu_{\tilde{M}}(x), \mu_{\tilde{N}}(y)\right) \tag{5}$$

As a special case, the result of the summation operator is also a fuzzy number as follows

$$\mu_{\tilde{M}+\tilde{N}}(z) = \sup_{z=x+y} \min\left(\mu_{\tilde{M}}(x), \mu_{\tilde{N}}(y)\right) \tag{6}$$

2.2. Ranking fuzzy numbers

Ranking fuzzy numbers seems a necessary procedure in decision making when alternatives are fuzzy numbers. Various methods for ranking fuzzy subsets have been proposed. Yager in [21] introduced a function for ranking fuzzy subsets in unit interval which is based on the integral of mean of the α -cuts. Jain in [22], Baldwin and Guild in [23] were also suggested methods for ordering fuzzy subsets in the unit interval. Ibanez and Munoz in [24] have developed a subjective approach for ranking fuzzy numbers. In this chapter, we use the subjective approach, as introduced in [24].

Ibanez and Munoz in [24] defined the following number as the average index for $\tilde{M}$

$$V_P(\tilde{M}) = \int_Y f_{\tilde{M}}(\alpha)dP(\alpha) \qquad \forall \tilde{M} \in U \tag{7}$$

where Y is a subset of the unit interval and P is a probability distribution on Y .

The definition of $f_{\tilde{M}}$ could be subjective for decision maker. In [24] the following definition for $f_{\tilde{M}}$ has been suggested, in which the parameter λ determines the optimism-pessimism degree of the decision maker

$$f_{\tilde{M}}^{\lambda} : Y \rightarrow R, \quad f_{\tilde{M}}^{\lambda}(\alpha) = \lambda b_{\alpha} + (1-\lambda)a_{\alpha} \tag{8}$$

where $\lambda \in [0,1]$ and $M_{\alpha} = \left[a_{\alpha}, b_{\alpha}\right]$.

When an optimistic decision maker ($\lambda = 1$) wants to choose the greatest value, the upper extreme of the interval (b_{α}) would be chosen, i.e. he prefers to choose the greatest possible value. A pessimism person ($\lambda = 0$) on the opposite prefers to decide on the lower extreme of the interval (a_{α}).

Using $V_P(\cdot)$ in (7), the ordering relations between fuzzy numbers $\tilde{A}$ and $\tilde{B}$ can be given as

$$\tilde{A} \leq \tilde{B} \Leftrightarrow V_P(\tilde{A}) \leq V_P(\tilde{B}) \qquad \forall \tilde{A}, \tilde{B} \in U \tag{9}$$

The fuzzy number $\tilde{A}$ is not preferred to $\tilde{B}$, if and only if their average index is the same

$$\tilde{A} \approx \tilde{B} \Leftrightarrow V_P(\tilde{A}) = V_P(\tilde{B}) \qquad \forall \tilde{A}, \tilde{B} \in U \tag{10}$$

For convenience, let $Y=[0,1]$ and P is the Lebesgue measure on [0, 1], as used in [24]. Then, the following $V_P(\tilde{M})$ is derived for a specific λ

$$V_P(\tilde{M})=\int_0^1\left(\lambda b_\alpha+(1-\lambda)a_\alpha\right)d\alpha \tag{11}$$

As a special case, the average index for triangular fuzzy numbers can be given as follows

$$V_L^\lambda=b\lambda+(a-\frac{1}{2}b) \tag{12}$$

where, a triangular fuzzy number with notation $\tilde{M}=T(a,c,b)$ is used as shown in Figure 1.

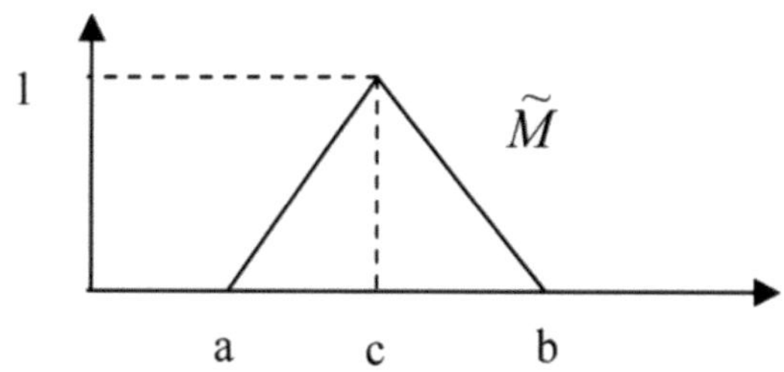

Figure 1. Triangular fuzzy number, $\tilde{M}=T(a,c,b)$.

3. Preference relations

3.1. Fuzzy preference relations

Preference relation is one of the most regular tools for stating decision maker's preferences. In the process of making decision, individuals are asked to give their preferences over an alternative, which is based on their comparison according to one's desire. Various kinds of preference relations have been developed including multiplicative preference relation [25], fuzzy preference relation [26], linguistic preference relation [27] and intuitionistic preference relation [28].

For a decision making situation, let $X=\{x_1,x_2,\ldots,x_n\}$ be a discrete set of alternatives. A preference relation P on the set X is defined by a function $\mu_P=X\times X\rightarrow D$, where D is the domain of representation of preference degrees provided by the decision maker for each pair of alternatives. In many situations, due to lack of information about the problems, the goals, constraints and consequences are not precisely known. Because of these uncertainties, fuzzy set theory allows a more flexible framework to express the preferences. Fuzzy preferences show the fuzziness of the decision maker's preferences. In [26], fuzzy preference relation is defined as follows.

Definition 5: A fuzzy preference relation R on the set X is defined as a matrix $R=\left(r_{ij}\right)_{n\times n}$ with some properties given as

$$\begin{cases} r_{ij} + r_{ji} = 1, \\ r_{ij} \geq 0, \qquad\qquad \forall i,j = 1,2,\ldots,n \\ r_{ii} = 0.5, \end{cases} \tag{13}$$

where r_{ij} denotes the preference degree of the alternative x_i to x_j.

In particular, $r_{ij} = 0.5$ shows indifference between x_i and x_j. The case of $r_{ij} \geq 0.5$ indicates that x_i is preferred to x_j. As r_{ij} increases, the degree of preference gets larger. Also $r_{ij} = 1$ shows that x_i is absolutely preferred to x_j and vice versa.

Deriving priorities, that is the degrees of importance of alternatives, are the main aspect of preference relations. Quite a number of approaches have been developed to derive priorities from fuzzy preference relations. Lipovetsky and Michael-Conklin [29] introduced an optimization approach and an eigen-problem to produce robust priority estimation of a fuzzy preference relation. Xu [30] developed a weighted least square approach and an eigenvector method for priorities of fuzzy preference relations. Xu and Da [31] proposed the Least Deviation method to obtain the priority vector of a fuzzy preference relation and considered its properties. In this chapter, we use the Least Deviation method presented in [31] to derive priorities from fuzzy preference relation. The algorithm to derive priorities in [31] is mentioned briefly as follows.

Let $R = (r_{ij})_{n\times n}$ be a fuzzy preference relation and $W = (w_1, w_2, \ldots, w_n)$ be the priority vector which is going to be calculated such that w_i shows the degree of priority of alternative x_i, with the following properties

$$\sum_{i=1}^{n} w_i = 1;\ w_i \geq 0 \tag{14}$$

Let k be the number of iterations:

Step 1. Initialize the weight $W(0) = (w_1, w_2, \ldots, w_n)$, specify parameter $0 \leq \varepsilon \leq 1$ and let $k = 0$.

Step 2. Calculate the following term, where $h(r_{ij}) = 9^{(2r_{ij}-1)}$

$$\mu_i\left(w(k)\right) = \sum_{j=1}^{n} \left[h(r_{ij}) \left(\frac{w_j(k)}{w_i(k)} \right) - h(r_{ji}) \left(\frac{w_i(k)}{w_j(k)} \right) \right], \forall i \tag{15}$$

If $\left|\mu_i\left(w(k)\right)\right| \leq \varepsilon$ for all i, then update w by $w(k)$ and go to step 5, otherwise continue with step 3.

Step 3. $\left|\mu_m(w(k))\right| = \max_{i=1}^{n}\left\{\left|\mu_i(w(k))\right|\right\}$, calculate $T(k)$ and then $x_i(k)$ using

$$T(k) = \sqrt{\left(\sum_{j\neq m} h(r_{mj})\left(\frac{w_j(k)}{w_m(k)}\right)\right) \Big/ \left(\sum_{j\neq m} h(r_{jm})\left(\frac{w_m(k)}{w_j(k)}\right)\right)} \tag{16}$$

$$x_i(k) = \begin{cases} T(k)w_m(k) \\ w_i(k) \end{cases} \tag{17}$$

Step 4. Set $k = k+1$ and go to step 2.
Step 5. Obtain w as the priority vector.
Step 6. End.

3.2. Preference ordering of the alternatives

Individuals usually provide their preferences over the alternatives by preference ordering. Let's assume that an expert expresses his preferences on X by a preference ordering $O = \{o(x_1), o(x_2), \ldots, o(x_n)\}$. It is assumed that the lower the position of the alternative in the preference ordering, the more contentment for the individual. For instance, an expert states his preference ordering on $X = \{x_1, x_2, x_3\}$ by the following ordering; $o(x_1) = 2, o(x_2) = 1, o(x_3) = 3$. This means that the alternative x_2 is the best and alternative x_3 is the worst.

3.3. Transforming preference ordering into fuzzy preference relation

An alternative satisfies the decision maker according to its position in the ordering preference relation. Various mappings to transform the ordering preference relation into fuzzy preference relation have been proposed [32-34]. In [32] a crisp relation was introduced for assessing the fuzzy preference relation, where the preference between alternatives x_i and x_j depends only on the values of $o(x_i)$ and $o(x_j)$. In [33-34], the following relations was introduced to achieve fuzzy preference relation from an ordering preference

$$v(x_i) = 1 - \frac{o(x_i) - 1}{n - 1} \tag{18}$$

$$r_{ij} = \frac{1}{2}\left(1 + v(x_i) - v(x_j)\right) \tag{19}$$

where r_{ij} denotes the preference between alternatives x_i and x_j.

4. Fuzzy games

In a strategic game, there are n players and ns_i strategies for player i. Suppose X_i is the strategy set for player i defined as follows

$$X_i = \left\{x_{1i}, x_{2i}, \ldots, x_{ns_i i}\right\} \tag{20}$$

In pure strategy, let s_i denotes the strategy chosen by player i and s_{-i} denotes the strategies chosen by the other players. Then $\Pi_i(s_1, \ldots, s_i, \ldots, s_n) = \Pi_i(s_i, s_{-i})$ is the payoff achieved by player i. By definition of classical game theory, $(s_1^*, \ldots, s_n^*)$ is the pure strategy Nash equilibrium, if and only if [35]

$$\Pi_i(s_1^*, \ldots, s_i^*, \ldots, s_n^*) \geq \Pi_i(s_1^*, \ldots, s_i, \ldots, s_n^*) \quad \forall i \in \{1, \ldots, n\}, \forall s_i \in X_i \tag{21}$$

In mixed strategy, each player assigns a probability distribution $\sigma_i = \left(\sigma_{1i}, \ldots, \sigma_{ns_i i}\right) \in \Sigma_i$ over his strategies, where Σ_i determines all possible probability distributions for player i. Then, the expected payoff for player i is the real value $E_i\left(\sigma_1, \ldots, \sigma_n\right)$, where E denotes the expectation operator. In this regard, mixed strategy Nash equilibrium $(\sigma_1^*, \ldots, \sigma_n^*)$ is defined as follows

$$E_i\left(\sigma_1^*, \ldots, \sigma_i^*, \ldots, \sigma_n^*\right) \geq E_i\left(\sigma_1^*, \ldots, \sigma_i, \ldots, \sigma_n^*\right) \quad \forall i \in \{1, \ldots, n\}, \forall \sigma_i \in \Sigma_i \tag{22}$$

In real games, players must often make their decisions under unclear or fuzzy information. In this regard, there are several approaches for explaining games with fuzzy set theory. As discussed earlier, a game has three main components: a set of players, a set of strategies and a set of payoffs. The set of players is defined as fuzzy set when the concept of coalition in cooperative games is fuzzified. Butnariu in [17] proposed core and stable sets in fuzzy coalition games, and introduced a degree of participation for players in a coalition. Mares in [16] considered fuzzy core in fuzzy cooperative games, where possibility for each fuzzy coalition was considered as fuzzy interval, and an extension of the core in classic TU games. He discussed Shapely value in cooperative games with deterministic characteristics and fuzzy coalitions. In [9], the concept of fuzzy strategies has been introduced. It defines a strategy set consisting of fuzzy subspaces of strategy spaces and assigned a fuzzy payoff for each set of player's strategies. Hence, they have defined a fuzzy inference system (fuzzy If-Then rules). However, they have assumed some real values as strategies and have solved the games with common crisp methods. The first two steps seem rational for modeling any system according to its specifications, however the final step is not a reasonable fuzzy decision making approach to find NEs. Generally, since the final decision of a player is a number in real world problems, they can not adopt fuzzy strategies except when meaningful interpretations exist. However, we should remark that fuzzy strategies are constructive to model the games and calculate the payoffs.

Defining payoffs as fuzzy sets is reasonable in the following two main situations, and lead to fuzzy numbers as payoffs.

1. When there is not sufficient information about the payoffs and they are not inherently repeatable.
2. When calculation of payoffs is difficult or time-consuming. Hence, they are usually defined by fuzzy if-then rules.

In this chapter, we consider pure and mixed strategies in games with fuzzy numbers as payoffs. In this regard, in pure strategies, $\Pi_i(s_i, s_{-i})$ is considered as a fuzzy number. Also, in mixed strategies, the expected values are calculated based on the fuzzy extension principle and definitions 3 and 4, because the payoffs are fuzzy numbers.

5. Playing games with fuzzy numbers as payoffs

In classical game theory, a crisp payoff is either greater than or less than others. However, in fuzzy ones, there are uncertainties in comparing fuzzy payoffs. We model these uncertainties using fuzzy preference relation on the preference ordering of the expected values. In this regard, using ranking fuzzy numbers, a fuzzy preference relation reflecting the uncertainties in payoffs is constructed and then, the resultant priorities of payoffs are considered as the grades of being NE. This definition for the grade of being NE seems meaningful because if a player knows the opponent's strategy, he is satisfied with his strategy by the degree that this strategy has priority for him. The more priority the players get for each strategy, the more possible the strategy is the game's equilibrium. The proposed algorithm to find the grade of being NE for every mixed strategy is shown in Table 1.

1. $\forall i$, do steps 2-9.
2. For player i, determine the optimism-pessimism degree λ.
3. $\forall \sigma_{-i}$, do steps 4-9.
4. Fix the opponent's mixed strategy σ_{-i}.
5. $\forall \sigma_i$, calculate the average index of the resultant expected value according to (11).
6. Order the expected values according to their average index in descending order, according to (9).
7. Transform the ordering preference relation into fuzzy preference relation according to (18-19).
8. Derive priorities of expected values from the resultant fuzzy preference relation based on the least deviation method.
9. For every mixed strategy (σ_i, σ_{-i}), set its grade of being NE equal to its corresponding priorities.
10. For every mixed strategy (σ_i, σ_{-i}), set its overall grade of being NE equal to the minimum of its grade of being NE for each player.

Table 1. Algorithm to find the grade of being NE for each mixed strategy

6. Simulation results

6.1. Application in pure strategy

In this subsection, games with two players are considered because of their easier understanding. However, our algorithm can be implemented to more than two players. Consider a fuzzy game, as defined in Table 2. Player one and player two have three strategies, namely a_1, a_2, a_3 and b_1, b_2, b_3, respectively. Each cell includes two fuzzy triangular numbers as payoffs; the first for player one and the second for player two.

(Π_1,Π_2)	b_1	b_2	b_3
a_1	T(4,5,6),T(1,3,5)	T(5,6,7),T(2,3,4)	T(4,5,6),T(1.5,3,4.5)
a_2	T(2,3,4),T(0,1,2)	T(1,3,5),T(3,4,5)	T(2,3,4),T(1,3,5)
a_3	T(3,4,5),T(2,4,6)	T(3,5,7),T(1,3,5)	T(6,7,8),T(4,6,8)

Table 2. A sample game with fuzzy payoffs

Implementing the algorithm presented in Table 1, there can be several grades of being NE for the game corresponding to different λ s. Tables 3 and 4 present the priority of each payoff according to possible choices of λ .

	b_1	b_2	b_3
a_1	(0.7,0.08)	(0.7,0.7)	(0.22,0.22)
a_2	(0.08,0.08)	(0.08,0.7)	(0.08,0.22)
a_3	(0.22,0.22)	(0.22,0.08)	(0.7,0.7)

Table 3. Priorities of payoffs for $\lambda \leq 1/4$

	b_1	b_2	b_3
a_1	(0.7,0.7)	(0.7,0.08)	(0.22,0.22)
a_2	(0.08,0.08)	(0.08,0.7)	(0.08,0.22)
a_3	(0.22,0.22)	(0.22,0.08)	(0.7,0.7)

Table 4. Priorities of payoffs for $\lambda > 1/4$

The minimum priority of all payoffs in a cell is interpreted now as the grade of being Nash equilibrium of that cell. The results of the mentioned game is tabulated in Tables 5 and 6 for the possible choice of λ , respectively

	b_1	b_2	b_3
a_1	0.08	0.7	0.22
a_2	0.08	0.08	0.08
a_3	0.22	0.08	0.7

Table 5. Grades of being NE for $\lambda \leq 1/4$

	b_1	b_2	b_3
a_1	0.7	0.08	0.22
a_2	0.08	0.08	0.08
a_3	0.22	0.08	0.7

Table 6. Grades of being NE for $\lambda > 1/4$

The difference between the results of Tables 5 and 6 is meaningful. As the player 2 becomes optimistic, he prefers to choose the greatest possible outcome. Hence, between three alternatives $\{T(3,2), T(3,1), T(3,1.5)\}$, $T(3,2)$ has the most priority. But a pessimistic player prefers to choose alternative $T(3,1)$.

6.2. Application in mixed strategy

In this subsection, mixed strategies in bi-matrix games are considered. Consider a fuzzy bi-matrix game as defined in Table 7.

(Π_1, Π_2)	b_1	b_2
a_1	T(10,11,12),T(3,4,5)	T(3,4,5),T(0.5,2,3.5)
a_2	T(2,4,6),T(0,1,2)	T(6,7,8),T(2,4,6)

Table 7. A sample game with fuzzy payoffs

Using the proposed approach, every mixed strategy has a grade of being NE as shown in Figures 2 and 3 for each neutral player ($\lambda = 0.5$).

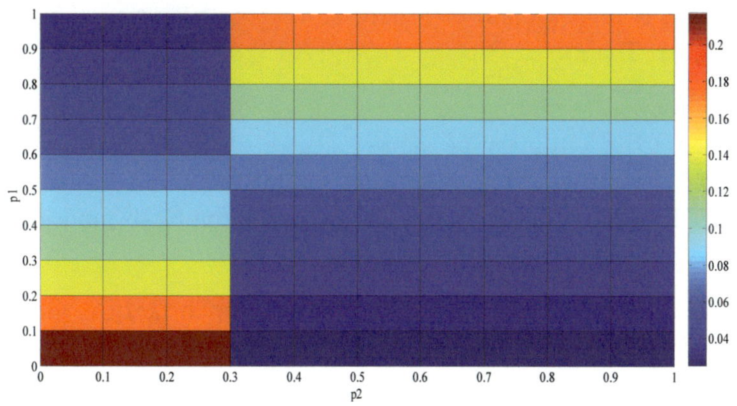

Figure 2. Grades of being NE for Player 1

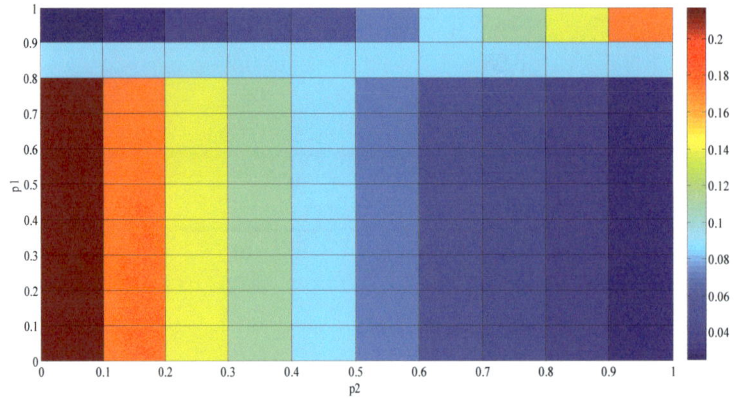

Figure 3. Grades of being NE for Player 2

Now, the minimum priorities for players can be interpreted as the overall grades of being NE. We remark that the minimum operator can be replaced by any other T-norms. The results of the game presented in Table 7, is shown in Figure 4.

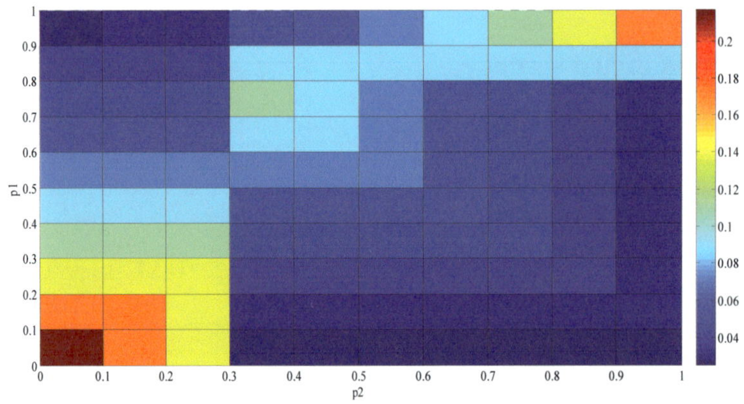

Figure 4. Grades of being NE for neutral players

Grade of being NE is high around two pure strategies, that is $(p_1 = (0,1), p_2 = (0,1))$ and $(p_1 = (1,0), p_2 = (1,0))$. The mixed strategy $(p_1 = (0.7,0.3), p_2 = (0.3,0.7))$ has also high grade of being NE.

In addition, the effect of optimism-pessimism degree of players in distribution of Nash grades is studied. For instance, if two players are optimistic ($\lambda = 1$), the Nash grades are the same as Figure 5. If two players are pessimistic ($\lambda = 0$), the Nash grades are the same as Figure 6.

The difference between the results of Figures 4 and 5 is important. As the player becomes optimistic, he prefers to choose the greatest possible outcome. Hence, payoff $T(4,1.5)$ gets more priority than payoff $T(4,1)$. In addition, Nash grades for neutral players are approximately the combination of Nash grades for optimistic and pessimistic players.

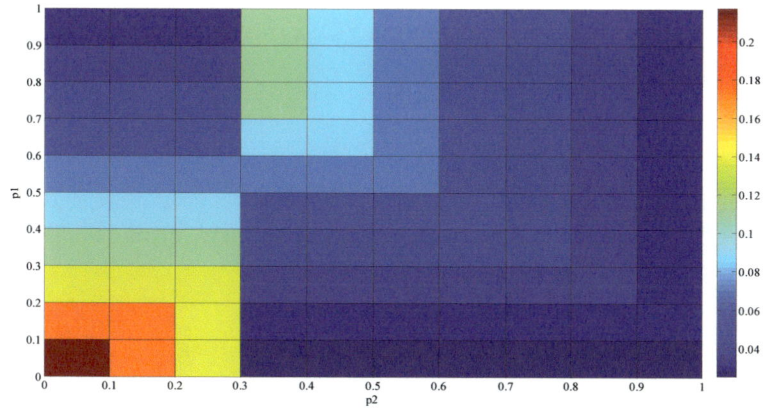

Figure 5. Grades of being NE for optimistic players

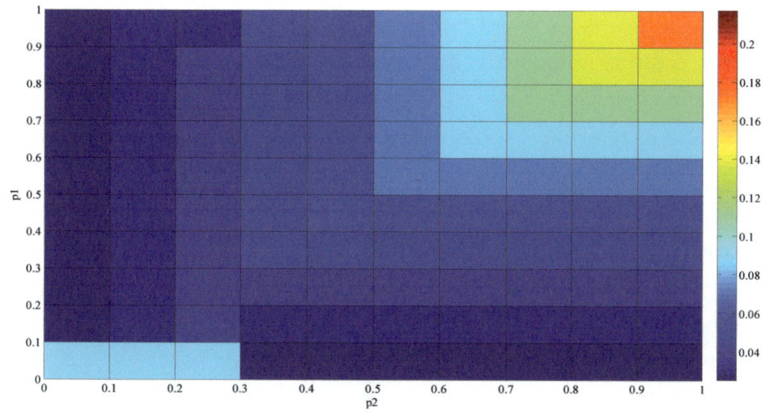

Figure 6. Grades of being NE for pessimistic players

7. Conclusion

In this chapter, some concepts in fuzzy sets including fuzzy numbers, fuzzy extension principle, ranking of fuzzy numbers and fuzzy preference relations were briefly introduced and consequently used to develop a new approach for practically analyzing the games with fuzzy numbers as payoffs. In this regard, a fuzzy preference relation was constructed on the preference ordering of payoffs using ranking fuzzy numbers. The priority of each payoff then was derived using the least deviation method. The priorities of payoffs were interpreted as the grades of being NE. We should remark that, in this chapter, we were not looking to show the pure and mixed strategy NEs, but rather we tried to assign a graded membership to each strategy to determine how much it is NE, i.e. it has the possibility for being NE.

Author details

Alireza Chakeri
Intelligent Systems Laboratory, Electrical Engineering Department, Sharif University of Technology, Tehran, Iran

Nasser Sadati
Intelligent Systems Laboratory, Electrical Engineering Department, Sharif University of Technology, Tehran, Iran
Electrical and Computer Engineering Department, The University of British Columbia, Vancouver, BC, Canada

Guy A. Dumont
Electrical and Computer Engineering Department, The University of British Columbia, Vancouver, BC, Canada

8. References

[1] Von Neumann J, Morgenstern O (1944) Theory of Games and Economic Behavior. New York: Wiley.
[2] Nash J (1950) Equilibrium Points in n-Person Games. Proc. Nat. Academy Sci. 36: 48-49.
[3] Nash J (1950) The Bargaining Problem. Econometrica. 18: 155–162.
[4] Nash J (1951) Non-Cooperative Games. Ann. Math. 54: 286–295.
[5] Fraser N. M, Hipel K.W (1984) Conflict Analysis: Models and Resolutions. New York: North-Holland.
[6] Harsanyi J. C (1967) Games with Incomplete Information Played by "Bayesian" Players, i-iii. Management Science. 14: 159-182.
[7] Zadeh L. A (1965) Fuzzy Sets. Inform. Contr. 8: 338–353.
[8] Butnariu D (1978) Fuzzy Games: A Description of the Concept. Fuzzy Sets and Systems. 1: 181-192.
[9] Garagic D, Cruz J. B (2003) an Approach to Fuzzy Non-Cooperative Nash Games. J. Optim. Theory Appl. 118: 475-491.
[10] Song Q, Kandel A (1999) A Fuzzy Approach to Strategic Games. IEEE Trans. Fuzzy Syst. 7: 634-642.
[11] Vijay V, Chandra S, Bector C. R (2005) Matrix Games with Fuzzy Goals and Fuzzy Payoffs. Omega. 33: 425-429.
[12] Maeda T (2003) On Characterization of Equilibrium Strategy of Two-Person Zero-Sum Games with Fuzzy Payoffs. Fuzzy Sets and Systems. 139: 283-296.
[13] Nishizaki I, Sakawa M (2000) Equilibrium Solutions in Multiobjective Bimatrix Games with Fuzzy Payoffs and Fuzzy Goals. Fuzzy Sets and Systems. 111: 99-116.
[14] Kim W. K, Leeb K. H (2001) Generalized Fuzzy Games and Fuzzy Equilibria. Fuzzy Sets and Systems. 122: 293-301.

[15] Li K. W, Karray F, Hipel K. W, Kilgour D. M (2001) Fuzzy Approach to the Game of Chicken. IEEE Trans on Fuzzy Syst. 9: 608-623.

[16] Mares M (2001) Fuzzy Cooperative Games. Heidelberg: Physica-Verlag.

[17] Butnariu D (1979) Solution Concepts for n-Person Fuzzy Games. Advances in Fuzzy Set Theory and Applications. 339-354.

[18] Zimmermann H. J (1996) Fuzzy Set Theory and its Applications. Kluwer Academic Publishers.

[19] Zadeh L.A (1975) The Concept of Linguistic Variable and Its Application to Approximate Reasoning. Inform. Sci. 8: 199-249.

[20] Yager R. R (1986) A Characterization of the Extension Principle. Fuzzy Sets and Systems. 18: 205-217.

[21] Yager R. R (1981) A Procedure for Ordering Fuzzy Subsets of the Unit Interval. Inform. Sci. 24: 143-161.

[22] Jain R (1977) A Procedure for Multiple Aspect Decision Making Using Fuzzy Sets. Int. J. Systems Sci. 8: 1-7.

[23] Baldwin J. F, Guild N. C. F (1979) Comparison of Fuzzy Sets on the Same Decision Space. Fuzzy Sets and Systems. 2: 213-231.

[24] Ibanez L. M. C, Munoz A. G (1989) A Subjective Approach for Ranking Fuzzy Numbers. Fuzzy Sets and Systems. 29: 145-153.

[25] Saaty T. L (1980) The Analytic Hierarchy Process. New York: McGraw-Hill.

[26] S.A. Orlovsky (1978) Decision-Making with a Fuzzy Preference Relation. Fuzzy Sets and Systems. 1: 155-167.

[27] Herrera F, Herrera-Viedma E, Verdegay J. L (1995) A Sequential Selection Process in Group Decision Making with Linguistic Assessment. Information Science. 223-239.

[28] Atanassov K (1986) Intuitionistic Fuzzy Sets. Fuzzy Sets and Systems. 87–96.

[29] Lipovetsky S, Michael Conklin W (2002) Robust Estimation of Priorities in the AHP. European Journal of Operational Research. 137: 110–122.

[30] Xu Z (2002) Two Methods for Priorities of Complementary Judgment Matrices_Weighted Least-Square Method and Eigenvector Method. Systems Engineering–Theory & Practice. 22: 71–75.

[31] Xu Z, Da Q (2005) A Least Deviation Method to Obtain the Priority Vector of a Fuzzy Preference Relation. European Journal of operational Research. 164: 206-216.

[32] Chiclana F, Herrera F, Herrera-Viedma E, Poyatos M. C (1996) A Classification Method of Alternatives for Multiple Preference Ordering Criteria Based on Fuzzy Majority. Journal of Fuzzy Mathematics.

[33] Dombi J (1995) A General Framework for the Utility-Based and Outranking Methods. Fuzzy Logic and Soft Computing. 202-208.

[34] Tanino T (1988) Fuzzy Preference Relations in Group Decision Making. Non-Conventional Preference Relations in decision making. Springer. 54-71.

[35] Fudenberg D, Tirole J (1991) Game Theory. MIT Press, Cambridge, MA.

Chapter 16

Game Theory as Psychological Investigation

Paul A. Wagner

Additional information is available at the end of the chapter

http://dx.doi.org/10.5772/53932

1. Introduction

Over the course of history mathematics and science have become increasingly entangled with one another. This has been especially true in the physical science wherein mathematical derivations have resulted in subsequent experimental pursuits. While the social sciences have relied on mathematical descriptions in recent generations it was not until the more general employment of game theory that mathematical modeling became itself a directive tool for subsequent evaluation as it had so become in the physical sciences, most prominently perhaps in high energy physics. This chapter shows how game theory forced upon the social sciences new avenues of investigation. It also shows that once those investigations were fully underway they in turn forced new considerations on the practice of game theory modeling. This boot-strapping dynamic between mathematical game theory and social science represents a novel turn in the relationship between the two disciplines bringing their relationship more into parallel alignment with what has long existed between physical science and mathematics.

Science and mathematics travel along coincidental paths. Just how coincidental is a source of perennial speculation and argument among sociologists, philosophers and historians of both science and mathematics. If the paths math and science travel are only coincidentally linked then neither serves as foundation or guide for the other. On the other hand, if the paths are destined to be linked in some fashion then the affiliation between the two is more than merely coincidental. Any answer to this query seems necessarily to raise the question of whether either, at least in its modern form, can exist without the contemporary development of the other. Any effort to address this issue inevitably prompts consideration of the ageless question of whether mathematics is invented or discovered.

Early explanations of the natural world can be put forward in strictly naturalistic terms common to the researcher's native tongue or in some cases scholarly language such as Latin or Greek in the West and Mandarin in the East. The science of antiquity was largely free of any necessary bondage to mathematics. Admittedly as soon as early cosmic observers in both the East and the West began tracking the heavens something akin to early arithmetic appeared as did some rudiments of geometry. Nonetheless, some descriptive sciences such as biology, physiology and psychology advanced for many centuries without necessary dependence on mathematics. Even well into the nineteenth and twentieth centuries scientists known as naturalists did amazing and properly described scientific work without much reliance on heavy mathematical machinery. Names that immediately come to mind are Charles Darwin and E. O. Wilson. Of course today the biological sciences are as dependent on the power of mathematics to aggregate the data of scientific thinking and organize inferential patterns as were the early cosmologists and contemporary high energy physicists [1]. Still the question remains: are the two paths necessarily linked?

Mathematician extraordinaire, G.H. Hardy spoke dismissively of his protégé, Norbert Weiner when asked what he thought about Weiner's applied accomplishments. Hardy said in no uncertain terms that Weiner was no mathematician! Weiner, Hardy complained, applied mathematics [2]. *Real* mathematicians Hardy declared, never leave the pristine purity of the number world itself. They don't use the beauty of mathematics to do the yeoman's work of science or engineering. A reporter asked Hardy why then should anyone do mathematics if, as it seems from what Hardy was saying, that it has no practical pay-off. Hardy's response was almost zen-like to anyone not a mathematician. Hardy said, "Because it's such a damn good sport! [3]"

Hardy's declaration makes clear that at least one noteworthy specialist in number theory sees no need to tie mathematics to anything in the sensuously detectable world of science or the daily life of ordinary humans. Mathematics or at least that domain of mathematics that Hardy was most interested in, depended in no way on the shifting sands of human observation and empirical science [4]. Contemporary science may need mathematics but there are areas of contemporary mathematics that can stand well apart from the fortunes and misfortunes of contemporary science.

To a mathematician like Hardy and logicians like Kurt Godel [5], Hao Wang [6] and Alfred Tarski [7], mathematics is something to be discovered. To these thinkers, mathematics could never be conceived as a mere game or a simple approach devised by shepherds for keeping track their flocks as some empirically-minded theorists have opined [8]. In any case however independent mathematics and science may be distinguished from one another as separate disciplines there can be no doubt that mathematics has often advanced the agenda of the various sciences. Not only has mathematics proved to be a wonderful tool for aggregating data and then organizing it in ways leading to prescriptive efficacy but inferences that appeared from various mathematical tinkering have often prompted fruitful scientific speculation. This of course has been especially true in higher energy physics and cosmology. For example, Paul Dirac's use of David Hilbert's infinite dimensional space led to unexpected and fruitful direction for empirically based theory. And, Dirac accurately predicted the existence of positrons on the basis of mathematical calculations alone [9]. And more recently this mathematical showing the way in empirical science was replicated again in Richard Feynman's sums-over-all-possible-histories.

In what follows, I will discuss how the coincidental crossing of paths in the social sciences of decision-making and mathematical game theory led to fruitful excavations in the psychology of decision-making and mental life, especially in the case of humans [10].

Blaise Pascal is generally recognized as the father of decision theory and judgment – making under conditions of uncertainty [11]. Shortly thereafter Thomas Bayes gave respectability to the practice of employing prior probabilities to continually adjust and update predictive calculations [12]. John von Neumann gets well – deserved credit for expanding this work into a theoretical approach for identifying strategies that increase the likelihood of decision-makers under conditions of uncertainty optimizing expected value (EV) in real world decision-making. Just as Pascal was attracted to the intrigue of investigating decision-making under conditions of uncertainty in playing dice games, in the 1920's von Neumann became enamored with investigating decision-making under conditions of uncertainty in poker where, in addition to the uncertainty of probabilistic distribution of card sets, each player employed differing bits of information of other's strategic style [12 p. 42].

Where Pascal relied on statistics alone for answers and Bayes opened the door to ready revision of planning estimates, von Neumann proposed a broader and more robust set of tactics for capturing relevant information explicitly for planning purposes. Von Neumann extended the reach of employable planning information into quantifiable estimates of the very mindsets of fellow players [12 p. 89-96]. The trouble was however that he didn't sufficiently appreciate the distinction between facts of the external world and facts of transient, human psychology.

Initially von Neumann was struck by the fact that unlike closed systems of transparency such as chess, checkers, Go, Tic Tac Toe and other like games, in poker there is a disproportionate distribution of information.

Information for von Neumann was all of one sort. It was a set of rule governed symbols. Once one knew what symbols to employ, formal structures made it possible to derive conclusions. The trouble is however, in poker, the disproportionate distribution of information also suffers from a differential reliability of evidence.

The epistemic challenges of securing reliable evidence are not all of one sort. Gaining knowledge of a somewhat static external world (at the macro-level of sensuous human experience) is different from gaining knowledge of the shifting sands of individual mental life. No player knows with certainty what cards other players hold. This is an epistemic challenge focusing on the facts of a momentarily static, external world. In contrast, no player knows what each other player anticipates his fellow players to do in a given set of circumstances. No player knows the strategy each other player may have in mind for playing this particular hand given the particular set of cards the player holds. These are both matters of transient human psychology about which far too little was known at the time of von Neumann's initial efforts at game theoretic modeling [13].

The first epistemic shortcoming referred to in the paragraph above is about the world as it exists external to the mind of any given player. The second and third epistemic shortcomings above referred to the psychological, more specifically, the mind set of other players. Nonetheless, depending on the previous betting patterns of each player in previous hands (and continually aggregating that information through the process of updating prior probabilities) there is important information available which together with a probabilistic assessment of the competitive strength of the hand one currently holds and an estimate of the strength of the hands held by others that *should* make strategizing more efficient than if one were to rely on the probabilities of certain card assortments and their distribution among the number of players in the game alone.

Von Neumann imagined using mathematical matrixes to sort and arrange probabilities of outcome to discount the EV of payoffs and determine player utility for each outcome in order to anticipate likely plays for each player [14]. With this mathematical tool in hand, along with reasonable estimates of the uncertainties referred to above, he imagined that an optimally effective winning strategy could be identified. Such a strategy could then lead to a generally profitable strategy over a run of playable hands. Of course, as the other authors of this volume will surely attest, the range of uncertainty in the game of poker is too vast to allow for the efficient application of game theoretic principles. This is still as true now as it was in von Neumann's time. It certainly is no surprise that von Neumann gave up his work on game theoretic reconstructions of poker playing strategies [14].

Von Neumann eventually moved beyond his playful distraction of subordinating poker to game theoretic principles. He along with the economist Oskar Morgenstern turned to human decision-making on the largest scale of systematic competitiveness (even though for illustrative purposes they limited examples to mostly two person games). Their large picture modeling emphasized zero sum and constant sum games. This emphasis probably was influenced by the post-war atmosphere, the increasing focus on central planning and strategic and tactical, Cold War worries. From these decision-models Neumann and Morgenstern produced a favored strategy for decision-making they called the maximin strategy [15].

Von Neumann and Morgenstern's maximin strategy became a competing paradigm to other schemes of classical economics for analyzing economic exchange and other forms of strategic behavior. This early work of von Neumann and Morgenstern has since become the classic statement of game theory in an ever expanding variety of game theoretic contexts to which all writers since have felt some obligation to acknowledge. The von Neumann and Morganstern term maximin is a bit of conceptual apparatus that is still in use though it has been shown that all maximin strategies can be reduced to a special case of minimaxing [16]. The maximin/minimax theory is also referred to in aggregate fashion as "satisficing strategy" [17]. For purposes of economy I will henceforth use the term satisficing to refer to either. Roughly speaking the strategic objective of minimaxing is to minimize risk while maximizing utility, to maximin, the strategy is to discount utility proportionately to manage acceptable risk. In short, both attempt to make the best of a risky and uncertain situation. Together minimax and maximin are strategies. Satisficing is the hedge fund strategy of decision theory.

The ambition to extend their early work to the largest scales of decision-making is evident in Morgenstern's later work [18] and also in von Neumann's work as Chair of the Atomic

Energy Commission [14]. The innovative mathematical applications of von Neumann and Morgenstern opened up vast new vistas of prescriptive decision theory [19] and, as will be demonstrated below, made possible new horizons in psychological and other descriptive social sciences.

Even before the computer and imaginative efforts in artificial intelligence and cognitive science, von Neumann and Morgenstern, along with those who shortly followed them, showed ways for illuminating much of the mystery of human mental life. To a brilliant mathematician like von Neumann the mathematics of game theory was not much of a challenge. Yet von Neumann's work in game theory remains his most lasting contribution to the sciences. Not only is game theory used in economics, international trade, military strategizing, and business operations at every level it is now also used to illuminate various evolutionary models in biology, anthropology, archaeology, sociology and psychology all in addition to economic theory [20 - 24].

As with any well thought out application of mathematical protocol, game theory depends upon a few simple and well-articulated base line assumptions. These assumptions address the nature of decision frames, their constitution and representation in formal models and the appropriate range of game theoretic applications.

At the time of von Neumann and Morgenstern's early work there were three basic assumptions common to economics and behavioral psychology. These assumptions were, first, humans are self – interested. Second, humans are rational. And third, humans are self-determining consumers. To these three, Von Neumann and Morgenstern added five more:

1. All outcomes can be known to varying degrees of certainty
2. Player information is often incomplete
3. Utilities (measures of one's relative gain) can be measured
4. The utilities of all outcomes when the other assumptions are met can be discounted and summarized in a single quantity (EV).
5. Some games are competitive (zero sum games), some are constant sum games and some constant sum games and non-zero sum games often favor dominant strategies whose equilibrium invites cooperation as an attractive strategy[25 p. 88-89; 26].

Game theory was employed almost immediately for strategizing by various governments around the world perhaps most notably in the United States at the government's quasi-private think tank, the Rand Corporation [15]. Decision-making models were driven less and less by simple cost/benefit analyses and weighted outcome averages and more on the basis of;

1. likelihood of outcomes
2. integrated with *perceived* utility and then…
3. derivation of an EV for each outcome

Of course, even with such integrated summaries at hand personal willingness to act on the basis of EV reflects the temperament and sense of responsibility of the decision maker [10, 13, 21, 27, 28, 29]. Witness for example the disproportionate risk the poor undertake to buy a lottery ticket when the EV to them is especially low. There is no secret about how little chance anyone has of winning a lottery [30 p.163 - 175]. Wealthy people can afford greater risk yet they tend to find the cost of lottery investment a poor buy. In contrast, many of the poor for whom the relative cost is so much more, imprudently absorb the risk time and again. This social phenomenon truly appears in sum to be a tax on ignorance. EV and human psychology are often in tension that tension illuminates both the social sciences and the challenges of exhaustively modeling best decision-making practices. Best to whom and under what circumstances remains a live challenge to normatively –driven, decision theorists and descriptively – driven social scientists alike [31]. How much individual agents understand and how much they are willing to satisfice to achieve acceptable results within a specified problem frame continues to be a subject of much uncertainty [32]. No mathematical theory can determine how risk aversive a decision maker might be or should be [33]. The mathematics can show only that integrated summaries spell out the discounted EV of available bets [34].

The enormity of this tension and the challenges it represents can perhaps be most vividly imagined in the context of the decision at Los Alamos during WW II to test detonate the first atom bomb. A major player in this decision was John Von Neumann [15]. In this case, von Neumann together with Stanislau Ulam was assigned the task of calculating the largest scale bet ever made in the history of humankind. Specifically they were assigned the task of figuring out the odds that when the first atom bomb exploded only the intended atoms would fission. In other words they were to calculate the odds that under the problem frame of detonation other surrounding atoms would not be drawn into an unrelenting process of nuclear destruction [35]. This calculation could not have been responsibly derived from formal probabilities alone or even some frequency study of previous atomic behavior under controlled conditions. At the very least subjective probabilities had to be utilized by the decision-makers and most likely they were utilized in some game theoretic fashion. Surely von Neumann's twenty years of game theory reflection influenced his calculations and subsequent recommendation to General Paul Griffiths. One can only wonder how small the odds of universal destruction ought to be in the mind of the great mathematician in order to recommend a properly satisficing go-ahead for the initial detonation.

While there is insufficient public information describing what happened at Los Alamos during that period it is nonetheless likely that game theory played a role [15]. In any case whatever reasoning was pursued by the mathematicians, generals and physicists at the time, game theory today could effectively model the potency of the actual, social and psychological forces presumably then at play [36]. Such modeling can prove illuminating in understanding ever more about the real process of human decision-making under conditions of uncertainty [37].

Throughout WW II game theory and Bayesian statistics both became more fully utilized at the highest levels of organizational decision-making [38 p. 185 – 220; 39 p.58-60]. At the same time game theory was becoming more widely embraced by academics and major decision-makers alike, it was also becoming apparent that there were problems to be solved in the utilization of these new mathematical tools in the ever increasing array of phenomena they were being used to model [39]. The focus in this brief chapter is game theory so there will be no further discussion of the problems some statisticians and other theorists found with the grounding or utilizing prior probabilities in Bayesian deployments. Such matters are important to the current topic but an adequate treatment would require far more attention than one can responsibly undertake in a single chapter.

In game theory, three major realms of difficulty emerged. First, settling issues of how risk aversive a given decision-maker *should be* in specific contexts proved increasingly intractable to any sort of formal analysis [40 - 41]. The more aggressive the efforts at formal analysis, the more evident it seemed that the psychology of human nature would in the end adjudicate matters of appropriate risk aversiveness for an individual or a group [42 p.94-110]. This realization was significantly responsible for prompting further psychological excavation into human motivation.

Second, less than a decade after publication of von Neumann and Morgenstern's *Theory of Games and Economic Behavior* [25], John Nash [43, 44], Richard Selton and John Harsanyi [45] showed that the psychological assumptions of classical economics and behavioral psychology led to unavoidable paradox. Expanding on von Neumann and Morgenstern, Nash alone and Selton and Harsanyi together demonstrated that given the current assumptions of game theory the most rational course of action for all players in any competitive game (zero sum or constant sum) was to follow a course of action that secured an equilibrium in EV among all players [46]. When every player recognizes an acceptable EV in a common strategy, that strategy dominates over all other choices for each and every player. Dominant strategies (whether mixed or pure) benefit everyone and impose no undue loss on anyone [47]. This platitude sounds too good to be true. If such dominant strategies could be identified for all human interactions game theory would have shown how peace and decorum can be realized throughout the world in every way [48-49]. All social problems would be reduced to simple puzzles in game theoretic modeling. Plato's philosopher kings would have been found but they would be mathematicians and not philosophers. But as Nash, Selton and Harsanyi all realized, much of the social world is not amenable to such modeling.

Somewhere in the Government-funded Rand Corporation think-tank, there emerged the puzzle of the most famous game in all of game theory, namely, the Prisoner's Dilemma (PD) [15]. Nash is often attributed with its first formulation but there seems to be some controversy over who the actual originator was [49]. Other names mentioned are Merrill Flood, Melvin Drescher and Albert Tucker [13 pp.5-6]. Nevertheless, that the PD became the source of much contested theoretical musings at Rand and then later in the specialist journals is undeniable [50].

The issue illustrated by the PD is that it leads to a paradox given the standard assumptions of classical economic and behavioral theory [49, 50]. The issue was not seen as a matter of psychology. Rather theorists simply noted that people might vary as to what they favor either individually at the moment or, over time (though those stipulations vanished as "time" has become a recognized issue in subsequent theorizing). The technical issue was that the classical model led to a disparity between securing optimal utility for each player despite the fact that one could derive a satisficing equilibrium that was demonstrably less than optimal.

Game theory was intended to show that by formalizing decision spaces and applying game theoretic principles, decision-makers could illuminate in every case the most rational course of action that by definition ought to produce the highest level of satisficing utility for each and for all.

Although some game theorists continue to deny that the PD ends in a paradox they represent a very small minority of thinkers [51]. This minority insists that the alleged paradox simply demonstrates that sometimes the only strategy open is not a minimax risk aversive strategy but rather a maximin strategy wherein all players wind up losing significantly but that a rationally derived equilibrium is achievable nonetheless and so game theory's value remains unimpeachable as classically portrayed [52]. This convoluted effort to sustain the distinction between maximin and minimax and derive a dominant strategy remains generally unconvincing to most [51]. Instead the paradox of the PD has become importantly informative in forcing theorists to reconsider the preferences of actual players in real world PD situations [33]. In doing so, game theory has led to the creation of a whole new dimension in the social sciences known now as preference theory [53].

Typically the scenario of the PD looks something like this. Two criminals are caught shortly after a robbery they committed. Authorities place each suspect in a room separate from the other and offer each criminal individually a deal to ease the authorities' way to a conviction for either one or both the suspects. For convenience, name the criminals Donald (row) and Rosie (column). Each is told that if he or she testifies against the other (defects) then the other person will spend ten years in prison and the defector will be released (0 years in prison). Counsel for each cautions their respective clients that if each defects then it is likely that the court will put each of them away for five years. Counselors' for each advise additionally that if each of the accused remains silent (cooperates) the authorities already have enough evidence to put away each of them for two years on a related but far lesser charge. Neither Donald or Rosie nor, their respective counselors, may consult with the other or with the other's counsel. Finally, the authorities' demand that an immediate response to their offer. Assuming each person is self-interested and rational, the choice seems obvious as illustrated in the matrix below.

		Defects	Cooperate
Donald	Defects	5, 5 (I)	0, 10 (II)
	Cooperate	10, 0 (III)	2, 2 (IV)

Table 1. Rosie

If Donald defects and Rosie cooperates by remaining silent then Donald loses nothing (cell II). The same possibility is true for Rosie if she defects and Donald cooperates by remaining silent (cell III). Neither Donald nor Rosie know what the other will do. However, by stipulated assumption, they each "know" that people are rational and self-interested and moreover there is no reason to think their counterparts in this situation are any different than they themselves and other rational people. Each is also aware that the other criminal may be thinking the same way about them and so the only way to avoid getting suckered into a ten year incarceration is to defect and expect that the other will too (cell I). In this case each will lose five years of life. Given the assumptions of both rationality and self-interest this determines a dominate strategy that is to say, an evident "right" choice. Yet something seems strongly counterintuitive here.

Clearly there is no likely outcome that optimizes Expected Value (EV) for either of the criminals. The EV for both, as noted above under the current set of assumptions is an EV of ten lost years for the two together (five years for each). Yet in cell four there is a payoff of four lost life-years for the pair (two years for each). This is clearly the best option for the pair. But again, given the standard assumptions of classical von Neumann and Morgenstern game theory, there is no way to get there [54]. The PD truly represents a paradoxical situation as most scholars allege and thus von Neumann and Morgenstern's game theoretic satisficing strategy is a poor tool for identifying a dominant decision-making strategy under such situations [55].

If the von Neumann and Morgenstern assumptions about modeling are changed, then the preferred equilibrium can be derived [52, 56]. For example, if Donald and Rosie *know,* to some high degree of certainty, that each has a strong sense of loyalty and *prefers* to demonstrate that preference above nearly any other rewarding outcome. The problem frame shifts significantly [57]. In addition, if each player values (assigns high utility) the well-being of the other at least as much as each values him or herself, then each can count on the other's silence. This new bit of surmizeable knowledge shifts the problem space even further. With the new assumptions in hand, and the problem context re-framed as a result, the dominant strategy is now the strategy that secures the EV in cell II. Of course now it seems that securing the best decisions in life is no longer a simple mathematical problem but is dependent on accurate psychological observations as well. The mathematical modeling led to the need for empirical investigation if applications to the world were to be truly productive. Coincidentally, the empirical investigations of psychology and other social sciences now had a robust and productive new direction to explore thanks to the revelations of mathematical game theory efforts at application.

But how does one ever know – even to a reasonable degree of certainty – that people in general or another person in particular embraces noble or at least cooperative preferences in a steadfast way? What counts as a *reasonable* degree of certainty? As with von Neumann and Ulam's recommendation regarding detonation of the first atomic bomb such considerations in the end are normative as more a matter of philosophy than empirical science or mathematics. Yet the empirical social sciences can speak to the likelihood that people in general address certain ranges of problems in fairly predictable ways and this gives each player more information under conditions of uncertainty than a fly – by – the pants guess. In addition, each player may be able to augment that situation by personal knowledge of the other player. This Bayesian updating of subjective probabilities improves the betting odds of each player at every step of the way. This is no small accomplishment. As Ken Binmore so wisely observes, "Only in a small world, in which you can always look before you leap, is it possible to consider everything that might be relevant to the decisions you take [41 p.139]

The only way to get a better grasp on human "problem-framing" tendencies is to do scientific work into the actual preferences of human beings. Psychologists and behavioral economists must measure the variability of sustaining preferential strength under various conditions. For game theory to be efficient in PD - like situations adaptive correction of previously assigned, prior probabilities, that prove inconsistent with the material facts of social interaction how much more need to be known about human psychology than people like von Neumann and Morgenstern thought necessary.

Two games devised for more deeply mining human preferences and problem-framing practices are the Ultimatum Game and its variant, the Dictator Game. Each game creates an empirically measureable construct in which people have to make a choice often at some sacrifice to self and with no immediately apparent reward. The evidence overwhelmingly demonstrates that people will sacrifice to punish a perceived unfair, non – cooperative defection [40]. Whether or not people engage in such apparently altruistic behaviors intentionally because of being hard-wired to cooperate and punish defection by evolution is yet to be determined. What no longer needs determining is the fact that people act on other than simple self – interest when framing a problem space. Other natural and wide spread human preferencing habits are now being identified and taken into account in developing more exhaustive game theoretic models and decision –processing strategies [45]. Without going into further detail, suffice to say that the general consensus among empirical researchers is that the PD in addition to games such as the Ultimatum Game and the Dictator Game demonstrate that humans have a tendency to act somewhat altruistically and are sensitive to violations of what they take to be fair play in practices of distributive justice [46]. It is interesting to note as well that empirical studies based on such games show that cross-culturally people are likely to accept some cost or sacrifice willingly to punish a player (whether familiar to the player or not) who makes little attempt to be cooperative with others [36].

When the constraint of self – interest is loosened and weighted appropriately along with other identified preferences, a more comprehensive picture of players' desired payoffs and likely strategies emerges making equilibrium strategies more evident as the dominant strategy for each player [50]. Moreover, as Bayesian revisions are made in light of assessments of the other players' weighted preferences and developing information set, the problem frame suggestive of an appropriate game theoretic matrix can be more approximately fitted to the real world state of affairs defining an applicable game [55]. This is particularly true when the real world is in flux or when an unlimited round of plays in the future is anticipated by the players. For example, branching tree decision models developed to reveal these sorts of anticipatory decision factors as well as other evolving relevant determinants of outcome that become evident as the sequence of subsequent decisions prompt further review to sustain equilibrium. This sort of application is effectively illustrated through the employment of the Beauty Game.

The Beauty game is a variant of the strategy recommended by the character of John Nash in the movie A Beautiful Mind. The Beauty Game shows that identical game structures can be nested within a larger game. While each problem space looks similar in character when taken one by one, when the nested structure is taken into consideration as a whole, the structure of the problem frame itself may shift into a different sort of game. Through use of a decision tree the nesting of games is revealed. For example, as I show in my discussion of the Beauty Game elsewhere [58] a cooperative non-zero sum game can quickly shift into a zero sum competitive problem frame. When the set of nested games are seized upon and reconstructed as fragments of a much larger problem frame identifying time and shifting identifiable preferences as factors realigning possible pay – offs in a serial fashion. Equilibrium achieved early on can be destabilized when the sequence of rounds can be shown to have a determinate limit. Thus backwards induction may reveal earlier distributions of EV are not as satisficing to all as may have been first perceived. (No matter how the game is played, not everyone can have the one uncontroversial "most beautiful girl")[58].

Beyond the evident paradox in the PD referred to above, the third difficulty that emerged for game theory was that psychological preference theory was impoverished in the 1950's [13; p. 1-6]. There was little for game theorists to draw upon when attempting to apply game theory to the actual world of daily life and especially in the more mundane aspects of daily living. Many real life situations seem to fold into PD matrix distributions. If so and if people recognized paradox after paradox then cooperative problem-solving efforts would run aground so often that human rationality would become generally discredited. The fact that humans somehow seem to navigate their way through the many PD situations they face as a matter of routine suggests on the other hand that there was no linear processing going on but rather just random or emotionally charged and unpredictable acts leading people through the PD problem frames. Or again, under the classical model people inevitably face making decisions that look optimally rational on the surface except the uncontroversial dominant strategy delivers less EV than could evidently be achieved.

Too many PD situations as in the Beauty Game, left an abundance of goodies on the table. Empirical researchers surmised these goodies could be distributed in an optimally efficient fashion. (The Beauty Game represents a further difficulty because the reward of the one most beautiful girl cannot be distributed and satisficing strategies do not work as long as the unit of reward she represents stays unimpeachably intact. In the movie, A Beautiful Mind, all the suitors satisfice but, only at the cost of removing the most beautiful girl from the pool of EV. And as I explain in the Beauty Game, the master strategist by removing himself from the first round of play may have masterminded a super game which remains satisficing to all only so long as no one catches on to his ploy to secure the greatest EV for himself in a discrete subsequent round of play in which only he and Beauty are left. [58]) Classically derived, dominant strategies appear on the surface to be insufficiently potent. For game theory to be of value in the nearly ubiquitous realm of PD situations more must be learned about actual human preferencing practices [55].

Psychology and the other social sciences had to develop insights that mathematical modeling by itself could not achieve. Mathematicians had to await the insights of social scientists studying preferencing to advance their own efforts to improve game theoretic modeling. Once again the coincidentally entangling paths of mathematics and science became evident with neither discipline wholly dependent upon the other. Instead game theoretic applications required mathematics and social science each to boot-strap further progress on results achieved by the other.

Thanks to the game theoretician's need for further insight into human preferencing empirical investigation into human motivational action by social scientists of every stripe accelerated. The result of the subsequent growth into human preferencing studies and related heuristical practices is that game theory was able to extend its range of apt application throughout the social sciences extensively and then ranging into fields such as business, finance, evolution and population genetics [31]. Together game theory and psychology excavated into mental life more than previous behavioral stipulations and methodologies would have ever allowed [13].

As noted above, a range of PD games confront people every day. Game theory is not an esoteric study for parlor mathematicians alone. This merging of the mathematical and the social sciences has enormous value in understanding the practical life of most humans every day. Consider when you are driving on a fast-moving but heavily congested freeway and you and another driver (player) want to change into the same lane in what seems to be a relatively rare opportune moment. This is a zero sum game at the moment when the problem space is defined narrowly in terms of one player's success guarantees another player's loss. But in the real world of driving, different drivers have different purposes and different heuristical practices they are likely to employ. How much information can a deliberative player have about the real situation he encounters? How should the different fragments of information be weighted in near spontaneous fashion? How best should the game theoretic driver weight her own EV in securing a move into the fast lane? What are the probable outcomes of any move on the part of each driver?

Once one moves from the narrow defining of the problem frame to formulating a game matrix that ostensively captures real world possibilities as well as identifying an optimal strategy. The lane changing situation is clearly a PD environment. However, practical application of PD technology constitutes a severely time – discounted, problem frame. Consequently, for all practical purposes game theoretic strategizing under such constraints is ill-advised. Drivers probably do best what they are already inclined to do and rely on Kahnemann describes as System 1 thinking [40]. Having acknowledged the prescriptive poverty of game theoretic thinking in such practical applications it may still be profitable for social scientists to observe large numbers of driver behavior under such circumstances and model such behavior to reveal whatever can be known about such driving strategies in general. Indeed, if researchers learn more about statistical patterns common to such contexts and survey driver preferencing patterns perhaps safer freeways, automobile safety additions and driving instructions might be derived.

Employing a more exhaustive set of assumptions allowing for individual preferencing avoids the paradox of the PD [55]. Of course, the challenge then becomes for social scientists to excavate into human mental life and identify relevant information about the preferences of human actors.

The practical advantage of game theory to decision – makers is obvious; less intuition and flying by the seat of one's pants and better positioning for placing prudent bets (securing EV) in the future. The evolving advantage of game theory to social scientists is that game theory draws attention to the range and weighting of human motivations in the course of decision-making more than any previous methodology in the social sciences [31].

When the overly restrictive stipulation of self – interest is set aside as the sole human motivation, rationality can still be preserved and shown to be as robust as in classical modeling [59]. And, in PD cases as illustrated in table 1, access to cell IV with its greater EV for the pair, looms compellingly as the evident choice for all players. When additional motivations, appropriately weighted, are considered along with an appropriately weighted value for self – interest, a once formidable paradox gives way to a decision procedure more sensitive to general welfare benefit. The satisficing of appreciably finite contests can serve aptly for generalizing welfare benefit on increasingly larger scale with ever more generalized descriptions of human nature provided by the social sciences.

Vilfredo Pareto described a special type of normative equilibrium [60]. In Pareto optimality, there is an ideal point such that if anything changed for any player than there would be a decrease in the distribution of optimal benefaction for each and every player [60]. In the PD illustrated in table 1 above, there is an evident Nash equilibrium that is satisficing on von Neumann and Morgenstern's principles. But, the inaccessibility of optimal EV for the pair, evident in the more mutually benefitting, cell IV, makes such an equilibrium seem alarmingly deficient [32].

Securing the optimal EV for the pair and which also secures an optimal EV for each player - seems the way to go. Any decision procedure which fails to point the way to such a solution seems prima facie deeply flawed. So rather than insist on treating every player as self – interested, in situations where the game can be defined in non-zero sum terms (including constant sum games)or in cases of coalition benefit, it is clearly better to seek Pareto optimality or coalition well – being over individual benefit (as suits the specific problem frame and respective information sets)[61].

Seeking Pareto optimality requires abandoning the standard classic assumption of self-interest in light of empirically based psychological studies and subsequent generalizations. More than ever before advances in evolutionary psychology and behavioral economics in particular are revealing much about players expected and hoped for utilities in life [61, 62].

Specifically motivations have been identified that empirical research has shown sometimes trump the constraint of self – interest [63]. This insight was pivotal as the paths subsequently pursued by mathematical game theory and social science merged.

Game theory demonstrated that identifying relevant human motivations is necessary if decision theory is to be increasingly informative both descriptively and prescriptively. In addition, neuroscientists found in these enriched models of cognitive function an apparent supervening upon neurological activity. In addition, game theory made social theorists increasingly aware of the range of actual players and player interests in both intra-cultural and intercultural contexts. Real world concerns with accessing Rousseau's fabled "will of the people" have been invigorated. Renewed attention to lotteries [64] and voting practices has led to revisions of Condorcet's Jury Theorem [65] and the grim realities posed by Arrow's paradox [66]. For anything to count as a science mathematical management of data and mathematical predictions seemed essential [1]

Quality control theorists and operations researchers both were similarly extending the range of their vision from abstract modeling to modeling with real human players in mind. For example, the physicist W. Edward Deming [67] in prescribing holistic approaches to quality control advised senior decision – makers to imagine the sport of crew. Each member of a team of eight and a coxswain may be excellent in her own right. Yet this team of excellent athletes may lose to a team of less talented athletes who had learned to row together more efficiently. Coordination and cooperation were proving to be demonstrably valuable assets in addition to the excellencies of individual participants. Despite the human tendency to want to show off individual excellence mathematical modeling was showing the way to establishing empirical evidence for the fact that cooperation and coordination are necessary to improve EV humans could reasonably come to expect.

Despite rhetoric to the contrary, folk psychology for most of the twentieth century in business and in other practical pursuits still seemed to imagine human competition as did Homer of Greek antiquity. Homer imagined armies fighting in the shadow of their respective heroes Ajax and Achilles. Find the vulnerable heel of the champion and the entire community loses its sense of conviction or so the folk psychology of the time attests. Mathematical modeling and game theory are showing that humans like all other herd species are much more likely to approach equilibrium in cooperative efforts by becoming more rationally astute in aggregating expectations and utilities in a fashion that optimizes EV for all players in the games of life.

A skull crew may have Hercules, Samson and six other super athletes but if each super athlete is violently working hard in his own way, the skull may jerk about in modestly productive thrusts as compared to an eight comprising lesser individual athletes but exhibiting greater team coordination. There is in this example an analogical metaphor with game theory's Pareto optimality. Cooperative play was increasingly becoming evident as a means for extracting the optimal utility from individual effort. From non-zero and cooperative sum games to total quality management and team sports and many PD contexts as well organizational proficiency could be deliberately increased as a result of deliberatively choreographing coalitional activity.

Games are being devised to reveal the potency of preferences such as variances of altruism that support the productive value of ever greater networks of cooperation [24]. Some of these games have become quite famous for revealing non-self-interested preferences in actual decision – making contexts. For example, versions of the Ultimatum Game and the Dictator Game (a conceptual derivative of the Ultimatum Game) demonstrate that people act altruistically at times. Such games have also revealed that there are a variety of different types of altruism ranging from reciprocal altruism to pure altruism [68]. Applications of game theoretic thinking to experimental situations have also shown that money (a supposed veritable token of self-interest under the classical schemes of von Neumann and Morgenstern). For example, one experiment found that able-bodied people were willing to help a stranger move a couch into a house when simply asked. In contrast, seemingly similar individuals would refuse to help if the solicitation involved a monetary reward of less than twenty dollars [68]. An offer of what some regarded as paltry was not only non-motivating but it had a dampening effect on response to requests for help. People seem to have some altruistic preference to help another in need. They also seem to calculate the value of money which increases their range of choices in life, against some value they assign to their labor we may have parochially referred to in the past as personal dignity.

Another startling advance in the social sciences stimulated by applications of game theoretic modeling techniques include the experimentally derived observation that in actual decision-making contexts, preferencing is not always transitive [69]. Imagine a fellow, Jones, whose prioritized preferences in life are ordered as follows: wine (w), women (f) and song (s) and getting close to his creator through prayer (p). So, whenever he is considering the utility of some act or, practice, we have following general order of his preferences as follows:

$$w \geq f \geq s \geq p.$$

Unfortunately, Jones has enjoyed way too much wine over the years. So now, Jones' doctor informs him that regardless of what he does from here on out, Jones will shortly die. The doctor further advises Jones that if he avoids alcohol, he will experience some remission of pain and extend his life somewhat as well. Under such circumstances imagine Jones quits drinking. If preferences are always transitive we should expect Jones will then turn to chasing women more often, singing a bit more and then finally praying a wee bit more as well. The remission of pain and the foregoing of immediate death, afford Jones a chance to enjoy his preferences in life a bit more than his continuing previous behavior would have permitted. So, in light of this consideration Jones eliminates his enjoyment of wine as a preference altogether. His EV for each of the three remaining preferences would increase in ranking under conditions of intransitivity. However, in the real world of sensuously experienced activity it is easy to imagine Jones "turns his life around". Jones may now prefer the activity of prayer above all else. His life may now be driven by dramatically and a transitively revised set of preferences such as:

$$p \geq s \geq f; \; \textit{or}, \; p \geq f \geq s.$$

Any ordering of the preferences will do to show that with the elimination of a preference it is not at all obvious that rank order of preferences transitively follows. Rather than go through all the permutations it is sufficient simply to show here the plausibility of the transitivity of preferences in actual life situations as problem frames change [70].

Further game theoretic-based experiments are showing the intensity of human motivations in the decision-making process. Concepts such as reputation, disgust and honor are now recognized as central in increasingly apt models of human decision-making. Game theorists have shown that by weighting preferences appropriately prescriptive values including moral commitments can be retained throughout a decision making procedure of nearly any social sort [69].

To illustrate the normatively sustaining points referred to in the paragraph above consider a simple two player zero sum game. Imagine two players. Call them "F" and "E". Both are confined in a prison-like situation and both are starving. A bit of food is given to F. The food is enough for F to survive another day or so but only if he consumes it all himself. E is given nothing. E will surely starve unless F gives him the food. But if F gives E his food then F will starve. This seems like a straight-forward zero sum PD. There is no equilibrium on the face of it. F and E presumably both want to live. The player who has the food should consume it assuming he is rationally self – interested.

This scenario is in fact an actual case as described by Ellie Weisel in his autobiography [71]. F is Ellie's father and E is Ellie himself. Ellie loved his father but he was desperately starving. If his father gave him the food Ellie would survive a day or more. If his father insisted that Ellie take the food and Ellie refused to take the food out of love for his father and the food then sat uneaten, then both would die. With these considerations in mind the game no longer seems so straight – forward. How much should a father's love count in making such a decision? How about the father's sense of honor, reputation or his own disgust at the thought that he might eat his way to his own son's destruction?

Even setting aside these moral preferences there may be other preferences relevant but not strictly moral in and of themselves. Consider time for example. Time is nearly always a factor deserving of some weighted consideration. A pay – off of ten dollars today is likely to be discounted differently than a pay – off of ten dollars ten years from now [50]. Not only is inflation likely to decrease the real dollar value over such a length of time but the recipient's socio-economic status may shift substantially as well. Ten years from now the recipient may have gone from rags to riches and ten dollars isn't a very impressive figure anymore. Moreover in one's new status in life seeking to recover the measly amount of ten dollars may seem a bother or below one's social status. In short, one consideration Father must address is the benefit of strengthening or sustaining the health of himself or his son over time. The caloric expenditure necessary to sustain the father for a day may be sufficient for the son to care for both over two or three days.

An evolutionary theorist may frame this PD by stipulating that it is a constant sum game rather than a zero sum game with father and son both equally committed to passing along their similar genes. In this case many of the moral considerations disappear and are replaced by economic considerations of evolutionary value. If the son's prospects for survival seem better than the father's any paradox fades into an equilibrium achieved in behalf of the survival of the community genes rather than either gene-carrier.

There is little to warrant thinking that sufficient background information can ever be sufficient for sorting through such destabilizing trauma [70]. This does not mean that game theory or other mathematical models of things such as swarm intelligence are of little value [72]. Rather, the point is that the coincidental crossing of paths between science and mathematics gives us increasing and well-earned confidence that we can increasingly manage world events but never to the point that mathematical certainty will account for all the vicissitudes of nature [72]. Supervenient applications of mathematical structures to the world of experience assist in scientifically understanding the world but never to the point of subordinating natural processes to our mathematical modeling [36].

The ambitions of Morgenstern [74] in rendering all human social activity to transparency through game theoretic modeling will never be fulfilled. With each new advance in game theoretic models of human behavior and biology, new understandings of the human social and cognitive dynamic emerges. However, while game theoretic modeling has become an indispensable device both in learning more about human action and normatively by matching more fully human action with human intent, the excavating efforts of such research are unlikely to ever be complete. Human nature and the environment are too complex. And, evolution never lets matters rest as they are.

By accommodating human preferencing and intent, game theoretic developments help humans improve their interactions with the world in more predictable and systematic fashion. Before game theory and modern decision theory, moral values and other social commitments were honored in speech but too often forgotten or over looked by well-meaning decision-makers who forgot these commitments once the seemingly more quantifiable elements of dollars and cents made an easier cost/benefit analysis [75]. Game and decision theory have shown how utilities of any sort can be heuristically quantified to preserve more conscientious decision-making and presumably achieve a better world for all [48,49,58,70].

Author details

Paul A. Wagner

Institute for Logic and Cognitive Studies, University of Houston – Clear Lake, Houston, Texas, USA

Acknowledgement

Funding for this work was through the gracious award to me by the University of Houston – Clear Lake to serve a year as one of its five distinguished University Fellows.

2. References

[1] Chaitlin, G. (2012) Proving Darwinism: Making Biology Mathematical. New York: Pantheon.

[2] Weiner, N. (1964) I Am a Mathematician. Cambridge, MA. MIT.
[3] Hardy, G. (1940) A Mathematician's Apology. New York: Cambridge.
[4] Resnik M. (1987) Choices: An Introduction to Decision Theory. Minneapolis: University of Minnesota.
[5] Wang H. (1987) Reflections on Kurt Godel. Cambridge, MA. MIT.
[6] Wang H. (1993) Popular Lectures on Mathematical Logic. New York Dover.
[7] Feferman A. & Feferman S. Alfred Tarski: Life and Logic. New York: Cambridge. 2008.
[8] Boyer, C. & Merzbach U. A History of Mathematics. New York: Wiley: 1968.
[9] Deutsch D. Beginning of Infinity. New York: Viking: 2011.
[10] Kanazawa, S. The Intelligence Paradox. New York: Wiley: 2012.
[11] Hacking I. The Emergence of Probability. A Philosophical Study of Early Ideas about Probability, Induction and Statistical Inference. New York: Cambridge: 1984.
[12] Hacking, I. The Taming of Choice. New York: Cambridge: 1990.
[13] Maestripieri, D. Games Primates Play. New York: Basic Books: 2012.
[14] Heims, S. John von Neumann and Norbert Weiner: From Mathematics to the Technologies of Life and Death. Cambridge, MA. MIT: 1980.
[15] Poundstone W. Prisoner's Dilemma; John von Neumann, Game Theory and the Puzzle of the Bomb. New York: Anchor: 1992.
[16] Rosenthal E. The Complete Idiot's Guide to Game Theory. New York: Alpha Pub: 2011.
[17] Slote M Beyond Optimizing: A Study of Rational Choice. Cambridge, MA. Harvard: 1989.
[18] Morgenstern O. The Question of National Defense. New York: Random House: 1959
[19] Barash D. The Survival Game: How Game Theory Explains the Biology of Cooperation and Competition. New York: Time Books: 2003.
[20] Skyms B. Evolution of the Social Contract. New York: Cambridge: 1996.
[21] Harford T. The Logic of Life. New York: Random House:2006.
[22] Sigmund K. The Calculus of Selfishness. Princeton, NJ. Princeton University. pp. 104-123.
[23] Glimcher, P. Decisions, Uncertainty and the Brain. Cambridge, MA. MIT: 2003.
[24] Dugatkin, L. Cooperation Among Animals: An Evolutionary Perspective. New York: Oxford: 1997.
[25] Morgenstern O. & von Neumann J. Theory of Games and Economic Behavior. Princeton: Princeton University: 1944.
[26] Binmore K. Playing for Real. New York: Oxford: 2007.
[27] Schelling T. Choices and Consequences. Cambridge, MA. Harvard: 1985.
[28] Wu J. & Axelrod R. "How to Cope with Noise in a Reiterated Prisoner's Dilemma." Journal of Conflict Resolution. 39, (1) pp. 183 – 189: 1995.
[29] Kahnemann D. & Tversky A. "Loss aversion in Reckless Choices: a Reference Dependence Model" IN Kahnemann D. & Tversky, A. eds, Choices, Values and Frames. pp. 143 – 159. 2000.
[30] Weirich, P. Collective Rationality: Equilibrium in Cooperative Games. New York: Oxford: 1998.

[31] Gintis H. The Bounds of Reason. Game Theory and the Unification of the Behavioral Sciences. Princeton: Princeton University: 2011.

[32] Binmore K. Natural Justice. New York: Oxford: 2005.

[33] Simon, H. Theories of Decision-making in Economics and Behavioral Science" American Economic Review Vol. 49, No. 3, pp. 253 – 283: 1959.

[34] Binmore K. & Samuelson L. "Evolutionary Stabilityin Repeated Games Played by Finite Automata" Journal of Economic Theory, 57 (3), 122-131. 1992.

[35] Ulam, S. Adventures of a Mathematician. Berkeley: University of California: 1976.

[36] Nowak M. Supercooperators: Altruism,Evolution and Why We Need Others to Succeed. New York: Free Press: 2011.

[37] Bermudez J. Decision Theory and Rationality. New York: Oxford: 2009.

[38] Leonard R. Von Neumann, Morgenstern and the Creation of Games from Chess to Social Science. New York: Cambridge.. pp.185-220: 2010

[39] McGrayne S. The Theory that Just Won't Die. New Haven: Yale: 2011.

[40] Kahnemann, D. Thinking: Slow and Fast. New York: Farrar, Srauss &Giroux: 2011.

[41] Binmore K. Rational Decisions. Princeton, New Jersey: Princeton. 2009. pp.58-60.

[42] Gilboa I. Theory of Decision Under Uncertainty. New York: Cambridge. pp.94-110. 2009.

[43] Nash, J. Equilibrium Points in N-person Games". Proceedings of the National Academy of Sciences. 36, (4) 1950. pp. 48-49.

[44] Nash J. "Non-cooperative Games" Annals of Mathematics. Vol. 54, No. 2, 1951.

[45] Harsanyi J. & Selton R. A General Theory of Equilibrium Selection in Games. Cambridge: MIT: 1988.

[46] Raiffa H. Decision Analysis. Reading, MA.: Addison-Wesley: 1968.

[47] Meyerson, R. Game Theory: Analysis of Conflict. Cambridge, MA.: Harvard: 1997

[48] Sen, A. Choices, Welfare and Measurement. Cambridge, MA.: Harvard: 1997.

[49] Tempkin, L. Inequality. New York: Oxford.

[50] Binmore, K. A Very Short Introduction to Game Theory. New York: Oxford: 2007.

[51] Roth A. & Kagel J. Handbook of Experimental Game Theory. Princeton: Princeton University: 1995.

[52] Weirich P. Realistic Decision Theory: Rules for Nonideal Agents in Nonideal Circumstances. New York: Oxford: 2004.

[53] Weirich, P. Equilibrium and Rationality: Game Theory Revised by Decision Rules. New York: Cambridge: 2004

[54] McClennan E. Rationality and Dynamic Choice. Cambridge: Cambridge University: 1990

[55] Hausman D. Preference Choice and Welfare. New York: Cambridge: 2012

[56] Moulin, H. Axioms of Cooperative Decision-making. New York: Cambridge: 1988.

[57] Gilboa, I. Rational Choice. Cambridge, MA. MIT: 2010.

[58] Wagner P. Socio-sexual Education: a Practical Study in Formal Thinking and Teachable Moments. Sex Education : 11, (2). Pp. 193 204: 2011.

[59] Luce R. & Raiffa H. Games and Decisions. New York: Wiley: 1957

[60] Lotov A., Bushenkov V., & Kemenev, G. Interactive Decision Maps: Approximation and Visualization of Pareto Frontiers. Dordrecht, THE NETHERLANDS: Kluwer. Pp. 24 -34; 69 – 74.

[61] Szpiro G. Numbers Rule: the Vexing Mathematics of Democracy from Plato to the Present, Princeton, New Jersey: Princeton: 2010.

[62] Wilson. E. The Social Conquest of Earth. New York: Liverbright: 2012

[63] Sober E. & Wilson D. Unto Others: the Evolution and Psychology of Unselfish Behavior. Cambridge, MA.: Harvard.

[64] Tuck R. Free Riding. Cambridge, MA. Harvard. 2008. [65] Poundstone W. (2008) Gaming the Vote. New York: Hill & Wang: 2008.

[65] Arrow K. Social Choice and Individual Values. New York: John Wiley: 1951.

[66] Walton M. The Deming Management Method. New York: Dodd:1986.

[67] Ariely D. Predictably Irrational: Hidden Forces that Shape our Reality. New York: Harper Collins: 2008.

[68] Tversky A. Slovic P. & Kahnemann D. "The Cause of Preference Reversal" In Lichenstein S. & Slovic P. eds., TheConstruction of Preferences. New York: Cambridge: 2009.

[69] Heath J. Following the Rules: Practical Reasoning and Deontic Constraint. New York: Oxford: 2008.

[70] Wiesel E. Night. New York: Farrar, Strauss & Giroux: 1972.

[71] Taleb N. The Black Swan. New York: Random House: 2007.

[72] Fisher L. The Perfect Swarm, the Rise of Complexity in Every Day Life. New York: Basic Books: 2008.

[73] Leonard R. Von Neumann, Morgenstern and the Creation of Games from from Chess to Social Science. New York: Cambridge: 2010.

[74] Bazerman M. & Tenbrunsel A. Blindspots. Princeton. Princeton University; 2011.